ADVANCES IN CHEMICAL ENGINEERING
Volume 8

CONTRIBUTORS TO THIS VOLUME

D. M. HIMMELBLAU
J. R. KITTRELL
N. R. KULOOR
R. KUMAR
C. E. LAPPLE
W. P. LEDET

# ADVANCES IN
# CHEMICAL ENGINEERING

*Edited by*

### THOMAS B. DREW

*Department of Chemical Engineering*
*Massachusetts Institute of Technology*
*Cambridge, Massachusetts*

### GILES R. COKELET

*Department of Chemical Engineering*
*Montana State University*
*Bozeman, Montana*

### JOHN W. HOOPES, JR.

*Atlas Chemical Industries, Inc.*
*Wilmington, Delaware*

### THEODORE VERMEULEN

*Department of Chemical Engineering*
*University of California*
*Berkeley, California*

## Volume 8

Academic Press · New York · London 1970

TP
145
.D7
1970
v.8

COPYRIGHT © 1970, BY ACADEMIC PRESS, INC.
ALL RIGHTS RESERVED
NO PART OF THIS BOOK MAY BE REPRODUCED IN ANY FORM,
BY PHOTOSTAT, MICROFILM, RETRIEVAL SYSTEM, OR ANY
OTHER MEANS, WITHOUT WRITTEN PERMISSION FROM
THE PUBLISHERS.

ACADEMIC PRESS, INC.
111 Fifth Avenue, New York, New York 10003

*United Kingdom Edition published by*
ACADEMIC PRESS, INC. (LONDON) LTD.
Berkeley Square House, London W1X 6BA

LIBRARY OF CONGRESS CATALOG CARD NUMBER: 56-6600

PRINTED IN THE UNITED STATES OF AMERICA

## CONTRIBUTORS TO VOLUME 8

D. M. HIMMELBLAU, *Department of Chemical Engineering, The University of Texas, Austin, Texas*

J. R. KITTRELL, *Chevron Research Company, Richmond, California*

N. R. KULOOR,[*] *Department of Chemical Engineering, Indian Institute of Science, Bangalore, India*

R. KUMAR, *Department of Chemical Engineering, Indian Institute of Science, Bangalore, India*

C. E. LAPPLE, *Stanford Research Institute, Menlo Park, California*

W. P. LEDET,[†] *Department of Chemical Engineering, The University of Texas, Austin, Texas*

---

[*] Deceased February 5, 1970.

[†] *Present address:* E. I. duPont de Nemours, Orange, Texas.

# CONTENTS

LIST OF CONTRIBUTORS . . . . . . . . . . . . . . . . . . . . . . . . v
CONTENTS OF PREVIOUS VOLUMES . . . . . . . . . . . . . . . . . . . ix

## Electrostatic Phenomena with Particulates
### C. E. Lapple

| | | |
|---|---|---|
| I. | Introduction . . . . . . . . . . . . . . . . . . . . . . . . . . . | 2 |
| II. | Electrostatic Fundamentals . . . . . . . . . . . . . . . . . . . | 2 |
| III. | Effects of Electrostatic Charges . . . . . . . . . . . . . . . . | 7 |
| IV. | Electrostatic Dissemination Techniques . . . . . . . . . . . . | 38 |
| V. | Methods of Charging or Discharging Particles . . . . . . . . | 43 |
| VI. | Charge Measurement Techniques . . . . . . . . . . . . . . . | 75 |
| | Appendix. Conversion Factors for Commonly Used Units; Universal Constants and Defined Values; and Properties of Standard Air . . . . . . . . . . . . . . . . . . . . . . . . . . | 84 |
| | Nomenclature . . . . . . . . . . . . . . . . . . . . . . . . . | 88 |
| | References . . . . . . . . . . . . . . . . . . . . . . . . . . | 91 |

## Mathematical Modeling of Chemical Reactions
### J. R. Kittrell

| | | |
|---|---|---|
| I. | Introduction . . . . . . . . . . . . . . . . . . . . . . . . . . | 98 |
| II. | Linearly Reducible Models . . . . . . . . . . . . . . . . . . | 102 |
| III. | Parameter Estimation . . . . . . . . . . . . . . . . . . . . . | 110 |
| IV. | Tests of Model Adequacy . . . . . . . . . . . . . . . . . . . | 131 |
| V. | Use of Diagnostic Parameters . . . . . . . . . . . . . . . . | 142 |
| VI. | Empirical Modeling Techniques . . . . . . . . . . . . . . . | 154 |
| VII. | Experimental Designs for Modeling . . . . . . . . . . . . . | 168 |
| | Nomenclature . . . . . . . . . . . . . . . . . . . . . . . . . | 178 |
| | References . . . . . . . . . . . . . . . . . . . . . . . . . . | 181 |

## Decomposition Procedures for the Solving of Large Scale Systems
### W. P. Ledet and D. M. Himmelblau

| | | |
|---|---|---|
| I. | Introduction . . . . . . . . . . . . . . . . . . . . . . . . . . | 186 |
| II. | Information Flow in Process Models . . . . . . . . . . . . . | 188 |
| III. | Finding an Output Set . . . . . . . . . . . . . . . . . . . . | 196 |
| IV. | Partitioning of the System Equations . . . . . . . . . . . . | 198 |
| V. | Identifying Disjoint Subsystems . . . . . . . . . . . . . . . | 209 |
| VI. | Tearing . . . . . . . . . . . . . . . . . . . . . . . . . . . . | 211 |
| VII. | Comparison of Precedence Ordering Techniques . . . . . . . | 222 |

| | | |
|---|---|---:|
| VIII. | Two Examples | 226 |
| | Appendix A. Computer Program for Precedence Ordering | 237 |
| | Appendix B. Reordered Occurrence Matrix of the Hanford N-Reactor System | 252 |
| | Nomenclature | 253 |
| | References | 253 |

## The Formation of Bubbles and Drops
### R. Kumar and N. R. Kuloor

| | | |
|---|---|---:|
| I. | Introduction | 256 |
| II. | Measurement of Bubble Volume | 257 |
| III. | Influence of Various Factors on Bubble Size | 265 |
| IV. | Bubble Formation under Constant Flow Conditions | 277 |
| V. | Bubble Formation under Constant Pressure Conditions | 304 |
| VI. | Bubble Formation in Non-Newtonian Fluids | 316 |
| VII. | Bubble Formation in Gas Fluidized Beds | 318 |
| VIII. | The Effect of Orifice Geometry on Bubble Size | 321 |
| IX. | The Influence of Orifice Orientation on Bubble Formation | 324 |
| X. | The Influence of Continuous Phase Velocity on Bubble Size | 332 |
| XI. | Formation of Bubbles at Multiple Orificed Plates | 333 |
| XII. | Formation of Drops | 334 |
| XIII. | Drop Formation in Non-Newtonian Fluids | 343 |
| XIV. | Drop Formation from Vertically Oriented Orifices | 346 |
| XV. | Spraying and Atomization | 347 |
| XVI. | Unified Model for Drop and Bubble Formation | 350 |
| XVII. | Bubble and Drop Size in Stirred Vessels | 354 |
| XVIII. | Bubble Formation under Intermediate Conditions and at Sintered Disks | 356 |
| XIX. | Concluding Remarks | 362 |
| | Nomenclature | 363 |
| | References | 365 |

| | |
|---|---:|
| Author Index | 369 |
| Subject Index | 377 |

# CONTENTS OF PREVIOUS VOLUMES

## Volume 1

Boiling of Liquids
*J. W. Westwater*

Non-Newtonian Technology: Fluid Mechanics, Mixing, and Heat Transfer
*A. B. Metzner*

Theory of Diffusion
*R. Byron Bird*

Turbulence in Thermal and Material Transport
*J. B. Opfell and B. H. Sage*

Mechanically Aided Liquid Extraction
*Robert E. Treybal*

The Automatic Computer in the Control and Planning of Manufacturing Operations
*Robert W. Schrage*

Ionizing Radiation Applied to Chemical Processes and to Food and Drug Processing
*Ernest J. Henley and Nathaniel F. Barr*

AUTHOR INDEX—SUBJECT INDEX

## Volume 2

Boiling of Liquids
*J. W. Westwater*

Automatic Process Control
*Ernest F. Johnson*

Treatment and Disposal of Wastes in Nuclear Chemical Technology
*Bernard Manowitz*

High Vacuum Technology
*George A. Sofer and Harold C. Weingartner*

Separation by Adsorption Methods
  *Theodore Vermeulen*

Mixing of Solids
  *Sherman S. Weidenbaum*

AUTHOR INDEX—SUBJECT INDEX

## Volume 3

Crystallization from Solution
  *C. S. Grove, Jr., Robert V. Jelinek, and Herbert M. Schoen*

High Temperature Technology
  *F. Alan Ferguson and Russell C. Phillips*

Mixing and Agitation
  *Daniel Hyman*

Design of Packed Catalytic Reactors
  *John Beek*

Optimization Methods
  *Douglass J. Wilde*

AUTHOR INDEX—SUBJECT INDEX

## Volume 4

Mass-Transfer and Interfacial Phenomena
  *J. T. Davies*

Drop Phenomena Affecting Liquid Extraction
  *R. C. Kintner*

Patterns of Flow in Chemical Process Vessels
  *Octave Levenspiel and Kenneth B. Bischoff*

Properties of Cocurrent Gas-Liquid Flow
  *Donald S. Scott*

A General Program for Computing Multistage Vapor-Liquid Processes
  *D. N. Hanson and G. F. Somerville*

AUTHOR INDEX—SUBJECT INDEX

## Volume 5

Flame Processes—Theoretical and Experimental
*J. F. Wehner*

Bifunctional Catalysts
*J. H. Sinfelt*

Heat Conduction or Diffusion with Change of Phase
*S. G. Bankoff*

The Flow of Liquids in Thin Films
*George D. Fulford*

Segregation in Liquid-Liquid Dispersions and Its Effect on Chemical Reactions
*K. Rietema*

AUTHOR INDEX—SUBJECT INDEX

## Volume 6

Diffusion-Controlled Bubble Growth
*S. G. Bankoff*

Evaporative Convection
*John C. Berg, Andreas Acrivos, and Michel Boudart*

Dynamics of Microbial Cell Populations
*H. M. Tsuchiya, A. G. Fredrickson, and R. Aris*

Direct Contact Heat Transfer between Immiscible Liquids
*Samuel Sideman*

Hydrodynamic Resistance of Particles at Small Reynolds Numbers
*Howard Brenner*

AUTHOR INDEX—SUBJECT INDEX

## Volume 7

Ignition and Combustion of Solid Rocket Propellants
*Robert S. Brown, Ralph Anderson, and Larry J. Shannon*

Gas-Liquid-Particle Operations in Chemical Reaction Engineering
*Knud Østergaard*

Thermodynamics of Fluid-Phase Equilibria at High Pressures
*J. M. Prausnitz*

The Burn-Out Phenomenon in Forced-Convection Boiling
*Robert V. Macbeth*

Gas-Liquid Dispersions
*William Resnick and Benjamin Gal-Or*

AUTHOR INDEX—SUBJECT INDEX

# ELECTROSTATIC PHENOMENA WITH PARTICULATES

## C. E. Lapple

Stanford Research Institute
Menlo Park, California

| | |
|---|---:|
| I. Introduction | 2 |
| II. Electrostatic Fundamentals | 2 |
| III. Effects of Electrostatic Charges | 7 |
|     A. On Material Properties | 7 |
|     B. On Containment or Bulk Storage | 8 |
|     C. On Aerosol Properties | 10 |
|     D. On Coalescence | 25 |
|     E. On Evaporation | 28 |
|     F. On Deposition | 28 |
|     G. On Adhesion | 30 |
|     H. On Biological Processes | 38 |
| IV. Electrostatic Dissemination Techniques | 38 |
|     A. Electrostatic Atomization of Liquids | 38 |
|     B. Electrostatic Dissemination of Powders | 43 |
| V. Methods of Charging or Discharging Particles | 43 |
|     A. Maximum Stable Charge | 43 |
|     B. Contact Charging | 46 |
|     C. Induction or Field Charging | 47 |
|     D. Diffusion Charging | 51 |
|     E. Thermionic Emission Charging | 53 |
|     F. Charging by Interface Alteration | 55 |
|     G. Charging by Phase Change | 73 |
|     H. Other Methods of Charging | 74 |
| VI. Charge Measurement Techniques | 75 |
| Appendix. Conversion Factors for Commonly Used Units; Universal Constants and Defined Values; and Properties of Standard Air | 84 |
| Nomenclature | 88 |
| References | 91 |

## I. Introduction

Industrial operations, particularly chemical and metallurgical, deal extensively with particulate systems, for example, powders, aerosols, fluidized beds, ores, paints, and insecticides. In such systems, electrostatic phenomena can play many important roles: (1) they can markedly affect the behavior or properties of the materials being handled, as in conveying, grinding, and screening; (2) they can be utilized as direct processing means as in electrostatic printing, atomization, minerals separation, or aerosol precipitation; and (3) they may be used to control, modify, or trigger other processes, both detrimentally and beneficially, as in setting off dust explosions, in causing adhesion of films and fibers, and in the controlling of spray deposition.

In the past, many of the electrostatic phenomena have been treated as unavoidable nuisances, and many processing problems have been blamed rather casually on electrostatic influences. Only in recent years has the scope of the role of electrostatics and its potential for application in processing begun to be more fully understood.

It is the purpose of this chapter to summarize knowledge dealing with the role of electrostatic phenomena in particulate systems, especially in aerosols. A relatively comprehensive annotated bibliography by Blake and Lapple (B10) has surveyed the bulk of the literature through 1965. This chapter cites and summarizes the results of the more important or directly relevant work in this field.[1] However, before considering specific phenomena, it is desirable to review some of the fundamental principles in the field of electrostatics.

## II. Electrostatic Fundamentals

Bodies that acquire a net electric charge because of the loss or gain of an electron or molecular ion are said to be charged. The charge is said to be unipolar if all particles have the same sign of charge, and ambipolar if particles of both signs are involved.[2] The terms homopolar and heteropolar refer to the uniformity or nonuniformity of the magnitude of the charge on the component particles.

---

[1] Although aimed at different objectives, the processing problems of Edgewood Arsenal, for whom this study was originally conducted, are in many respects very similar to those of the process industries. While this chapter has been written with the process industries in mind, its origin has resulted in a somewhat greater emphasis on the aerosol particulate system than might otherwise have been the case.

[2] In some literature the terms "monopolar" and "bipolar" are used, but these may be mistakenly thought to indicate the magnitude of the charge, especially when referring to ions.

In the presence of an electrostatic field, all materials become polarized; that is, they acquire a dipole resulting from orientation of their molecules by the field or displacement of their free electrons. This dipole tends to counteract the imposed electrostatic field and is directly related to the dielectric constant of the material. In normal gases (dielectric constant = 1), the polarization effect is negligible. It is a maximum in conductors, in which the polarization is actually the result of migration of free electrons rather than being simply an orientation of molecules. In most materials (always with conductors), the polarization will disappear when the field is removed. In some materials, called electrets, the dipoles induced by an electrostatic field will remain even after the field is removed, although they may disappear after long periods of time. In other materials (e.g., ferroelectrics), stable dipoles tend to form spontaneously, even in the absence of electrostatic fields. Polarized bodies will have an electrostatic force field around them because of the dipoles but will not have a net charge unless it is acquired by other means. Polarization processes may play a significant role in some processes (e.g., electrostatic atomization) and are considered in those instances. The bulk of this chapter, however, is concerned with the phenomena related to particles with net electrical charges.

The magnitude of charge on a particle can be expressed in any one of several ways: total charge ($\mathcal{Q}_p$), surface charge density ($\mathcal{Q}_{ps}$), specific particle surface gradient or field ($\mathcal{E}_{ps}$), charge-to-mass ratio ($R_c$), or specific particle potential ($E_{ps}$). These are related by the following identities:

$$\mathcal{E}_{ps} = \frac{2E_{ps}}{D_p} = \frac{\mathcal{Q}_{ps}}{\varepsilon\delta} = \frac{\mathcal{Q}_p}{\pi\varepsilon\delta D_p^2} = \frac{\rho_p D_p R_c}{6\varepsilon\delta} \qquad (1)$$

where $D_p$ is the diameter of particle; $\rho_p$, the density of particle; $\varepsilon$, the permittivity of free space, a universal constant; and $\delta$, the dielectric constant of the surrounding medium. Of these terms, $\mathcal{Q}_p$ and $R_c$ are always physically real quantities. The rest of the terms have real physical significance under certain circumstances. For example, $\mathcal{E}_{ps}$ is the potential gradient or field intensity existing on the outside surface of a particle only if the charge is distributed uniformly around the surface of the particle. However, in any case the terms $\mathcal{E}_{ps}$, $\mathcal{Q}_{ps}$, and $E_{ps}$, are defined in terms of $\mathcal{Q}_p$, or $R_c$ by Eq. (1), regardless of their physical meaning.

Electrostatic phenomena are a function of particle properties as well as particle charge. Of particle properties, the one most important and subject to the widest variation is particle size.[3] Consequently, it would be most con-

---

[3] Particle size will always be expressed in terms of diameter rather than radius in this chapter. In all developments in this chapter, the particles will be assumed to be spherical. Consequently, when applying the results to other than spherical particles, that equivalent spherical diameter must be used which would correspond to the phenomenon involved.

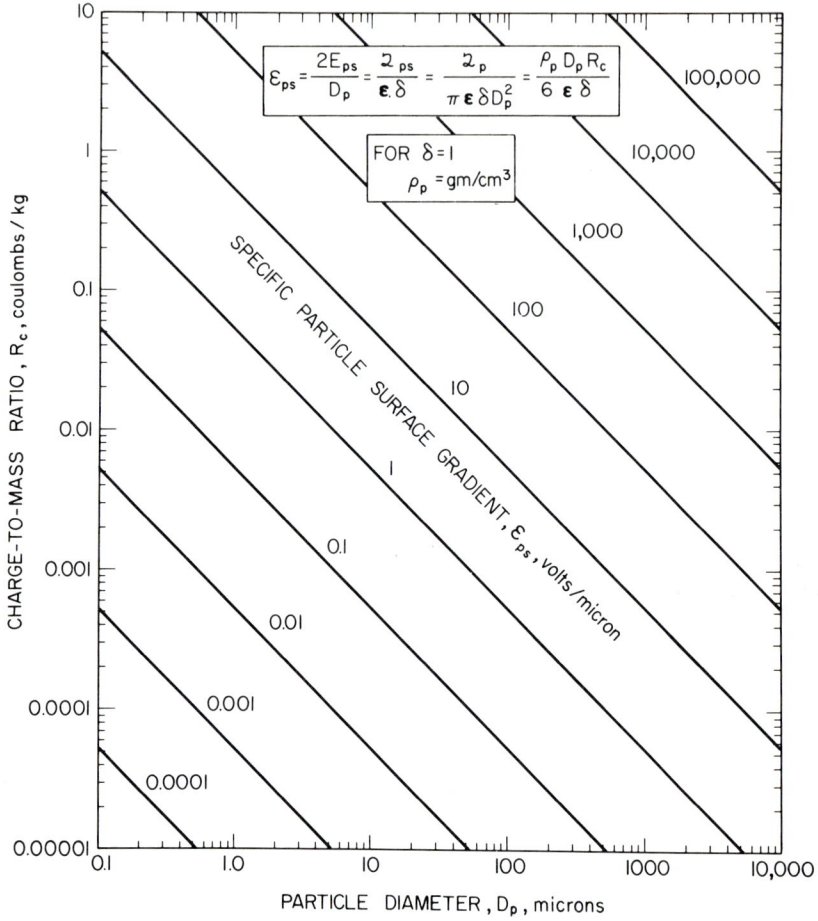

Fig. 1a. Relationship between specific particle surface gradient and charge-to-mass ratio for various particle diameters.

venient to express particle charge in that combination with particle size which most nearly is a direct measure of the phenomenon. While this combination varies with the phenomenon, it is believed that the term $\mathscr{E}_{ps}$ is a direct measure of the most frequently encountered phenomena. For this reason all particle charges are reported in terms of $\mathscr{E}_{ps}$ in this chapter, usually using the units of volts/micron. Since particle charge is often measured in terms of charge-to-mass ratio, $R_c$, the relationship between $R_c$ and $\mathscr{E}_{ps}$ is shown in Fig. 1a as a function of particle diameter.

In order to provide a better means for visualizing the physical significance of both the specific particle surface gradient, $\mathscr{E}_{ps}$, and the specific particle

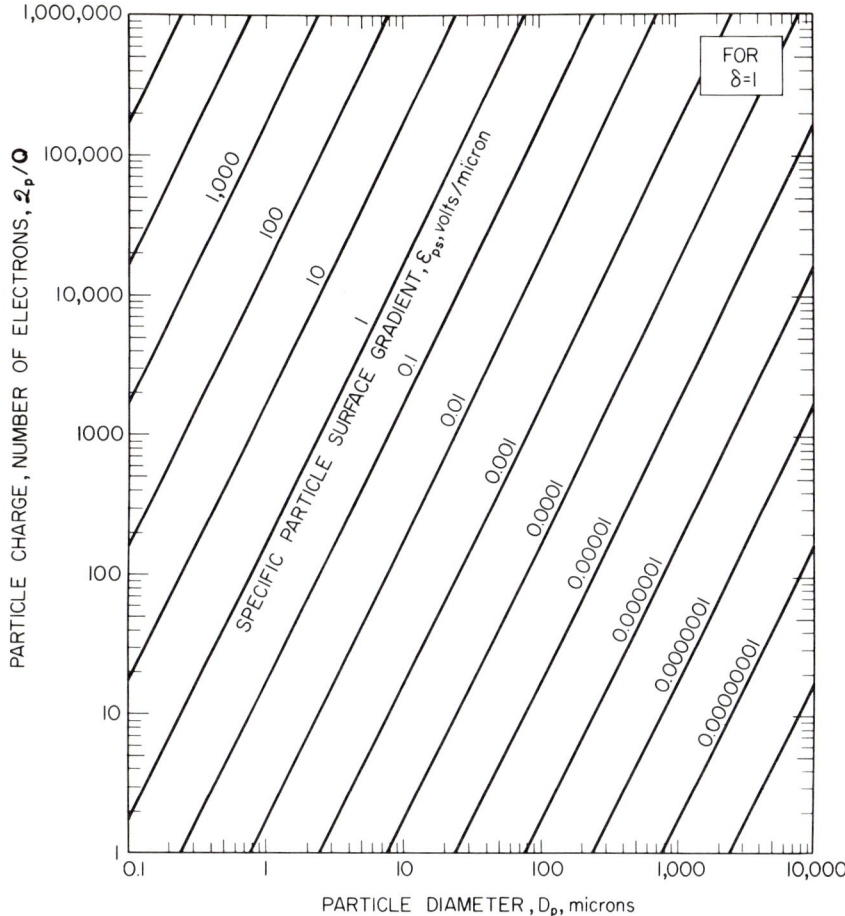

FIG. 1b. Relationship between specific particle surface gradient and number of electrons on particle for various particle diameters.

potential, $E_{ps}$, Figs. 1b and 1c give the particle charge, in number of electrons, for various particle sizes corresponding to various levels of $\mathscr{E}_{ps}$ and $E_{ps}$. Figures 1a–1c are simply graphical portrayals of Eq. (1).

In this chapter all equations are written in a form for which any consistent system of units may be used. The table of nomenclature gives illustrative units in the MKSA system. By definition,

$$F = \mathscr{E} \mathscr{Q} \qquad (2)$$

where $F$ is the force acting on a point charge $\mathscr{Q}$ situated in an electrostatic field of intensity $\mathscr{E}$. In the rationalized form, the basic electrostatic force

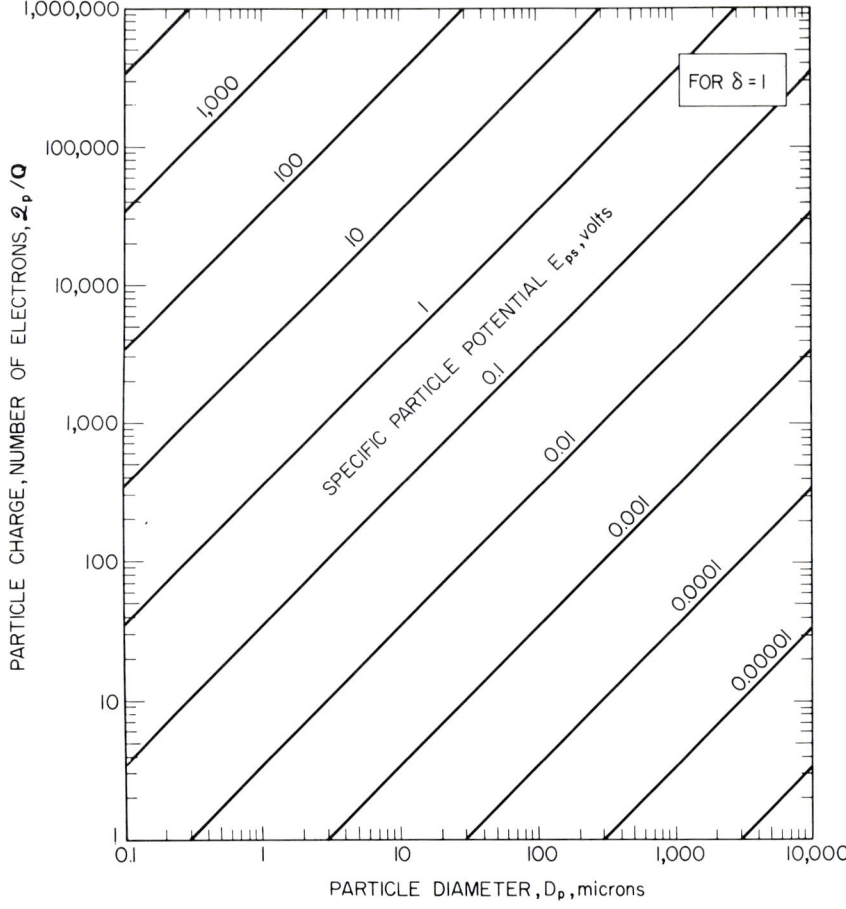

FIG. 1c. Relationship between specific particle potential and number of electrons on particle for various particle diameters.

equation (Coulomb's law) is

$$F = \mathcal{Q}_1 \mathcal{Q}_2 / 4\pi\varepsilon\delta L^2 \tag{3}$$

where $F$ is the force between two point charges, $\mathcal{Q}_1$ and $\mathcal{Q}_2$, distance $L$ apart in a medium of dielectric constant $\delta$.

Davies (D2) has given a very brief review of some of the recent literature related to electrical properties of aerosols.

## III. Effects of Electrostatic Charges

### A. On Material Properties

The presence of an electrostatic field subjects a material to a stress. Such fields can arise from the presence of free charges in the material or from the action of an external field. A number of authors (Hogan, H12, H13; Graf, G8; Doyle, Moffett, and Vonnegut, D6; and others) give the outward pressure on a liquid sphere with a uniformly distributed surface charge as

$$p_e = \frac{\varepsilon \mathscr{E}_{ps}^2}{2} \quad (4)$$

which assumes that the sphere is in a medium with unit dielectric constant. Thus a value of $\mathscr{E}_{ps}$ of 1 V/micron, would correspond to an outward electrostatic pressure of 44 dyn/cm² or 0.00064 psi, which is a comparatively small pressure.

For comparison, the constraining pressure due to surface tension, $\sigma$, is given by

$$p_\sigma = \frac{4\sigma}{D_p} \quad (5)$$

For a 1-micron-diam water particle, $p_\sigma$ will be $2.88 \times 10^6$ dyn/cm² or 42 psi.

The effect of electric charge on the vapor pressure of a liquid is given by the classic equation of Thomson as quoted by Gallily and Ailam (G1) and Vonnegut, Moffett, Sliney, and Doyle (V3)

$$\ln\left(\frac{p_p}{p_\infty}\right) = \frac{M}{\rho_p RT}\left[\frac{4\sigma}{D_p} - \frac{\varepsilon \mathscr{E}_{ps}^2}{2}\right] = \frac{\rho_g}{\rho_p}\left(\frac{p_\sigma - p_e}{p_T}\right) \quad (6)$$

where $p_p$ is the equilibrium vapor pressure from a drop of diameter $D_p$ of a liquid of density $\rho_p$, whose vapor pressure from a flat surface under the same total pressure as that of the fluid surrounding the drop would be $p_\infty$ at the same temperature. The density $\rho_g$ is that at the absolute pressure $p_T$ and temperature $T$ of the pure vapor of the liquid. The terms $p_\sigma$ and $p_e$ are the pressures due to surface tension and surface charges as given by Eqs. (5) and (4), respectively. This equation is derived for a conducting liquid in a medium, devoid of external force fields, with a dielectric constant of unity. Surface tension results in an inward pressure on the liquid, which is partly counteracted by the presence of a charge at the surface. Thus, surface tension will cause an increase in vapor pressure but only because of the curved surface. This effect does not normally become significant until the drop diameter is less than 1 micron. The effect of a charge on the surface is to reduce vapor pressure and hence counteract the effect of surface tension. Gallily and Ailam

(G1) have also derived a similar modified equation for the case of a drop of conducting liquid surrounded by a gaseous suspension of both neutral and charged particles, enclosed in a spherical container.

From the illustrative value given for $p_e$ above it is apparent that electrostatic effects on vapor pressure will be negligible unless the particle charge level is orders of magnitude greater than 1 V/micron.

The discussion above has considered the effect of free charges and surface tension on vapor pressure. By exposing a polarizable liquid surface to an electrostatic field, similar effects can result because of the surface charges created by polarization. Hogan (H12, H13) considers these effects, including the effect due to electroconstriction. Although the effect of a free surface charge is always to exert an outward pressure, an inward pressure can exist if a sufficiently large tangential field intensity exists at the surface of a polarizable liquid. In the absence of tangential stresses, polarization effects can be allowed for by multiplying the right-hand side of Eq. (4) by a factor that is a function of the properties of the material. As indicated by Schultz and Wiech (S3), this factor may range from 0.55 for nonpolar unassociated liquids to unity for polar highly associated liquids.

There is also the possibility of having surface tension affected directly by the presence of an electrostatic field. To some extent this will be a matter of definition since the outward pressure due to a surface charge could be defined as an apparent effect on surface tension. Hurd, Schmid, and Snavely (H15) measured the surface tension of water and water solutions when fields up to 0.7 V/micron were applied across the air–solution interface. The results showed a reduction in surface tension of less than 1%. These data must not be considered conclusive, however, because insufficient details are reported to permit assessment of the exact nature of the electrostatic field applied or of the validity of a number of corrections that had to be applied but were reported to be very large and difficult to apply.

Electrostatic charges or force fields could also influence other properties such as viscosity and freezing point, but no attempt has been made to extend the review to these areas.

B. On Containment or Bulk Storage

In storing a precharged powdered material, one must consider the rate and manner in which electric charge may be lost. This charge may be lost by conduction through the powder to the container and then either by conduction to the surrounding supports or by corona to the atmosphere.

It can be readily shown that for a volume of unipolar and uniformly distributed charged particles bounded by a conductive surface equidistant from a center of symmetry [either a plane (flat volume); a line (cylindrical volume);

or a point (spherical volume)], the following describes the rate at which the volume will lose its charge

$$t = \varepsilon \delta_v \mathscr{R}_0 \ln(\mathscr{Q}_0/\mathscr{Q}) \tag{7}$$

where $t$ is the time for the total charge within the volume to drop from $\mathscr{Q}_0$ to $\mathscr{Q}$ for a material of resistivity $\mathscr{R}_0$ and dielectric constant $\delta_v$. If the powder is ambipolar, Eq. (7) will still apply if $\mathscr{Q}_0$ and $\mathscr{Q}$ are considered as net charges. This is essentially a generalized form of the analysis given by Shaffer (S5). Values of $\delta_v \mathscr{R}_0$ corresponding to various rates of charge loss are shown in Table I.

TABLE I

VALUE OF $\delta_v \mathscr{R}_0$ CORRESPONDING TO VARIOUS RELAXATION TIMES

| Relaxation time[a] $t_R$ | $\delta_v \mathscr{R}_0$ ($\Omega$)(cm) |
|---|---|
| 1 msec | $1.129 \times 10^{10}$ |
| 1 sec | $1.129 \times 10^{13}$ |
| 1 min | $6.77 \times 10^{14}$ |
| 1 hr | $4.06 \times 10^{16}$ |
| 1 day | $9.75 \times 10^{17}$ |
| 1 year | $3.56 \times 10^{20}$ |

[a] Relaxation time is defined as the time for the charge to drop by a factor of $e$, i.e., for $\ln(\mathscr{Q}_0/\mathscr{Q}) = 1$, following the terminology used by Vonnegut et al. (V3) for the group $\varepsilon \delta_v \mathscr{R}_0$ in a study of another but related phenomenon.

From these values it is apparent that very few materials could retain a significant fraction of their original charge for many days. Most materials could retain their charge for only a matter of minutes or less. Although Eq. (7) is derived for a specific type of geometry, it should also be applicable to give the correct order of magnitude for the time required for a deposited particle to lose or acquire charge in the absence of a superimposed field. In that case $\mathscr{R}_0$ would include any surface resistivity.

An assemblage of even lightly charged particles can result in very high electrostatic field intensities at the surface of the assemblage. It is readily shown that for a volume of unipolar and uniformly distributed charged particles bounded by a surface equidistant from a center of symmetry, the field intensity at the outer edge of the volume is given by

$$\mathscr{E} = \frac{6c_p(V/A)\mathscr{E}_{ps}}{\rho_p D_p} = \frac{6m\mathscr{E}_{ps}}{A\rho_p D_p} = \text{(for spherical assemblage)} \frac{c_p D \mathscr{E}_{ps}}{\rho_p D_p} \quad (8)$$

where $c_p$ is the mass concentration of particles of average size $D_p$, average density $\rho_p$, and average net charge corresponding to $\mathscr{E}_{ps}$ in a volume $V$ bounded by a surface of area $A$. The total mass of particles in the volume $V$ is $m$, and $D$ is the diameter of the spherical volume. This is also a generalized treatment of the analysis given by Shaffer (S5).

For powders, $c_p/\rho_p$ is approximately one-half. If we further assume that the outer edge of the volume will go into corona when $\mathscr{E}$ exceeds 3 V/micron, which is the normal electrical breakdown strength of a flat surface in air, Table II shows the maximum size containers that can be used for various net levels of particle charging without having the charge lost by corona.

TABLE II

ILLUSTRATIVE VALUES OF MAXIMUM ALLOWABLE CONTAINER DIAMETER FOR VARIOUS PARTICLE CHARGE LEVELS
(for $\mathscr{E} \leq 3$ V/micron; $c_p/\rho_p = 0.5$)

| Specific particle surface gradient, $\mathscr{E}_{ps}$, V/micron | Maximum allowable container diameter $D/D_p$ |
|---|---|
| 1 | 6 |
| 0.1 | 60 |
| 0.01 | 600 |
| 0.001 | 6000 |

Thus, even at the very low particle charge levels of 0.001 V/micron (corresponding to an average charge of 0.2 electron on a 1-micron particle), the container cannot exceed 6000 times the particle diameter or 6 cm for a 10-micron particle.

The above confirms the conclusions reached by Shaffer (S5) that unipolar *precharging* of powders is impractical because of considerations of charge loss during storage.

C. ON AEROSOL PROPERTIES

A cloud is essentially an aerosol with mobile boundaries. The stability of a cloud involves two distinct aspects. Because of Coulomb, image, and diffusion forces, particles in a given volume will flocculate. This will result in a lower number concentration but the same mass concentration of particles (provided no loss occurs by diffusion to bounding walls or by settling out under the action of gravity). This can occur even with uncharged particles (by either Brownian motion or turbulence flocculation). If the cloud contains a net charge of one sign, the cloud will tend to expand because of the net repulsive forces set up within the cloud. This will result in a decrease in both number and mass concentration and will be superimposed on flocculation effects. The following discusses the role of these phenomena in determining the behavior and properties of an aerosol (or cloud).

## 1. Cloud Expansion and Cloud Properties

For a cloud of unipolar and uniformly distributed charged particles bounded by a surface equidistant from a center of symmetry, the rate of cloud expansion due to the repulsion of the charged particles within the cloud is given by

$$\left[\frac{-dc_p}{c_p\,dt}\right] = \left[\frac{-dn_p}{n_p\,dt}\right] = \left[\frac{2k_C\,\varepsilon\,\delta_v\,c_p\,\mathscr{E}_{ps}^2}{\mu\rho_p}\right] = \frac{1}{t_{sd}} \tag{9}$$

where $c_p$ is the mass concentration of particles at time $t$; $n_p$ is the number concentration at the time $t$; $\mu$ is the gas viscosity; $\rho_p$ is the particle density; $k_C$ is the Stokes–Cunningham factor (a function of particle diameter and mean free path of gas molecules), and $\mathscr{E}_{ps}$ is the specific particle surface gradient. The specific dilution time, $t_{sd}$, is defined by the above equation. This and all subsequent equations, while derived for a unipolar cloud, will still hold for an ambipolar cloud if $\mathscr{E}_{ps}$ is defined in terms of the net charge. This is essentially a generalized version of the analysis presented by Foster (F1) and mentioned by Zebel (Z2). If $\delta_v$ is independent of $c_p$, Eq. (9) can be integrated to

$$t = t_{sd}[1 - (c_p/c_{p0})] \tag{10}$$

where $t$ is the time for the cloud to expand from a concentration $c_{p0}$ to a concentration $c_p$. Thus the specific dilution time, $t_{sd}$, is actually the hypothetical time required for a cloud to expand from an infinite concentration of particles to a given concentration $c_p$. For most practical purposes, $t_{sd}$ will also be identical with the time after initial formation for a cloud to drop to a concentration $c_p$.

The radial outward velocity of the edge of the cloud is given by

$$u_R = \frac{V/A}{t_{sd}} = \text{(for spherical cloud)} \frac{D}{6t_{sd}} \tag{11}$$

where $V$ is the cloud volume, $A$ the total area of the edge of the cloud, and $D$ the diameter of a spherical cloud. The field intensity at the edge of the cloud is given by an equation identical to Eq. (8).

A plot of Eq. (9) is given in Fig. 2 for air at 25°C taking $\rho_p = 1$ gm/cm³, $\delta_v = 1$, and $k_C = 1$. It is apparent from Fig. 2 that concentrations of the order of 100–1000 µg/liter could not exist for a matter of more than a few seconds if $\mathscr{E}_{ps}$ is greater than 1 V/micron. The above analysis allows only for cloud expansion due to the net charge on the particles. In practice, expansion rates will be even greater because of any added convective dilutions due to temperature or wind velocity gradients. It is, therefore, apparent that net charges greater than that corresponding to $\mathscr{E}_{ps} = 1$ V/micron cannot be tolerated from considerations of cloud stability for applications requiring aerosol concentrations greater than 100 µg/liter.

It is interesting to note from Eq. (11), that a 60-ft-diam field cloud of concentration 100 µg/liter with a charge corresponding to $\mathscr{E}_{ps} = 1$ V/micron

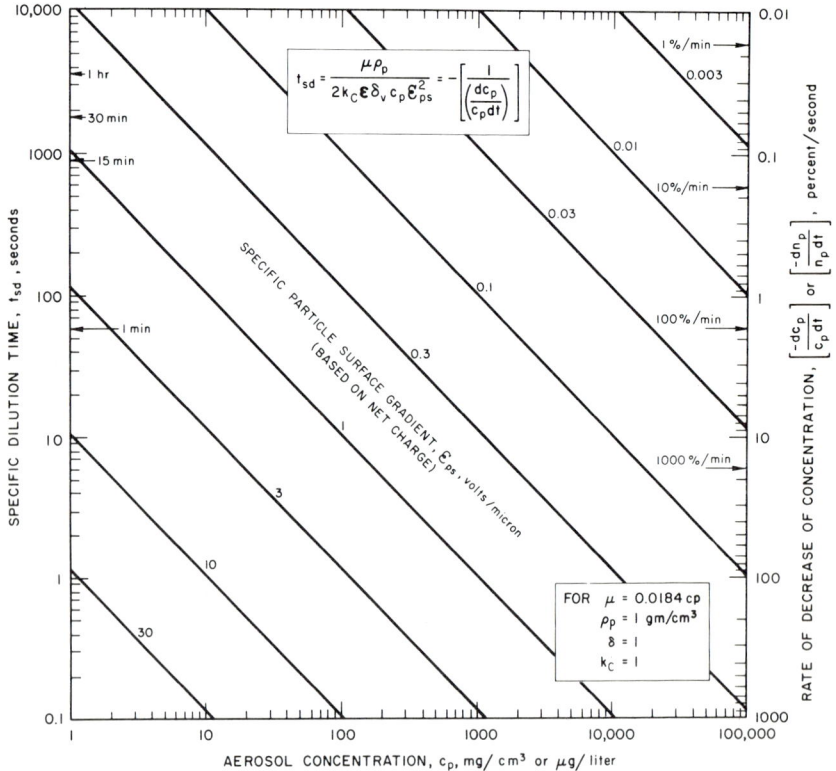

FIG. 2. Rate of expansion of charged clouds.

would be expanding at a radial velocity of 1 ft/sec. Prior to that point, the cloud would be expanding at a greater velocity.

Fuchs (F4, pp. 102–105) also points out that, for a unipolar aerosol, the rate of concentration change, as given by Eq. (9) or (10), is an intrinsic property of the aerosol, depending on neither size nor shape of cloud. Some of the other relationships [Eqs. (8) and (11)], however, are subject to the cloud symmetry specified above. Fuchs also points out that in an ambipolar cloud the particles of minority sign tend to be driven toward the center or core where the cloud approaches neutrality asymptotically. The outer edge of the cloud tends to become unipolar with the sign of the initially most prevalent charge.

Equation (8) is applicable to clouds of particles as well as powders. From Eq. (8) it is readily shown that even a lightly charged cloud would have a field intensity in excess of 10 V/cm (or 0.001 V/micron) at the outer edge. (For example, $\mathscr{E}$ will be 15 V/cm, when $\rho_p = 1$ gm/cm$^3$; $c_p = 1000$ μg/liter; $D_p = 10$ microns; $D = 50$ ft; and $\mathscr{E}_{ps} = 0.001$ V/micron.)

It is also interesting to note that natural atmospheric or terrestrial electrostatic fields are of the order of 1 V/cm (Schonland, S1) in the regions near the ground, with the earth's surface usually negative relative to the atmosphere. These fields are small compared with those resulting from even a lightly charged cloud. Although the small terrestrial fields may be negligible from the standpoint of aerosol dissemination, they may be significant in other atmospheric phenomena.

There is a possibility that a cloud might lose charge by corona during its early concentrated stage. For example, if it is assumed that the maximum field intensity at the outer edge cannot exceed 3 V/micron (the normal breakdown strength of air) without having corona ensue, then Eq. (8) would predict the relationships shown in Table III for a net particle charge corre-

TABLE III

Illustrative Values of Maximum Permissible Cloud Diameter for Various Particle Concentrations
(for $\mathscr{E}_{ps} = 1$ V/micron; $\rho_p = 1$ gm/cm$^3$; $D_p = 10$ microns)

| Particle concentration in cloud, $c_p$ | | Maximum permissible cloud diameter, $D$ (without exceeding field intensity of 3 V/micron) (ft) |
|---|---|---|
| (μg/liter) | (lb/lb air) | |
| 10$^6$ | 1 | 0.1 |
| 10$^5$ | 0.1 | 1 |
| 10$^4$ | 0.01 | 10 |
| 10$^3$ | 0.001 | 100 |

sponding to 1 V/micron, a particle density of 1 gm/cm³, and a particle diameter of 10 microns.

This would imply that the concentration of the initially disseminated cloud would be limited by its initial size as indicated above if no charge is to be lost owing to corona. However, it is not certain that corona would actualy ensue, especially where the cloud is expanding simultaneously. Even if there were some corona, the lost ions might be recaptured by the expanding cloud. Consequently, the above simply represents a conservative estimate of possible limitations on the initial cloud or aerosol jet.

## 2. *Optimum Charge*

In such applications as insecticide or pesticide spraying, riot control, or chemical warfare, it is the objective to expose objects within a given area to either an incapacitating or lethal dosage of an agent, as the case may warrant. If a cloud expands to cover a ground area $A_b$ in time $t$, then any object exposed at the edge of this area will have received a critical dosage at a later time $t_a$ provided

$$\int_t^{t_a} c \, dt \geq (ct)_c \tag{12}$$

where $(ct)_c$ is the critical dosage usually referred to as a *ct* value. All other objects within this area will already have received at least a critical dosage in time $t_a$. If it is further assumed that the number of target objects exposed is proportional to the area $A_b$ and that the area $A_b$ is proportional to the $\tfrac{2}{3}$ power of the cloud volume at any time, it can be shown that there is an optimum value of $\mathscr{E}_{ps}$, which will result in a maximum number of casualties for any given time $t_a$ and total mass of agent released. The optimum is given by

$$\mathscr{E}_{ps_{opt}} = \left[ \frac{\mu \rho_p}{2 k_C \, \varepsilon \, \delta(ct)_c} \right]^{1/2} \tag{13}$$

The actual number of casualties will, of course, depend on both the value of $t_a$ desired and the mass of agent disseminated. Assuming a value of 10 (μg/liter)(min) for $(ct)_c$, the optimum value of $\mathscr{E}_{ps}$ will be approximately 1 V/micron. This is, of course, a simplified analysis and various further considerations should be included such as allowances for target objects seeking shelter or protection, variation of $(ct)_c$ values with exposure time and particle charging, nature of agent dispersion prior to dissemination, etc. This also assumes that cloud expansion is governed solely by the electrostatic charge of the cloud and makes no allowance for other factors such as atmospheric turbulence.

## 3. Aerosol Stability

Kunkel (K9) has derived expressions for the collision rate of charged particles allowing for both Coulomb attraction forces and ionic image forces. His final expressions, however, are not readily applied and he concludes only that the rate of growth of a particle over a period of several seconds is small if

$$\frac{|\mathcal{Q}_p \mathcal{Q}_{pa}| n_p}{D_p/2} < 10^{10} \tag{14}$$

where $\mathcal{Q}_p$ is the charge of the particle in electrons; $\mathcal{Q}_{pa}$ is the average charge of other particles of one sign; $n_p$ is the number concentration of particles, particles/centimeter$^3$; and $D_p$ is the average particle diameter in microns. Assuming an equal magnitude of charge on all particles, this can be written in the form

$$\mathscr{E}_{ps}^2 c_p / \rho_p < 100 \tag{15}$$

where $\mathscr{E}_{ps}$ is in volts/micron; $c_p$ is in micrograms/liter; $\rho_p$ is in grams/centimeter$^3$. Thus for $\rho_p = 1$ gm/cm$^3$ and $c_p = 1000$ μg/liter, $\mathscr{E}_{ps}$ must be less than 0.3 V/micron. Thus one would conclude from Kunkel that flocculation due to charge would be small in a period of minutes if $\mathscr{E}_{ps}$ were less than 0.01 V/micron.

A more thorough analysis of flocculation is given by Fuchs (F3) and by Zebel (Z1, Z2), both of whom arrived at essentially the same result. The flocculation rate can be written in the form

$$-(dn_p/dt)/n_p = k_{em} K_S n_p \tag{16}$$

where $K_S$ is the flocculation coefficient in the absence of charge and $k_{em}$ is a correction factor to allow for both particle charge and particle mean free path. Fuchs and Zebel obtained for $k_{em}$

$$k_{em} = \frac{N_q}{e^{N_q} - 1 + N_{me}} \tag{17}$$

where

$$N_q = \frac{\mathcal{Q}_{p1} \mathcal{Q}_{p2}}{2\pi\varepsilon\delta k T (D_{p1} + D_{p2})} \tag{18}$$

$$N_{me} = N_m N_q e^{N_q} \tag{19}$$

$$N_m = \left[\frac{4(\mathcal{Q}_{p1} + \mathcal{Q}_{p2})}{(D_{p1} + D_{p2})}\right]\left[\left(\frac{4}{3kT}\right)\left(\frac{m_{p1} m_{p2}}{m_{p1} + m_{p2}}\right)\right]^{1/2} \tag{20}$$

$N_q$ is a parameter to allow for charge effects and $N_m$ is a parameter to allow for slip between particles (as distinguished from slip between molecules) in the collision between two particles of different sizes $D_{p1}$ and $D_{p2}$, diffusion coefficients $\mathscr{D}_{p1}$ and $\mathscr{D}_{p2}$, and masses $m_{p1}$ and $m_{p2}$. The term $k_{em}$ is the collision rate correction factor, which measures the combined effect of both of these factors, and **k** is the Boltzmann constant. Both Fuchs and Zebel independently obtained the $N_q$ term in Eq. (17). The $N_{me}$ term is a refinement taken from Zebel.

When dealing with a monodisperse homopolar aerosol (i.e., one in which all particles have the same size and magnitude of charge), the coefficient $K_S$ takes the well-known form

$$K_s = 4k_C \mathbf{k} T/3\mu \tag{21}$$

and

$$N_q = \frac{\pm \mathscr{Q}_p^2}{4\pi\varepsilon\delta \mathbf{k} T D_p} = \frac{\pm \pi\varepsilon\delta D_p^3 \mathscr{E}_{ps}^2}{4\mathbf{k}T} \tag{22}$$

$$N_m = \left(\frac{16 k_C^2 \rho_p \mathbf{k} T}{81 \pi \mu^2 D_p}\right)^{1/2} \tag{23}$$

where the plus sign in $N_q$ applies where unipolar charges are involved (all particles the same sign and repelling each other), and the minus sign applies to the ambipolar case where the particles attract. The term $N_m$ will always be negligible for $D_p > 1$ micron.

The following special cases of Eq. (17) should be noted:

(a) Particles uncharged:

$$k_{em} = \frac{1}{1 + N_m} \tag{24}$$

(b) Large ambipolar charge:

$$k_{em} = |N_q| \tag{25}$$

(c) Large unipolar charge:

$$k_{em} = \frac{N_q}{e^{N_q}(1 + N_q N_m)} \tag{26}$$

In the derivations of the above equations, image forces (which are likely to be small compared to the Coulomb forces) were neglected and every collision was assumed to result in adhesion. Although some have questioned the validity of the latter assumption, there is no evidence to indicate that this is not, at least nominally, the case when dealing with particles smaller than 10 microns colliding at low velocities. While this subject of adhesion has never been investigated systematically, the bulk of all the data on aerosol flocculation and deposition would imply that for these conditions adhesion is nominally complete. Zebel (Z2) also extends the above analysis to computer calculations for actual distributions.

In order to illustrate the quantitative implications of the above equations, Figs. 3–5 have been prepared. Figures 3 and 4 represent a plot of Eq. (16) using Eq. (21) to define $K_S$. For air at 25°C and $k_C = 1$, $K_S$ is $2.98 \times 10^{-10}$ cm³/sec. Figure 3 shows flocculation rate in terms of number concentrations while Fig. 4 shows it in terms of mass concentration. It should be remembered that for the flocculation considered here, the mass concentration will always remain constant; only the number concentrations will change with time because of flocculation. Any change due to cloud expansion, which will cause mass concentrations to change, is a separate effect that would be superimposed on the above. If $k_{em}$ in Figs. 3 and 4 is taken as unity, the figures will essentially give a direct representation of flocculation rates in the absence of electrical charges. It will be noted that at a concentration of $10^6$ particles/cm³, the flocculation rate is only about 2%/min. The common statement that an aerosol must have a concentration of less than $10^6$ particles/cm³ in order to be stable simply means that such an aerosol will be stable for a period of many minutes without significant flocculation.

Figure 5 shows the effect of particle charge on flocculation rate. It is based on Eq. (17), basing $N_q$ on Eq. (22) and for $N_m = 0$. This figure shows that a unipolar charge causes a very rapid reduction in $k_{em}$, and hence in flocculation rate, while the rate of increase in flocculation rate due to ambipolar charge is not as rapid. Figure 5 shows only the relative effect of charge on flocculation rate for various charge levels and particle sizes. It implies a very large relative effect for large particles but tells nothing about the absolute rate directly, since the particle flocculation rate decreases rapidly with particle size for a given mass concentration. The absolute effect can be obtained by considering Figs. 4 and 5 simultaneously.

For ambipolar charging, $k_{em}$ can be approximated by

$$k_{em} = 1 + N_q \tag{27}$$

This will overestimate the value of $k_{em}$ by as much as 30% in the range of $N_q$ between 1 and 4 when $N_m$ is neglected. The difference between this approximation for $k_{em}$ and the exact expression as given by Eq. (17) is best seen by comparing the dashed line in Fig. 5 with the solid line for ambipolar charges at $\mathscr{E}_{ps} = 0.001$. The same spread in the curves would exist at all other values of $\mathscr{E}_{ps}$.

Thus, within this precision, it is apparent from Eq. (16) that the flocculation rate is an additive function of ordinary diffusion or Smoluchowski flocculation and electrostatic attraction. The ordinary diffusion rate is given by Eq. (16) for $k_{em} = 1$, whereas the flocculation rate due to electrostatic

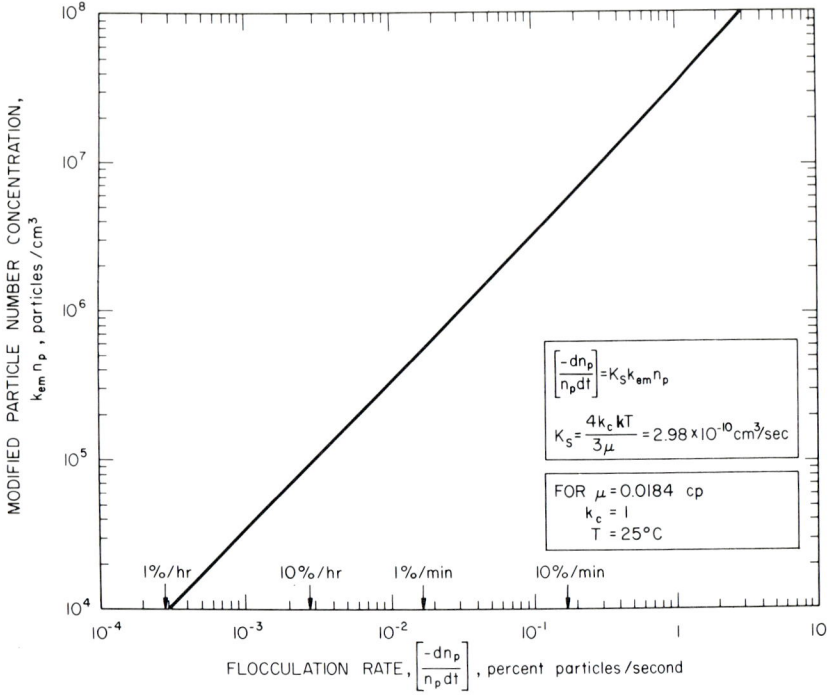

Fig. 3. Effect of particle number concentration on flocculation rate.

effects alone for an ambipolar homopolar monodisperse aerosol is given by

$$\left[\frac{-dn_p}{n_p\,dt}\right]_e = \frac{2k_C \varepsilon \delta c_p \mathscr{E}_{ps}^2}{\mu \rho_p} \tag{28}$$

This result is obtained by substituting the values of $K_S$ and $N_q$ from Eqs. (21) and (22), respectively, in Eq. (16) for $k_{em} = N_q$. It will be noted that this equation is identical to Eq. (9). It should be remembered, however, that Eq. (9) is the expression for the rate of change of concentration due to cloud expansion from net unipolar charging while Eq. (38) gives the rate of flocculation of particles within a given highly charged ambipolar cloud. Nevertheless, this identity permits one to also use part of Fig. 2 to evaluate Eq. (28). It is interesting to note from Fig. 2 that for a concentration of 1000 μg/liter and a particle charge corresponding to $\mathscr{E}_{ps} = 0.3$ V/micron, the flocculation rate is 10 %/sec, which agrees reasonably well with the value obtained from Kunkel [Eq. (15)]. Equation (28) or Fig. 2, however, offers a more quantitative basis for assessing flocculation rate.

For nonconducting equal spheres with equal surface distribution of charge, the following equation can be derived, which will give either (1) the time for

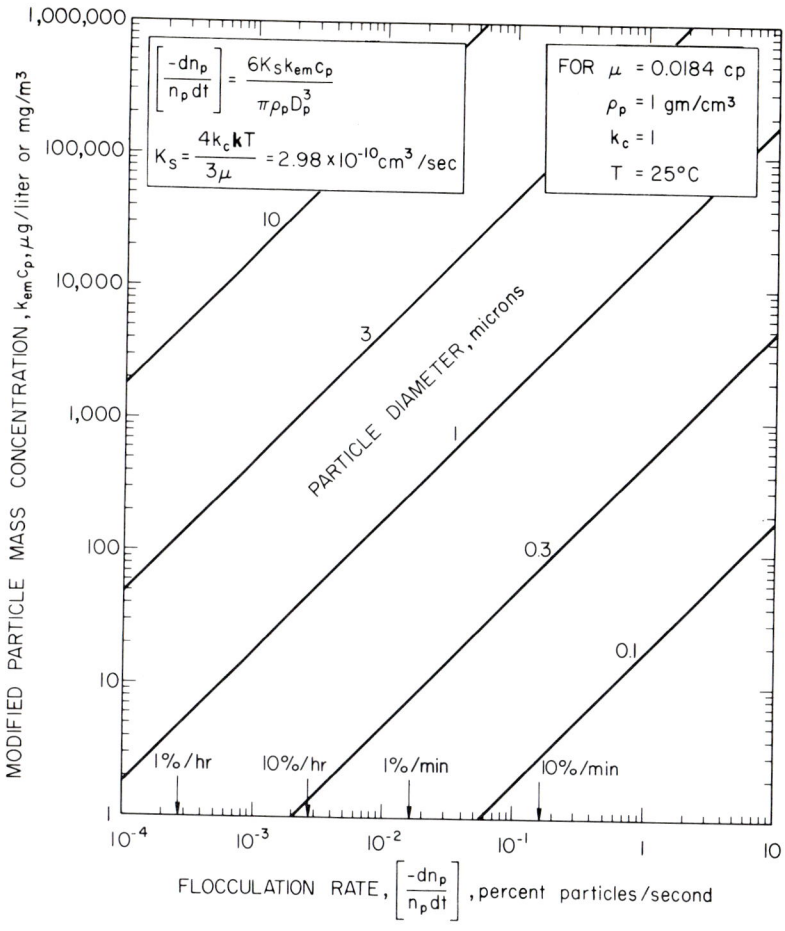

Fig. 4. Effect of particle mass concentration on flocculation rate.

particles of opposite charge to coalesce if they are initially a distance $L$ apart from center to center; (2) the time for particles of the same sign of charge that are initially almost in contact to be repelled to a distance $L$ from center to center

$$t = \left[\frac{2\mu}{k_C \varepsilon \delta \mathscr{E}_{ps}^2}\right]\left[\left(\frac{L}{D_p}\right)^3 - 1\right] \tag{29}$$

In deriving this equation it was assumed that particle acceleration is a negligible factor, which is true if $t$ is large compared to the specific stopping time, $t_{ps}$.

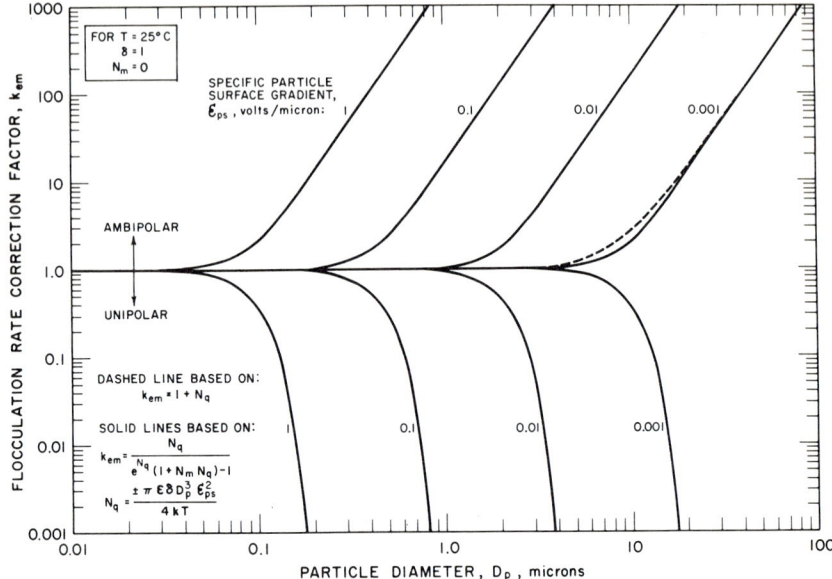

FIG. 5. Effect of particle charge on flocculation rate.

The relationship between particle spacing and mass concentration for particles in a cubical array is given by

$$\left(\frac{L}{D_p}\right) = \left(\frac{\pi \rho_p}{6 c_p}\right)^{1/3} \qquad (30)$$

Table IV indicates the magnitude of particle spacings corresponding to various concentrations assuming a cubical array and a particle density of 1 gm/cm³.

If we calculate the mass concentration, $c_p$, which would correspond to a cubic array of particles with a center–center distance of $L$, Eq. (29) can be written in terms of concentration to yield

$$t = \left(\frac{2\pi}{3}\right)\left[\frac{\mu \rho_p}{2 k_C \varepsilon \delta c_p \mathscr{E}_{ps}^2}\right] \qquad (31)$$

for the case where $L$ is large compared to $D_p$. Thus $t$ is the time for a pair of particles to expand to or coalesce from a spacing corresponding to a concentration $c_p$. As might be expected, this is of the same form as Eq. (9) or (28). The difference is in the proportionality constant, which reflects the fact that Eq. (31) deals with individual particles on a specified path rather than a statistical migration between random particles.

## TABLE IV

### Spacing between Particles for Various Particle Concentrations
(for cubical array and $\rho_p = 1$ gm/cm$^3$)

| Particle concentration $c_p$ | Particle spacing in cubical array $(L/D_p)$ |
|---|---|
| 32.8 lb/ft$^3$ or 0.524 gm/cm$^3$ | 1.000 (particles touching) |
| 1 lb/ft$^3$ gas | 3.20 |
| 1 lb/lb gas or 1 gm/gm gas | 7.62[a] |
| 1 mg/cm$^3$ | 8.06 |
| 1 grain/ft$^3$ | 61.2 |
| 1 gm/meter$^3$ or 1000 μg/liter | 80.6 |
| 1 mg/ft$^3$ | 246.0 |
| 1 grain/1000 ft$^3$ | 612.0 |
| 1 μg/liter or 1 mg/meter$^3$ | 806.0 |

[a] Based on air at 1 atm, 25°C.

## 4. Floc Properties

Having considered the factors that govern the extent or rate of flocculation in an aerosol, it is also important to consider the effects of flocculation on the properties of an aerosol. Consider a large sphere of diameter $D_{p2}$, with its surface covered with small particles of diameter $D_{p1}$. If the adhering particles are no more than one layer thick, it can be shown as a good first approximation that

$$n_{p1} = \left(\frac{\pi}{k_d}\right)\left(\frac{D_{p2}}{D_{p1}}\right)^2 \tag{32}$$

$$D_{pf} = D_{p2}\left[1 + 2\left(\frac{D_{p1}}{D_{p2}}\right)\right] \tag{33}$$

$$D_{pSfa} = D_{p2}\left[1 + \left(\frac{\pi - 2k_d}{2k_d}\right)\left(\frac{D_{p1}}{D_{p2}}\right)\right] \tag{34)[3a]}$$

where $n_{p1}$ is the number of $D_{p1}$ particles adhering to the surface; $D_{pf}$ is the floc diameter; $D_{pSfa}$ is the apparent size of the floc (defined as that diameter of particle having the same density as the individual particle that will have the same Stokes' settling velocity as the floc); and $k_d$ is a distribution factor. The above will be valid only if $n_{p1}$ is very large so that a large fraction of the surface will not be bare, and $D_{p2}$ is very large compared to $D_{p1}$. If the particles

---

[3a] Assumes all particles have the same density.

are arranged on a tight square array, $k_d = 1$. If the adhering small particles are more widely spaced, $k_d$ will be greater than 1. Thus, from Eq. (34) it is apparent that if the large particle is ten times as large as the small ones, the apparent size $D_{pSfa}$ will be only 6 % larger than $D_{p2}$ for an assumed tight square array. If fewer small particles adhere than would correspond to a tight square array, $k_d$ could become greater than $\pi/2$. This would make $D_{pSfa}$ less than $D_{p2}$ and the adhering small particles would be exercising a buoyancy effect on the larger particle (in general, the small particles would serve to reduce the mobility of the large particle). Data are not available by which to assess this effect quantitatively; in any event, the limiting effect will probably be small. The purpose of the above discussion is to point out that even where a large particle is entirely covered with small ones, its settling velocity (or its apparent size, $D_{pSfa}$) will not be increased by a significant factor. The following relates this phenomenon to the electrostatic flocculation between large and small particles.

Consider the deposition of small charged particles ($D_{p1}$) on a large particle ($D_{p2}$) of opposite charge. When the large particle is completely neutralized, the final floc will contain $n_{p1}$ small particles and have a mass, $m_{pf}$, given by

$$n_{p1} = \left(\frac{D_{p2}}{D_{p1}}\right)^2 \left(\frac{\mathscr{E}_{ps2}}{\mathscr{E}_{ps1}}\right) \tag{35}$$

$$m_{pf} = m_{p2}\left[1 + \left(\frac{D_{p1}}{D_{p2}}\right)\left(\frac{\mathscr{E}_{ps2}}{\mathscr{E}_{ps1}}\right)\right] \tag{36}[3a]$$

where $m_{p2}$ is the mass of the large particle, and $\mathscr{E}_{ps1}$ and $\mathscr{E}_{ps2}$ are the corresponding charge levels of each of the respective size of particles. If $\mathscr{E}_{ps1}$ is of the same order as $\mathscr{E}_{ps2}$ (i.e., if the particle charge is essentially proportional to particle surface), which is a likely case, it is apparent that $n_{p1}$ will be of the order $(D_{p2}/D_{p1})^2$. From the previous paragraphs and Eq. (32), it will be observed that this number is not greater than that needed to cover the surface of the large particle. Also, the total mass of material associated with the small particle needed to neutralize the large particle is small compared with the mass of the large particle. Thus, neutralizing a large particle will require a large number but little mass of small particles and will not have a significant effect on the settling velocity of the large particle. Thus, there will be little effect on the aerosol in terms of its settling velocity distribution on a mass basis. On a number basis (which might be of importance in biological warfare applications), the effect would be large.

If the small particles did not deposit as a coating but tended to form chains anchored to the large particle, the physical floc diameter would be effectively very much larger but its effective size from the standpoint of settling velocity would be less. This is a case where the small particles would undoubtedly serve to reduce average particle mobility. The greatest effect of this condition would be in connection with those deposition mechanisms that depend on physical interception. Chainlike particles should be deposited more efficiently in the alveoli of the lungs, but fewer such particles should reach the alveoli. It is not certain what the net effect on alveoli deposition would be, although the total deposition in the respiratory system should be increased.

For flocs of a few particles of the same nominal size, Kunkel (K10) has reported the data given in Table V.

TABLE V

DATA ON EFFECTIVE SIZE OF FLOCS[a]

| Floc arrangement | Number of particles in floc | Ratio $\left(\dfrac{\text{effective floc diameter}}{\text{discrete particle diameter}}\right)$ $(D_{psfa}/D_p)$ |
|---|---|---|
| Particles in line | 1 | 1.02 |
|  | 2 | 1.18 |
|  | 3 | 1.26 |
|  | 4 | 1.28 |
|  | 8 | 1.37 |
| Particles in a plane | 3 | 1.29 |
|  | 7 | 1.54 |
| Clustered in space | 6 | 1.59 |

[a] From Kunkel (K10).

From Kunkel's data it is apparent that a fair number of particles can adhere together without radically changing the effective size. Flocs formed as the result of attraction due to electrical charges cannot be composed of very many particles of the same size because of their tendency to neutralize each other. Where, because of low conductivity, actual neutralization is a slow process, the floc may consist of dipoles. While these dipoles are still capable of attracting particles, the force fields around such flocs become rapidly weaker owing to the net neutralization that results even if localized areas on the floc are both positively and negatively charged.

From the above it is concluded that flocculation due solely to electrostatic attraction cannot have too great an effect in increasing the effective particle size because of the limits inherent in such flocculation.

## 5. Particle Stability

As discussed previously, the presence of an electrostatic charge or force field results in a stress within the material. In the absence of polarization, this stress will always tend to rupture a particle. This tendency to rupture is resisted either by the action of surface tension or by the molecular bonds or tensile strength.

In the absence of polarization effects, if the material can flow, as in a drop, and if the stress is applied slowly enough, an equilibrium is established with surface tension. When any perturbation is applied to this system, vibrations are set up with frequencies given by

$$f^2 = \left(\frac{2n(n-1)}{\pi^2 \rho_p D_p^3}\right)\left[(n+2)\sigma - \frac{\mathscr{D}_p^2}{2\pi^2 \varepsilon D_p^3}\right] ; n \geq 2 \qquad (37)$$

where $f$ is the frequency of the mode of vibration for which the departure of the drop from sphericity is described by the surface zonal harmonic[4] of integral order $n$, and $\sigma$ is the surface tension. When the bracketed term becomes zero or negative, the system becomes unstable and drop breakup ensues. These relationships were first evaluated by Rayleigh (R3) and are summarized by Hogan (H12).

Substituting for $\mathscr{D}_p$ from Eq. (1), this breakup ensues when

$$\mathscr{E}_{ps}^2 > \frac{2(n+2)\sigma}{\varepsilon D_p} \qquad (38)$$

This has come to be known as the Rayleigh instability. The lowest admissible value of $n$ is 2, for which

$$\mathscr{E}_{ps}^2 > \frac{8\sigma}{\varepsilon D_p} \qquad (39)$$

It will be noted that when the outward pressure due to electrostatic stress [Eq. (4)] is made greater than the inward pressure due to surface tension [Eq. (5)], the same relationship as Eq. (39) is obtained. Hogan (H12) also reports that Schneider found that for liquid emitted from the end of a capillary only modes corresponding to odd values of $n$ are permitted. Hence, since 3 is the lowest admissible value of $n$ for this case

$$\mathscr{E}_{ps}^2 > 10\sigma/\varepsilon D_p \qquad (40)$$

---

[4] A surface zonal harmonic of the first kind of order $m$ and argument $z$ is the polynomial of the $m$'th degree in $z$ given by $(1/2^m m!)\,(d^m/dz^m)\,(z^2 - 1)^m$ and usually denoted by $P_m(z)$. In describing waves on a spherical surface these so-called Legendre polynomials fill the part played by trigonometric functions in describing the waves on a stretched string. In such applications $z$ is the ratio to the radius of the perpendicular distance of a point on the spherical surface to an equatorial plane.

Hogan (H12) also writes Eq. (40) in terms of the drop potential

$$E_{ps}^2 > 5\sigma D_p/2\varepsilon \qquad (41)$$

This is valid only if the drop is spherical and is remote from any other object. Actually the drop potential is likely to be sensitive to geometry and Eq. (41) may not represent a very valid conversion.

Values based on Eq. (39) are given in Fig. 6 in terms of $\mathscr{E}_{ps}$ and in Fig. 7 in terms of equivalent charge-to-mass ratio for various values of surface tension. Actually the maximum value of $\mathscr{E}_{ps}$ before breakup, as given by Eq. (39), is independent of pressure. The pressure term has been incorporated in Figs. 6 and 7 in order to permit direct comparison with later developments of instability due to corona. For present purposes, Figs. 6 and 7 will give a correct representation of Eq. (39) if all pressure terms (in the coordinates or in the parameter) are ignored or assumed to be unity.

Experimental data reported by Doyle, Moffett, and Vonnegut (D6) and, more indirectly, by Hendricks (H6) essentially check the order of charge predicted by Eq. (39).

When an electrostatic field is applied so rapidly that flow phenomena cannot occur (or in the case of a solid, which does not flow), breakup may not occur until the electrostatic stress exceeds the tensile strength of the liquid. Schultz and Branson (S2) and Schultz and Wiech (S3) claim that this is the case in their liquid atomization studies. For this case, at breakup,

$$\mathscr{E}_{ps}^2 > 2\tau_y/\varepsilon \qquad (42)$$

Since for liquids $\tau_y$ is of the order of $2 \times 10^9$ dyn/cm³, $\mathscr{E}_{ps}$ must be of the order of 7,000 V/micron by this mechanism. Schultz and Wiech suggest that filaments of liquid may first be formed by Rayleigh instability from which are torn finer drops by local electrostatic stresses exceeding the tensile strength.

The field strengths required for the mechanism suggested by Schultz and Wiech to occur are of the order of those that cause extensive field emission. Their experiments were conducted at high vacuum and such charge levels could not be stable at atmospheric conditions.

## D. On Coalescence

Vonnegut, Moffett, Sliney, and Doyle (V3), Berg, Fernish, and Gaukler (B7), and Lindblad (L7) have confirmed Rayleigh's (R2) original reports that coalescence rates of droplets can be radically increased by means of electrostatic charges. Vonnegut et al. worked with the coalescence of a jet of drops from a needle. They concluded that the field necessary to coalesce the drop increased with an increase in the relaxation time (product of resistivity and

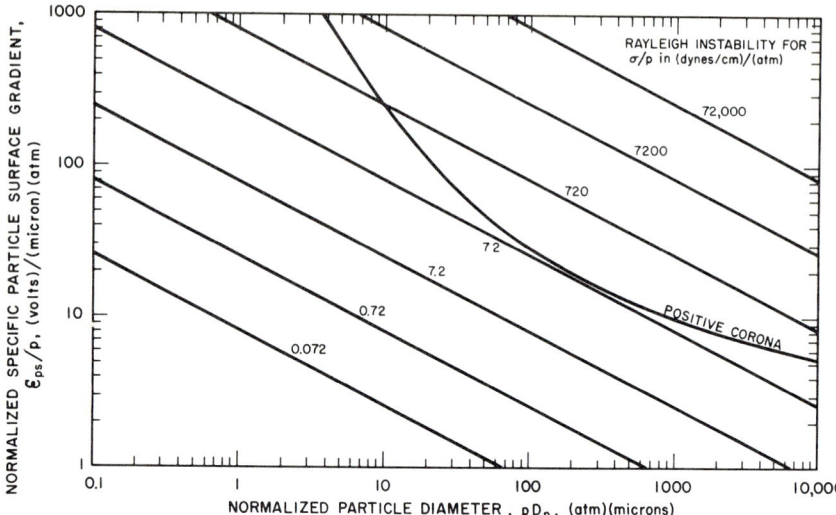

FIG. 6. Maximum stable charge on particles expressed as specific particle surface gradients.

permittivity) of the material. Lindblad, dealing with water drops, found that it took 200–1400 msec to coalesce the drops in the absence of charge, the time being about twice as great at 100 % relative humidity as at 0 %. This coalescence time decreased as voltage was applied across the drops, being 10–20 msec at 1 V. Berg et al. found essentially the same results at 1 V but extended the data to over 10 V with various liquids. At 10 V the coalescence time was only about ½ msec.

Berg, Fernish, and Gaukler (B7) indicated that their data extrapolate back to zero coalescence rate at zero voltage although they did not actually explore the region below 1 V. Presumably on the basis of this extrapolation, Berg (B3) concludes that charging of aerosol particles can increase retention in the respiratory system by many fold because of the increased rate of coalescence. An examination of Berg's data leads us to the conclusion that at zero voltage the coalescence time is considerably greater than 10 msec. However, it is difficult to consider the extrapolation to infinite coalescence time (i.e., zero coalescence) at zero voltage as anything more than a possibility. The data can be extrapolated back to confirm Lindblad's findings just as readily.

The available data on retention in the respiratory tract, as summarized by Mitchell (M6) indicate that upwards of 20% of inhaled aerosols are retained in the respiratory tract, approaching 100% for particles over 5 microns in diameter. The order of retention follows reasonably well what would be expected from the physics of the various deposition mechanisms assuming that once deposited a particle is retained by the surface. This would,

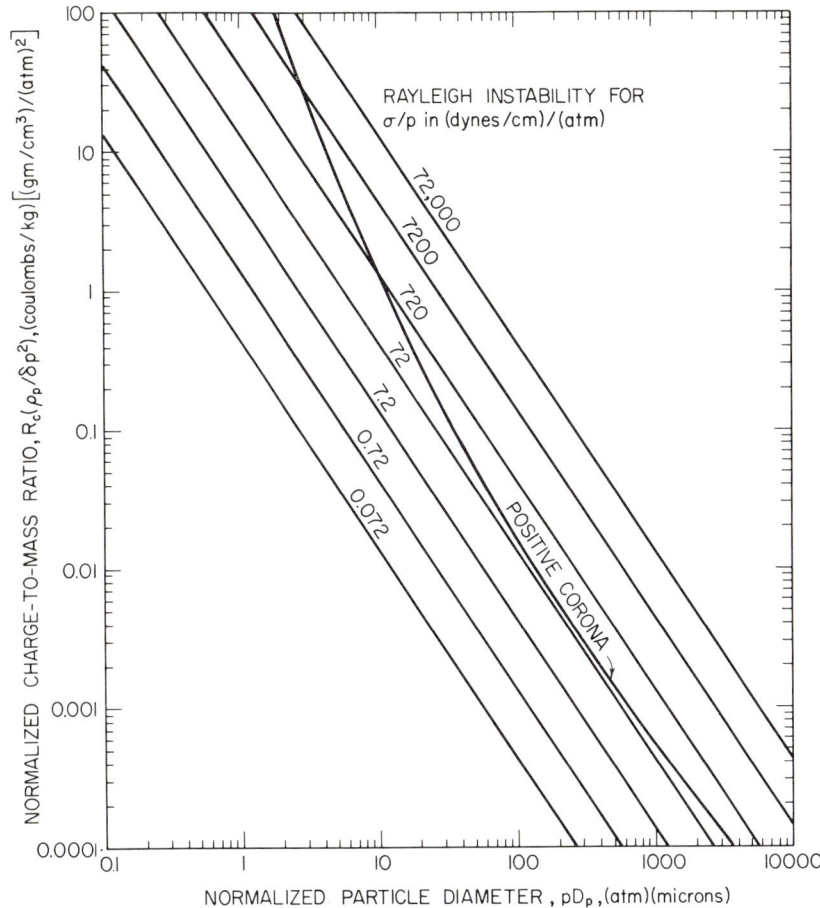

FIG. 7. Maximum stable charge on particles expressed as equivalent charge-to-mass ratio.

therefore, appear to refute Berg's claim that charging is necessary to secure aerosol retention. It can, of course, always be argued that some stray charges are usually present to cause coalescence. This cannot be readily refuted. However, if this is the case, the low level of charge needed would appear to be so universally present that there need be no concern about lack of coalescence under normal conditions.

The above comments on coalescence are aimed specifically at the area of deposition in the respiratory tract and are not intended for universal application. The same comments would probably apply to any case where relatively small particles (<10 microns) are deposited at relatively low velocities

(<1 ft/sec). For other conditions, involving more vigorous contact, it is known that rebound can occur. For example, it was indicated by Wright *et al.* (W6) that the extent of adhesion of 1-micron ammonium bromide crystals was reduced as deposition velocities exceeded 1 ft/sec.

### E. On Evaporation

It has previously been indicated that charges have comparatively little direct effect on the vapor pressure of a material until the particle diameter is substantially less than 1 micron and their effect is toward keeping the vapor pressure closer to that from a plane surface. Consequently, it may be concluded that charge will not significantly affect evaporation rate of a given droplet.

The data of Doyle *et al.* (D6) would indicate that a charged evaporating droplet cannot lose charge except with mechanical loss of material. As a droplet evaporates any charge remains until the charge concentration is high enough to result in Rayleigh instability [Eq. (39)]. At that point there is an emission of fine drops (less than 15 microns diam) from the parent droplet These drops are few in number (perhaps 10 from a 100-micron parent), represent a negligible part (say 3%) of the mass of the parent, but carry an appreciable part (say 30%) of the charge. This might be expected since the ejected fine drops presumably arise at the surface where the charges are concentrated. It seems likely that the ejected droplets have a charge level close to the maximum permitted by Rayleigh instability. In that case, they should almost immediately eject another series of drops as they in turn evaporate.

Thus while the presence of electrostatic charge will tend to hasten the evaporation process because of the extended surface provided by the ejected drops, this is probably not a major consideration since the initial stage of evaporation of a droplet is actually the one that limits the over-all rate. The concentration of charge within the drop, however, means that evaporation of a drop can ultimately yield very highly charged small particles of residual nonvolatile soluble or suspended contaminants.

### F. On Deposition

A number of investigators have indicated increased deposition rates due to electrostatic charges. LaMer *et al.* (L2) and Gillespie (see Richardson, R5) were concerned with aerosol filtration. Göhlich (G5), using corona charging of air suspensions of powders, reports two- to threefold increases in field deposition rates as a result of the charges. The greatest improvements, however, are reported by Vonnegut *et al.* (V3) who reported nine- to 83-fold increases in deposition on surfaces as the result of electrostatic charges. In this study they used both dc and ac corona charging of DOP (dioctylphthalate) particles. Although they indicate that an ac charged cloud should be neutral, they do not report any check of this point. With ac corona there is usually some net rectification.

Ambipolar charges can give increased deposition but such deposition depends for the most part on image effects. In this sense the rate of deposition will be comparable with the rate of flocculation of ambipolar aerosols. Improved deposition rates by such means are likely to be small.

With unipolar (or net unipolar) clouds, however, the rate of deposition will be essentially the same as the rate of expansion of the edge of the cloud. In fact, with very large conductors or grounded objects, because of image forces (which will reflect those of the entire cloud rather than just those of an individual particle), the rate of deposition would be expected to be just twice as fast as the rate of expansion of the edge of the cloud. Thus, since the deposition rate will be the product of velocity and concentration, one may use Eqs. (9) and (11) to predict the following deposition rate from a net unipolar charged cloud:

$$G_{de} = 2k_C \varepsilon \delta_v c_p^2 \mathscr{E}_{ps}^2 D/(3\mu\rho_p) \tag{43}$$

where $G_{de}$ is the mass deposition rate per unit of surface area exposed to the cloud. The factor of 2, to allow for image forces, has been incorporated in Eq. (43). Although $D$ was originally defined as cloud diameter, this assumes a free cloud. Therefore, in applying Eq. (43), $D$ will be essentially the apparent cloud size that the surface sees. For example, the top surface of the top leaves on a plant in the cloud will see the entire cloud. The under surfaces will see only that cloud between them and the next layer of leaves, so that $D$ for that case will be radically reduced. This results because one layer of leaves will electrically shield the adjacent lower layer from the cloud above. Similarly, the deposition on top of a man's head may be determined by the whole cloud diameter, whereas the deposition on his side will correspond to an effective diameter more nearly approximating his height. These are only approximations since the actual effects will depend on the field intensities as determined by the specific geometries involved. This can become an exceedingly complex estimate to make.

In order to provide some basis for comparison, the deposition rate per unit of surface area due to gravity settling is given by

$$G_{dg} = g(\rho_p - \rho)k_C c_p D_p^2/(18\mu) \tag{44}$$

The ratio $G_{de}/G_{dg}$ is, therefore,

$$\frac{G_{de}}{G_{dg}} = \frac{12\varepsilon\delta_v c_p \mathscr{E}_{ps}^2 D}{g\rho_p(\rho_p - \rho)D_p^2} \tag{45}$$

For $D = 3$ in., $c_p = 1000$ μg/liter, $\mathscr{E}_{ps} = 1$ V/micron, $D_p = 10$ microns, $\rho_p = 1$ gm/cm³, and $\delta_v = 1$, the ratio $G_{de}/G_{dg}$ will be 8.25. Thus, even for such a small exposed charged cloud, deposition by electrostatic means will be almost an order of magnitude greater than by gravity. The electrostatic mechanism will also result in deposition on the under side of the surface.

A detailed analysis of the various mechanisms of electrostatic deposition has been presented by Kraemer (K6), Kraemer and Johnstone (K7), and Dawkins (D3).

## G. On Adhesion

Corn (C7) has given a recent review of adhesive forces between particles in which he concludes that inadequate data on static electrification are available to arrive at any definite conclusion as to the relative role of electrostatics in particle adhesion. A detailed evaluation of the analysis presented by Russell (R9), however, sheds considerable light on the subject.

### 1. *Electrostatic Force between Charged Particles*

Charged particles will exert a force on each other that depends on the magnitude of the charge and on the distance between particles. This force is given by the simple Coulomb law [Eq. (2)] for point charges or for charged particles that are far apart compared with their diameter. When particles are brought close together, however, the nature of the force depends on the nature of the particle as well. On particles of perfect insulators the charges will remain in their respective locations on the particle regardless of the proximity of another charged particle. With conductors, on the other hand, the charges can shift because of induction effects. The charges on a conducting particle will move relative to the center of mass of that particle, toward those of the opposite sign on a proximate particle. This will result in a greater force of attraction between the particles than would be calculated from the simple Coulomb law without allowing for migration of the charges within the particles. Where the proximate particle has charges of the same sign, the charges will move further apart resulting in a lower repulsive force than indicated by Coulomb's law naively applied. The force fields around and the fields between spherical bodies have been extensively treated by Russell (R9). The force between particles can be expressed by a modified form of the Coulomb law. For conducting spheres of equal size and equal but opposite charge, the force of attraction is given by

$$F_A = k_{sA}\left[\frac{\mathcal{Q}_p^2}{4\pi\varepsilon\delta L^2}\right] = k_{sA}\left[\frac{\pi\varepsilon\delta\mathscr{E}_{ps}^2 D_p^2}{4(L/D_p)^2}\right] \quad (46)$$

where $\mathcal{Q}_p$ is the magnitude of charge on each particle; $L$ is the distance between centers; and $k_{sA}$ is a factor to allow for any inductive shifting of charges. This factor is a function of the relative distance between the particles $[L/D_p$ or $L_s/D_p$, where $L_s$ is the clearance between particles and is equal to $(L - D_p)]$.

The maximum field intensity around such particles will exist at the surface of each particle at the point of intersection of their line of centers and may be expressed as

$$\mathscr{E}_{max} = k_{s\mathscr{E}} \mathscr{E}_{ps} = k_{sE}(E/L_s) \tag{47}$$

where $\mathscr{E}_{ps}$ is a measure of particle charge as defined by Eq. (1), $E$ is the potential difference between the spheres, and $k_{s\mathscr{E}}$ and $k_{sE}$ are functions of $L/D_p$ or $L_s/D_p$.

The capacitance between such a pair of spheres is given by

$$\mathscr{C} = \mathcal{Q}_p/E = 2\pi\varepsilon k_{s\mathscr{C}} D_p \tag{48}$$

For conducting spheres of equal size and equal and like charge, the force of repulsion is given by

$$F_R = k_{sR}\left[\frac{\mathcal{Q}_p^2}{4\pi\varepsilon\delta L^2}\right] = k_{sR}\left[\frac{\pi\varepsilon\delta\mathscr{E}_{ps}^2 D_p^2}{4(L/D_p)^2}\right] \tag{49}$$

where $k_{sR}$ is a function of $L/D_p$ or $L_s/D_p$.

Table VI shows values of $k_{sA}$, $k_{sR}$, $k_{s\mathscr{E}}$, $k_{sE}$, and $k_{s\mathscr{C}}$ as given by Russell (R9) for various particle spacings.

Equations (46) and (49) would also hold for two equal spheres that are perfect insulators. If the charge is initially uniformly distributed within the sphere or on its surface, the value of $k_{sA}$ or $k_{sR}$ for such insulator spheres would be unity regardless of spacing since the charges cannot migrate. This, however, assumes that no polarization has taken place.

Equations (46) and (47) may be combined with the definition of Eq. (1) to yield

$$F_A = k_{F\mathscr{E}}\left[\frac{\pi\varepsilon\delta D_p^2 \mathscr{E}_{max}^2}{4}\right] \tag{50}$$

where

$$k_{F\mathscr{E}} = \left[\frac{k_{sA}}{k_{s\mathscr{E}}^2[1 + (L_s/D_p)]^2}\right] \tag{51}$$

This gives the relationship between the attractive force between two particles and the maximum surface field intensity. There is a limit to the magnitude that $\mathscr{E}_{max}$ can reach without incurring some type of discharge or emission. From data such as those of Earhart (E1) one would conclude that $\mathscr{E}_{max}$ cannot exceed some 100 to 200 V/micron before the particle would discharge

## TABLE VI

Factors for Forces, Field Intensity, and Capacity between Equal-Sized, Equally Charged, Conducting Spheres[a]

| $2L_s/D_p$ | $2L/D_p$ | $1+(L_s/D_p)$ | $k_{sA}$ | $k_{sR}$ | $k_{s\mathscr{E}}$ | $k_{sE}$ | $k_{s\mathscr{C}}$ | $k_{F\mathscr{E}}$ |
|---|---|---|---|---|---|---|---|---|
| $10^{-7}$ | $2.0_61$ | 1.000 | $5.77 \times 10^5$ | 0.615 | $2.140 \times 10^6$ | 1.000 | 4.6647 | $1.260 \times 10^{-7}$ |
| $10^{-6}$ | $2.0_51$ | 1.000 | $2.99 \times 10^4$ | 0.615 | $2.442 \times 10^5$ | 1.000 | 4.0891 | $5.01 \times 10^{-7}$ |
| $10^{-5}$ | $2.0_41$ | 1.000 | $4.05 \times 10^3$ | 0.615 | $2.845 \times 10^4$ | 1.000 | 3.5134 | $5.00 \times 10^{-6}$ |
| $10^{-4}$ | $2.0_31$ | 1.000 | $5.79 \times 10^2$ | 0.615 | $3.400 \times 10^3$ | 1.000 | 2.9378 | $5.01 \times 10^{-5}$ |
| $10^{-3}$ | $2.0_21$ | 1.000 | 91.1 | 0.615 | 423. | 1.000 | 2.3626 | $5.09 \times 10^{-4}$ |
| $10^{-2}$ | 2.01 | 1.005 | 15.60 | 0.618 | 56.1 | 1.003 | 1.7896 | $4.91 \times 10^{-3}$ |
| $10^{-1}$ | 2.10 | 1.050 | 3.355 | 0.643 | 8.38 | 1.034 | 1.2237 | $4.34 \times 10^{-2}$ |
| 0.5 | 2.5 | 1.250 | 1.507 | 0.757 | 2.640 | 1.173 | 0.8892 | 0.1381 |
| 1 | 3 | 1.500 | 1.213 | 0.850 | 1.770 | 1.359 | 0.7677 | 0.1722 |
| 2 | 4 | 2.000 | 1.072 | 0.935 | 1.320 | 1.770 | 0.6705 | 0.1538 |
| 5 | 7 | 3.500 | 1.012 | 0.990 | 1.080 | 3.151 | 0.5836 | 0.0709 |
| 8 | 10 | 5.000 | 1.004 | 0.996 | 1.033 | 4.604 | 0.5556 | 0.0378 |
| 10 | 12 | 6.000 | 1.003 | 0.997 | 1.026 | 5.587 | 0.5452 | 0.02645 |
| 100 | 102 | 51.000 | 1.000 | 1.000 | 1.000 | 50.51 | 0.5049 | 0.000384 |

[a] From Russell (R9).

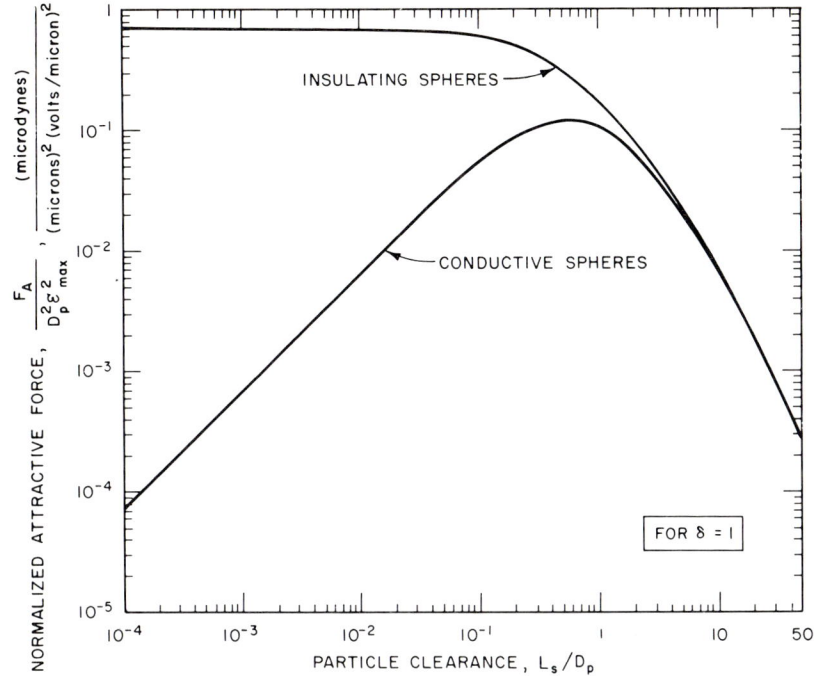

FIG. 8. Attractive force between equal-sized, equally and oppositely charged spheres at various particle clearances.

itself in air at 1 atm. Equation (50) is useful, therefore, for calculating the maximum possible force that can exist between two particles. The various factors grouped inside the last bracket are all a function of the clearance between particles and the value of this group, $k_{F\varepsilon}$, is given in Table VI. Since $k_{F\varepsilon}$ is a maximum at a clearance of approximately one particle radius, the maximum force of attraction will exist at this spacing. As the particles move closer together, the particles will lose charge so that the force of attraction will actually become less. This assumes, of course, that there is actually a limiting field intensity, a fact which is indicated but has not been proven.

Equation (50) may also be used to calculate the attractive force between equal sized, equally and oppositely charged spheres that are perfect insulators. For this condition, the value of $k_{sA}$ and of $k_{s\varepsilon}$ is taken as unity if polarization possibilities are neglected, and if the charge is initially uniformly distributed. Any polarization will tend toward an approach to the conductor condition (which basically represents a condition of infinite polarization). Figure 8 presents a plot of Eq. (50). Thus, if we assume that the maximum possible field intensity is 200 V/micron, the attractive force between equally but oppositely charged, conducting spheres of 1 micron cannot exceed 0.0048 dyn. This force would occur when the particles are approximately ½ micron apart provided that the particles had an initial charge at least corresponding to a value of $\mathscr{E}_{ps}$ of 113 V/micron [as calculated from Eq. (47)]. Any charge in excess of this would be lost as the particles approach each other. As the particles continue to come closer together than ½ micron, additional charge will be lost and the attractive force gradually drops to zero.

For the same conditions with insulating spheres the maximum attractive force would be 0.028 dyn and would occur when the particles are touching. It might be argued that because of localized regions of very high charge an even higher attractive force could result. This, however, is unlikely because of the loss of charge that would take place at those spots. Where $\mathscr{E}_{ps}$ is less than would correspond to the limiting critical field strength, however, such localized concentrations could result in higher attractive forces than would be calculated on the basis of the average $\mathscr{E}_{ps}$ value. It should be noted that with a perfect insulator there will also be a charge loss in the region of closest proximity. However, this will be confined to a relatively narrow region and will not be a major factor in the total charge left on the particles. The exact solution for this case would be quite complex.

Table VII gives both the electrostatic field intensity and force due to a point charge corresponding to one electron in magnitude. It is apparent from these values that elementary point charges do not exert significant forces upon each other until the distance between them approaches molecular dimensions. At this point, field intensities become so high that electron transfer or tunneling must occur for the case where the point charges are of opposite sign.

The presence of insulating films over the surface of charged conducting particles would tend to have relatively little effect, since the charges would

TABLE VII

FORCE AND FIELD INTENSITY DUE TO SINGLE ELECTRON POINT CHARGE

| Radial distance from point charge | | Field intensity due to point charge (V/cm) | Force on another point charge of 1 electron ($\mu$dyn) |
|---|---|---|---|
| (Å) | (micron) | | |
| 1 | 0.0001 | $1.437 \times 10^9$ | $2.300 \times 10^3$ |
| 10 | 0.001 | $1.437 \times 10^7$ | $2.300 \times 10^1$ |
| 100 | 0.01 | $1.437 \times 10^5$ | $2.300 \times 10^{-1}$ |
| 1,000 | 0.1 | $1.437 \times 10^3$ | $2.300 \times 10^{-3}$ |
| 10,000 | 1 | $1.437 \times 10^1$ | $2.300 \times 10^{-5}$ |

still be free to migrate and leak off at the point of closest proximity. They might, however, alter somewhat the critical value of field intensity at which charges leak off, since this field intensity may be a function of the surface material. It should also be noted that, since no materials are perfect insulators, the above discussion is strictly relative to the time periods under consideration. The distinction between insulators and conductors can be drawn on the basis of the relaxation time of the material relative to the time period of importance as discussed previously.

From the above discussion it is apparent that while charge can cause particles to come together, it cannot be a significant factor in determining the force of adhesion with *conductive* particles. With perfect insulators, very high charge levels (corresponding to $\mathscr{E}_{ps} > 100$ V/micron) can result in an effective force of adhesion.

## 2. Comparison with Other Forces

In order to illustrate further the magnitude of electrostatic forces in relation to other forces, Fig. 9 has been prepared. In this figure the forces due to various mechanisms have been depicted as a function of particle size. The equations that are plotted in Fig. 9 are given in Table VIII. In order to simplify the presentation, the various forces have been presented for specific conditions as indicated in the legend. An attempt has been made to select conditions for which each particular mechanism will exert a maximum effect. In doing this within the bounds of realism, an oversimplification may have resulted. Hence before considering the figure in general some specific explanations are in order.

Some of the forces presented are actually pseudo- or equivalent forces. This is true in the case of diffusion. The force formula given is the equivalent force that would give a particle the same migration velocity toward a surface as it experiences by diffusion when in close proximity to the surface.

Of adhesive forces other than electrostatic, only van der Waal's forces have been presented. Jordan (J1), in discussing the work of Bradley (B11) and others, indicates that the force of attraction between two particles of diameter $D_{p1}$ and $D_{p2}$ may be expressed as

$$F_A = K_W D_{p1} D_{p2}/(D_{p1} + D_{p2}) \qquad (52)$$

where $K_W$ is a constant of the material with a magnitude of the order of 200 dyn/cm (0.2 N/meter). For equal-sized particles this becomes

$$F_A = (K_W/2) D_p \qquad (53)$$

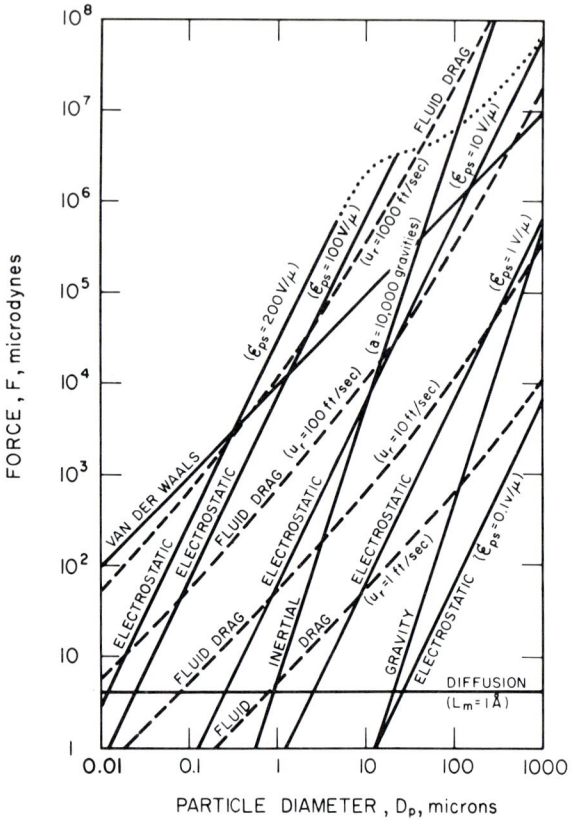

FIG. 9. Comparison of forces arising from various mechanisms.

Corn (C7), however, points out that adsorbed films or gas layers may exist between the particles and may alter the nature of the adhesive force. Corn also indicates that various investigators have derived a value of the order of $4\pi\sigma$ for $K_W$ when adsorbed liquid films are involved. Bradley (B11) has derived an identical expression where $\sigma$ is defined as the surface energy of the solid. All of these forms yield values of $K_W$ of the same general magnitude. There are, however, other reports (Fuchs, F4, pp. 363, 373) that indicate adhesive forces between particles as much as two orders of magnitude smaller than these.

For the forces due to fluid drag (shear and pressure), the maximum velocity has arbitrarily been taken as 1000 ft/sec. This, being essentially sonic velocity in room air, is the largest velocity that will normally be realized. However, for all velocities over say 300 ft/sec, compressibility effects will become significant. These are not allowed for in the equations and to this

## TABLE VIII

### Equations Plotted in Fig. 9

| Type force | Force equation | Values of constants and physical properties used in preparing plot |
|---|---|---|
| Earth's gravity, less buoyancy of fluid | $F = \pi g \rho_p D_p^3 [(\rho_p - \rho)/\rho_p]/6$ | $\rho_p = 1.00$ gm/cm$^3$ (density of particle)<br>$\rho = 1.185 \times 10^{-3}$ gm/cm$^3$ (density of air) |
| Inertial force to give acceleration $a$ | $F = \pi a \rho_p D_p^3/6$ | $\rho_p = 1.00$ gm/cm$^3$<br>$a = 10^4$ gravities |
| Total fluid drag at relative velocity $u_r$ | $F = \pi \rho C_D u_r^2 D_p^2 / 8$<br>where<br>$C_D = \psi(N_{Re_p})$<br>$= 24/N_{Re_p}$<br>if $N_{Re_p} < 0.2$<br>$= 0.44$ if $N_{Re_p} > 1000$ | Air at 25°C, 1 atm;<br>$\rho = 1.185 \times 10^{-3}$ gm/cm$^3$<br>$\mu = 0.0184$ cp<br>$u_r = 1, 10, 100,$ and $1000$ ft/sec (as indicated on curves) |
| Diffusion—equivalent force causing migration over distance $L_m$ | $F = kT/L_m$ | $T = 25°C$ (298°K)<br>$L_m = 1$Å ($10^{-8}$ cm) |
| van der Waals attraction between two equal spheres | $F = \tfrac{1}{2} K_W D_p$ | $K_W = 200$ dyn/cm |
| Electrostatic attraction of two equal spherical insulators | $F = \dfrac{\pi \varepsilon \delta \mathscr{E}_{ps}^2 D_p^2}{4(L/D_p)^2}$ | $\delta = 1$<br>$(L/D_p) = 1$ (particles touching)<br>$\mathscr{E}_{ps} = 0.1, 1, 10, 100$ and $200$ V/micron (as indicated on curves) |

extent the values plotted are fictitious, although still of the correct order of magnitude. The curves for forces due to fluid drag are based on data for the drag coefficient, $C_D$, for spherical particles in infinite space. In the case of electrostatic forces, as will be shown later, the limiting surface field strength that a particle can tolerate without ionizing air becomes less as size increases. This limit is indicated in a very approximate way by the dotted curve in Fig. 9, although such limits are not actually known with any precision.

It will be noted in Fig. 9 that diffusional forces are small compared with other forces for particles smaller than 1000 microns in diameter. Inertial forces can become significant with particles larger than 10 microns. Fluid drag forces can be of the same general order as van der Waal's forces at these assumed high velocity levels. Electrostatic forces can be quite large but only if relatively high charge levels are involved. If the charge level is as low as would correspond to $\mathscr{E}_{ps} = 1$ V/micron, the forces begin to approach those due to the earth's gravitational field.

It is doubtful whether electrostatic charges can play a significant role in adhesion under any but the most unusual circumstances. Although high force levels are shown in Fig. 9, there is considerable question as to whether the field intensities indicated can be maintained for any but the best insulators. It should also be remembered that conductive particles can show no adhesion due to electrostatic charges unless they are covered with an insulating film, and even then the magnitude of the possible force is very much less than with insulating particles. The subjects of switchgear arcing and electrical contact wear are areas that have not been covered in this study and may contain some relevant data.

### H. On Biological Processes

No attempt has been made in this chapter to cover the biological or medical effects of electrostatic charges. It should be noted, however, that Gordieyeff (G6) gives a bibliography of forty-two references, a large number of which deal with the medical aspects.

## IV. Electrostatic Dissemination Techniques

Electrostatic forces can be utilized for the direct dissemination of materials, both liquid and solid. The dissemination of liquids by this means (electrostatic atomization) has been studied extensively. The corresponding dissemination of powders is referred to in the literature but has not received much attention in depth. Dissemination of bulk solids (i.e., consolidated solids as distinguished from powders) by electrostatic means has not been studied and, although it may be technically possible, is probably not feasible.

### A. Electrostatic Atomization of Liquids

A large part of the interest in electrostatic atomization has arisen from two fields of interest: combustion of fuel oils and ion propulsion of rockets in space. Work in the latter area has been concerned primarily with atomization at high vacuum ($10^{-8}$ atm) as in the work of Cohen (C5), Hogan (H12, 13), Hendricks (H6), Schultz and Branson (S2), and Schultz and Wiech (S3). Graf (G8), Matthews and Mason (M3), Peskin and Raco (P3), Randall, Marshall, and Tschernitz (R1), and Vonnegut and Neubauer (V4) have been concerned primarily with atomization at atmospheric pressure. Most of the investigators claim that atomization is achieved as the result of Rayleigh instability. However, Schultz *et al.* claim that the ultimate atomization in their case is due to rupturing molecular bonds because of the high discharge rates involved in their tests.

The data in general seem to indicate that greater conductivity (lower resistivity) is conducive to better atomization. The failure to obtain atomization at very high resistivities in the type of apparatus used by Drozin (D7) and Randall *et al.* (R1) is easily understood since the source of high potential is far removed from the point of atomization. The poor atomization reported for liquids of very high conductivity is more difficult to understand, however. It also seems clear that there are two general modes of atomization. The first is due to Rayleigh instability, which results in filament formation and probably is capable of giving only relatively coarse droplets (> 10 microns). The other involves a more intensive atomization for which the mechanism is not quite clear. It may involve an intensification of the Rayleigh instability, corona, or breaking of molecular bonds as suggested by Schultz and Branson (S2). It must be concluded that, while guidelines have been established for electrostatic atomization, the principles are far from understood and that at present there is no means of predicting atomization performance either from the standpoint of the charge level or the size attained by the drops formed. The rest of this section is devoted to a brief resumé of the data generated during some of the more important investigations in this area.

The high vacuum work was aimed primarily at achieving relatively high charge-to-mass ratios. These investigations were essentially all made with flow of liquid out of a fine charged capillary or over a fine charged needle. Hendricks (H6), atomizing Octoil (dioctyl phthalate), obtained a range of charge-to-mass ratios of 0.01 to 5 C/kg when producing particles in the size range of 0.2–10 microns. His highest charge levels corresponded to $\mathscr{E}_{ps}$ values of about 80 V/micron. Schultz and Branson (S2) and Schultz and Wiech (S3) claim that by employing an electron emitter arranged to direct electrons at the liquid to be atomized, much higher charge levels (up to 300 C/kg) could be achieved. They report results of 80–130 C/kg when atomizing dioctyl phthalate from a fine positively charged needle point. They attribute the improvement over Hendricks' results to high charge emission rates made possible by the electron emitter, thereby overcoming liquid tensile strength rather than simply surface tension. Cohen (C5) obtained charge levels of 100–1000 C/kg when using an atomizing needle of either positive or negative polarity in the absence of any emitter. Cohen, however, did use a liquid (glycerol) doped with antimony chloride or sulfuric acid to make it conductive. Schultz and Branson (S2) report that for resistivities greater than $10^{14}$ Ω · cm, the liquid cannot be charged except by either liquid breakdown or corona. Neither Schultz and Branson nor Cohen report any data on particle size.

Hogan (H12) obtained $R_c$ values of 0.01–10 C/kg. With dibutyl phthalate, however, he reports a mode of atomization in which no liquid jets could be seen with a microscope and for which $R_c$ values of $10^5$ C/kg were measured, corresponding to singly ionized dimer molecules. No particles could be formed with silicone pump oil, whereas charges of 200–400 C/kg were obtained with suspensions of either Cab-O-Sil[4a] or carbon in glycerine.

In summary, the data on high vacuum atomization have in general dealt with charge levels much higher than those of interest in dissemination and have given little attention to the question of particle size. Hogan (H12) does give a good summary of theoretical considerations involved.

Peskin and Raco (P3) have given a theoretical analysis of both ultrasonic and electrostatic atomization from the point of view of liquid instability. They conclude that atomization with low frequency ac will require about twice the field strength as dc but that, by going to high frequency, lower fields are possible with conducting liquids. The value for the critical field for atomization given by these authors for a dc field is, however, smaller than that which would be calculated from Eq. (39) by a factor of $(1/32)^{1/2}$. This presumably reflects the simplified one-dimensional model used in their derivation.

Matthews and Mason (M3) consider the rupture of a large droplet due to aerodynamic forces both with and without the presence of an external electrostatic field. With the field present, the charge on the drop fragments was of the order of that which might be expected by induction from the field, and the charge also reversed sign with the field. The particle charge attained with the external field was also some two orders of magnitude larger than any charge attained in the absence of the field. This study was aimed primarily at explaining charge buildup in thunderstorms.

Graf (G8) measured the pressure buildup due to electrostatic charging of stationary drops. He found this pressure to be some 30–60% of the value given by Eq. (4), independent of polarity of charge or dielectric constant of the liquid. This factor, which he terms a charging factor, was a function of both geometry and liquid properties. It may also reflect polarization effects that are neglected in Eq. (4).

Graf (G8) also briefly investigated atomization. Charge-to-mass ratios of the order of $10^{-3}$ C/kg were obtained with oils having resistivities of $10^8$–$10^{12}$ $\Omega \cdot$ cm. No particle size data are given but he reports better atomization and higher charging rates as the resistivity of the liquid becomes less.

Drozin (D7) presents an analysis corresponding to Eq. (4) for a polarizable liquid in the absence of any surface charge. However, he expresses the electrostatic pressure in terms of the electric field inside the drop rather than

---

[4a] A very fine silica marketed by Cabot Corp.

in terms of that external to the drop. As a result, his theoretical analysis shows an apparent marked effect of dielectric constant of the liquid. If the conversion is made, dielectric constant is found to have relatively little effect as reported by most investigators. Using a buret-type atomizer, Drozin reports four stages of atomization as the field intensity is increased: (1) formation of large uniform drops with low or no fields; (2) formation of liquid filaments; (3) eruption of fine drops from the filaments; and (4) gradual shortening of filaments and replacement of them by a cloud of fine droplets. When a complete cloud of fine droplets only formed, it exhibited higher order Tyndall spectra, implying a relatively uniform size distribution. He also reports that, with a grounded electrode beneath his atomizer, voltages in excess of the condition resulting in a complete cloud caused a traversal of the above atomization stages in reverse order. Drozin claims that the fine particle cloud condition can only be achieved when the liquid is positively charged. It is not clear whether this represents his own observation or is a conclusion based on the data of Vonnegut and Neubauer (V4) who reported that a "smoke" could be obtained only with a positively charged liquid and only with liquids of very low conductivity (distilled water, lubrication oil, alcohol). Drozin also concludes that atomization is possible only if resistivity is between $10^5$ and $10^{13}$ $\Omega \cdot$ cm; liquids that are more conductive (such as salts and acids) or less conductive (such as xylene or carbon tetrachloride) could not be dispersed even with potentials up to 30 kV. Drozin also concludes that liquids with dipole moments smaller than $10^{-18}$ $dyn^{1/2} \cdot cm^2$ cannot be dispersed, in agreement with the conclusions of Straubel (S12).

Vonnegut, Moffett, Sliney, and Doyle (V3) in general confirm the modes of atomization reported by Drozin. In their case the drops were mostly greater than 100 microns except for the smoke or fine-particle cloud condition. They also report that the smoke was obtainable only with positive polarity at high voltage and that very conductive water (resistivity less than $10^3$ $\Omega \cdot$ cm) or poorly conductive materials (resistivity greater than $10^{13}$ $\Omega \cdot$ cm) would not form smoke. They also report that drop formation either becomes erratic or ceases when corona ensues.

Randall et al. (R1) examined some 60 liquids with a buret-type atomizer and positive dc voltage. They also concluded that atomization could not be achieved if the resistivity was greater than $10^{10}$ $\Omega \cdot$ cm and became much poorer if the resistivity was less than $10^5$ $\Omega \cdot$ cm. Increased viscosity resulted in a coarser spray, especially for viscosities above 20 cp. Measurements of size distribution are reported which indicate mass median diameters larger than 20 microns. An optimum voltage of 25–40 kV was found at which the mass median diameter was a minimum. They also ran limited tests with glass capillary atomizers (<200 microns i.d.) for which they estimated droplet sizes were less than 10 microns when atomization was achieved. They give a semiquantitative description of the type of atomization achieved with each type of atomizer at various voltages and with various liquids.

Shaffer (S5) also made some exploratory evaluations of the electrostatic atomization of dibutyl phthalate using a camel's hair brush for the atomizing nozzle. On a count basis 73% of the particles were smaller than 10 microns and the largest particle obtained was 40 microns. The energy input corresponded to 0.5 cal/g liquid atomized (0.00026 kWh/lb) and the charge level on the particles as atomized corresponded to a value of $\mathscr{E}_{ps}$ of the order of 3–5 V/micron. Current and flow rate measurements reported by Vonnegut and Neubauer (V4) would correspond to an energy input of 0.1 kWh/lb.

As a criterion for determining the size droplets formed by electrostatic atomization, Vonnegut and Neubauer (V4) postulated that the total energy of the system should be a minimum at equilibrium. On this basis they concluded that the most probable droplet diameter would be given by

$$D_p = (72\varepsilon V^2 \sigma / \mathscr{Q}^2)^{1/3} \tag{54}$$

where $V$ is the volume of liquid, having a surface tension $\sigma$ and charge $\mathscr{Q}$, that is atomized. This can also be written in terms of charge-to-mass ratio, $R_c$,

$$D_p = (72\varepsilon\sigma / \rho_p^2 R_c^2)^{1/3} \tag{55}$$

Both Hogan (H12) and Ryce (R10) have discussed this relationship. However, it has not been recognized that if Eq. (55) applies to an atomized jet it must also apply to a droplet that is evaporating. When this drop reaches a critical charge level, as given by Eq. (39), it will atomize as reported by Doyle *et al.* (D6). This critical charge level, expressing Eq. (39) in terms of charge-to-mass ratio, is given by

$$R_c^2 = [288\varepsilon\sigma / \rho_p^2 D_{p0}^3] \tag{56}$$

where $D_{p0}$ is the diameter of the evaporating drop before it atomizes. Since, if the whole drop atomizes, the final charge-to-mass ratio remains the same, combining Eqs. (55) and (56) will give the final drop diameter, $D_p$, in terms of the initial drop diameter, or

$$D_p = D_{p0}/4^{1/3} \tag{57}$$

This is equivalent to dividing the original particle into four equal pieces. From Doyle's data, this was obviously not the case since a few very small droplets of higher charge-to-mass ratio were ejected from the large droplet. One may conclude, therefore, that an atomizing system is not an equilibrium system and that the approach leading to Eq. (55) is not valid.

Dunskii and Kitaev (D9) describe a pneumatic atomizing nozzle in which there is an added provision for electrostatic charging with a potential of 1000 V and a power of 0.1 W. They report that the drops formed in the presence of the electrostatic field are two to four times smaller than with pneumatic dissemination alone and that more aerosol is deposited on the ceiling of the testing chamber when the field is applied. They claim that the charging is by induction rather than by corona.

## B. Electrostatic Dissemination of Powders

Electrostatics plays two separate and distinct roles in the dissemination of powders. It can be employed as an actual disseminating force to cause outward migration of the individual particles or it can be used simply as an adjunct to some other technique of dissemination, such as pneumatic dissemination. The former is entirely responsible for the transfer of powders in certain forms of electrostatic printing. Some of the problems and considerations involved have been discussed by Schultz and Wiech (S3) in connection with the use of powders for ion propulsion. A number of articles refer to dissemination of charged powders, but in most of these cases the dissemination process is pneumatic and electrostatic fields are employed primarily to obtain unipolar aerosols rather than for dissemination purposes. It is, however, likely that the electrostatic field will also influence the dissemination process, whether this was the intent or not. Göhlich (G5) discusses systems for pneumatic dissemination with simultaneous charging. Kitaev (K4) describes corona charging of powders and concludes from the usual charging equations that all particles will be almost fully charged within 0.01 sec. The nozzle arrangement described by Dunskii and Kitaev (D9) would appear to be equally applicable for dissemination of powders.

## V. Methods of Charging or Discharging Particles

### A. Maximum Stable Charge

Any particle in the atmosphere will gain or lose charge because of chance encounters with ions, photons, etc., but these events are usually so rare that charge changes due to them are negligible over periods of, at least, minutes to hours. Schonland (S1), for example, reports data for the atmosphere that would correspond to relaxation times, $t_R$, ranging from 10 min to 2 hr near ground level. He also indicates that small ion concentrations may range from 100 to 1000 ions/cm$^3$ while large ion concentrations may range from 200 to 80,000 ions/cm$^3$. These are all small compared to particulate concentrations of concern in agent dissemination. Chalmers (C1) reports fine weather relaxation times of 5–30 min. However, when the charge of the particle becomes sufficiently high it will either emit electrons or cause electrons to be generated in the surrounding atmosphere, which will serve to neutralize the charge on the particle. Actually, this can happen to a minor extent at all levels of charging but does not become significant until the charge level reaches a specific critical level. This is identical to the level at which a wire or other charged body will go into corona.

In the normal atmosphere if a large body is charged to such an extent that the outside surface field intensity (or $\mathscr{E}_{ps}$) exceeds approximately 30,000 V/cm (3 V/micron), corona will ensue. This critical field intensity may be somewhat higher with a negatively charged body than with a positively charged one. The critical field intensity is also known as the electrical breakdown strength of air. It is strictly valid only for a uniform field, as between flat plates,[5] but is still approximately true for large bodies outside of which the field intensity does not drop off too rapidly.

In order to achieve breakdown, electrons (either from the air or from the body) must be accelerated to a sufficient velocity to ionize the air and breed more electrons by any one of several processes. In an actual gas, however, some of the kinetic energy of the electrons is lost in collisions with air molecules without resulting in ionization. This combined effect has been expressed in terms of the Townsend ionization coefficient. As a body becomes smaller, its curvature increases and the electric field intensity drops off more rapidly with distance from the surface; consequently, to accelerate electrons a given amount, the body surface field intensity must be higher than for a flat surface. Actually, because of increased attenuation resulting from the increased distance that an electron must travel through air to achieve a given acceleration, the required surface intensity must increase even faster.

No attempts to calculate those limiting surface field intensities or gradients for small particles have been reported in the literature. However, point-to-plane corona has been studied in detail. Loeb (L8) has proposed a method for calculating positive corona threshold limits making use of established values of the Townsend first ionization coefficient. Loeb's threshold formula is

$$\zeta \xi \exp \int_{x_1}^{x_2} \alpha \, dx = 1 \tag{58}$$

where $\alpha$ is the first Townsend ionization coefficient; the integral

$$\int_{x_1}^{x_2} \alpha \, dx$$

is known as the Townsend integral. The exponential term represents the number of electrons (or ions) created in an avalanche originated by a single

---

[5] Actually, with flat plates, the first sign of breakdown is a spark and there is no initial corona.

electron. The term $\zeta$ represents the fraction of these electrons that produce a photon capable of ionizing the gas photoelectrically, and $\xi$ is the chance that such a photon will give rise to a photoelectron that can be the source of a new avalanche. The term $\alpha$ is a function of both pressure and field intensity.

Loeb et al. (L13) have shown that for the positive corona threshold the Townsend integral is essentially a constant having an average value of 10.2. Using this value for the integral, the field distribution relationships around a sphere, and the values of the ionization coefficient $\alpha$ (summary given by Loeb, L10, p. 676), the positive corona threshold limits for spherical particles have been calculated and are presented in Fig. 6 in terms of a normalized specific particle surface gradient, and in Fig. 7 in terms of charge-to-mass ratio. It will be noted that for $p = 1$ atm, the threshold charge limit on all sizes of liquid particles is determined by Rayleigh instability rather than by corona considerations, although for water ($\sigma = 72$ dyn/cm) the difference is small for particles of the order of 100 microns.

Negative corona is considerably more complicated and depends more on the emission properties of the discharge surface and of the gas. White (W4) gives a detailed discussion of corona mechanisms. In general, the negative corona threshold is higher than the positive.

In Fig. 6, it will be noted that at $p = 1$ atm for particle sizes less than 10 microns, the corona threshold rises rapidly. There is strong evidence that there is an upper limit to the specific particle surface gradient, $\mathscr{E}_{ps}$, of the order of 200 V/micron, and that the rapidly increasing values calculated for the corona threshold cannot be realized. Earhart (E1), for example, finds when surfaces are closer together than 2 microns that discharge occurs at a value of $\mathscr{E}_{ps}$ of 162 V/micron at atmospheric pressure and at even lower gradients for reduced pressures. Also, if the gradient reaches the order of $10^4$ V/micron, field emission from the surface can occur. There are also some other reports (Germer, G3; Kisliuk, K3) that suggest some type of field emission at gradients of the order of 100 V/micron. It has been suggested that these low apparent emission gradients may be the result of local intensification due to surface roughness (Robertson, Viney, and Warrington, R6).

It should be noted that the calculation of positive corona thresholds for spheres has been based on values of the Townsend integral for point-to-plane configurations. Thus, geometry may play an additional role that has not been allowed for. Also with particles smaller than 10 microns, the mean free path of electrons, molecules, and ions is being approached. If it is assumed that an electron or an ion can be accelerated without attenuation over a distance of its mean free path, then the specific particle surface gradient required to raise the energy level of the electron or ion to a potential of $E_i$ volts can be shown by a simple integration to be given by

$$\mathscr{E}_{ps} = (E_i/\lambda_m)[1 + (2\lambda_m/D_p)] \approx (E_i/\lambda_m) \tag{59}$$

where $\lambda_m$ is the mean free path. Since ionizing potentials are of the order of 20 V and mean free paths are of the order of 0.1 micron, it is clear that specific surface gradients of 200 V/micron can accelerate either electrons or ions to ionizing energy levels. It would seem, therefore, that specific surface gradients cannot exceed a value given by Eq. (59), if $E_i$ is defined as the ionization potential for the suspending gas and $\lambda_m$ is taken as the mean free path of a gas molecule (or positive ion). It is interesting to note that this level checks the order of gradient reported by Earhart for surfaces that were very close together. This subject, however, requires further study.

### B. Contact Charging

When two bodies are in contact with one another, a potential difference may be set up between the two bodies. This is known as a contact potential and is most commonly applied to conductors or semiconductors. If the bodies are physically separated, some of the charge segregation resulting from this contact potential will remain, causing the bodies to be oppositely charged; the rest may leak back during the separating process. The magnitude of this charge can be estimated from a knowledge of the contact potential and the capacitance of the system before separation.

As previously discussed, for two equally sized, equally and oppositely charged conducting spheres, the capacitance is given by

$$\mathscr{C} = \mathscr{Q}_p/E = 2\pi k_{s\mathscr{C}} \varepsilon D_p \tag{48}$$

where $k_{s\mathscr{C}}$ is a function of the ratio $(L_s/D_p)$; $L_s$ is the clearance between the spheres; $\mathscr{Q}_p$ is the charge on each sphere; and $E$ is the potential difference between the spheres. Values of $k_{s\mathscr{C}}$ are given in Table VI if $L_s$ is known. From the definition of $\mathscr{E}_{ps}$ [Eq. (1)], Eq. (48) becomes

$$\mathscr{E}_{ps} = 2k_{s\mathscr{C}} E/D_p \tag{60}$$

It is also easily shown that the corresponding equations for a charged conducting sphere near a grounded conducting plane are given by

$$\mathscr{C} = \mathscr{Q}_p/E = 4\pi k_{s\mathscr{C}} \varepsilon D_p \tag{61}$$

$$\mathscr{E}_{ps} = 4k_{s\mathscr{C}} E/D_p \tag{62}$$

where $\mathscr{Q}_p$ is the charge on the sphere; $E$ is the potential difference between the sphere and the plane; and $\mathscr{E}_{ps}$ is the average gradient corresponding to $\mathscr{Q}_p$ on the sphere. The value of $k_{s\mathscr{C}}$ may also be obtained from Table VI if $L_s$ is defined as twice the distance between the plane and the edge of the sphere. Thus, if $E$ is the contact potential acting over a distance [$L_s$ for two equal spheres, $(L_s/2)$ for the sphere and plane], the charge can be calculated from the above equations if the separation distance is known. Actually, the effective distance over which $E$ acts will be of the order of 1 Å (or a portion of one molecular diameter). Fortunately, an examination of Table VI shows that $k_{s\mathscr{E}}$ is very insensitive to $L_s$ and that this distance need not be known accurately to obtain $k_{s\mathscr{E}}$. Also some charge may leak back if $\mathscr{E}_{max}$ between the bodies is greater than the breakdown strength. For larger clearances it was previously indicated that this breakdown would probably correspond to a value of $\mathscr{E}_{max}$ of some 200 V/micron. However, for the extremely small clearances here involved, it is more likely that back discharge either will not occur until the potential difference between the bodies exceeds the ionization potential or will proceed by the electron tunneling effect.

As the bodies are separated the potential difference will increase. However, the potential difference at any separation will be directly proportional to the respective value of $k_{s\mathscr{E}}$. Since this will not vary markedly, the potential difference between the bodies will not increase very much as they are separated. Therefore, since contact potentials are normally considerably below ionization potentials, it is unlikely that back discharge will actually occur owing to air ionization. However, from Table VI and Eq. (47) it is also apparent that the maximum surface gradient may be of the order of $10^8$ V/cm (if $E = 1V$ and separation distance is 1Å), which should be enough to cause field emission and discharge by this means.

Thus the amount of back discharge cannot be readily calculated. It would seem unlikely, however, that the potential difference would drop below the contact potential as the bodies are separated. This would mean that the actual charge leaking off would be only a small fraction of that originally present.

Assuming a value of $k_{s\mathscr{E}}$ of 2 (which is a likely value from Table VI), it can be shown from Eq. (60) that $\mathscr{E}_{ps}$ will be of the order of 4 V/micron for a 1-micron-diam particle if the contact potential is 1 V. It is thus apparent that substantial charges can be generated by simple contact potential.

The contact potential represents the difference between the work function of the two metals. Hence, the particle charge will be positive on that body which has the lower work function and negative on that body having the higher work function.

Cho (C2) has calculated charging by contact potentials essentially on the basis described above, assuming a contact potential of 1 V for the case of an aluminum particle in contact with a steel plate. Arabadzhi (A1) reports a contact potential between ice and water of 0.15 V, although the abstract for another article by Arabadzhi (A2) reports this value as 1.5 V.

## C. Induction or Field Charging

It is possible to charge a particle by placing it in an externally generated field. This external field will cause either electrons or ions that are present in the surrounding space or on a contacting surface to flow to the body. This is sometimes known as induction charging.

Two specific cases will be considered. The first involves a conductive particle resting on a charged plate. It has been shown (Cho, C2) that the average field at the surface of a spherical particle resting on a flat plate is $(\pi^2/6)$ times that at the surface of the charging plate. Thus, assuming that the particle is maintained on the surface long enough to reach this equilibrium condition, the particle will reach a charge level given by

$$\mathscr{E}_{ps} = (\pi^2/6)\mathscr{E}_0 = 1.65\mathscr{E}_0 \tag{63}$$

where $\mathscr{E}_0$ is the field at the surface of the base plate (or the field between two plates with the particle resting on one of them). In actual fact the repulsive forces may cause the particle to leave the plate surface before the equilibrium charge is reached. Cho (C2) has investigated this method of charging experimentally using aluminum, iron, nickel, molybdenum, and magnetite particles, and essentially confirms Eq. (63) for particles in the size range of 0.3–15 microns at field strengths, $\mathscr{E}_0$, up to 2.5 V/micron. In the case of aluminum he found that when the plate was positive the aluminum particles received a slightly higher charge than when the plate was negative. The difference was of the order predictable for charging due to contact potential. He also found that conductor and semiconductor particles would continuously migrate back and forth between the charging plates. Insulator particles did not react immediately to the application of a potential across the charging plates, but some did hop off the charging plate after a period of time. This was attributed to the small but finite conductivity of such particles. Such particles did not migrate back and forth between the plates. Nonconductors can only be polarized in such a field. Fields around polarized particles are usually an order of magnitude lower than those resulting from charged particles.

The second case of field charging that is considered is that of charging a particle located in a uniform field in the presence of a net concentration of unipolar gas ions. This case has been extensively treated by Deutsch (D4), Rohmann (R7), Ladenburg (L1), Pauthenier and Moreau-Hanot (P1) Mierdel (M5), and Cochet (C3). A good summary is given by White (W4). Under these circumstances, a particle will always reach a specific limiting charge, but the time required to reach this limit will depend on the concentration and mobility of the gas ions. The most general relationship is that given by Cochet (C3)

$$\left(\frac{\mathscr{E}_{ps}}{\mathscr{E}_{ps\max}}\right) = \frac{(\beta_i n_i \mathbf{Q}/4\varepsilon)t}{1 + (\beta_i n_i \mathbf{Q}/4\varepsilon)t} \tag{64}$$

$$\mathscr{E}_{ps\max} = \left\{\left[1 + \left(\frac{2\lambda_i}{D_p}\right)\right]^2 + \left[\frac{2}{1 + (2\lambda_i/D_p)}\right]\left[\frac{\delta_p - \delta}{\delta_p + 2\delta}\right]\right\}\frac{\mathscr{E}_0}{\delta} \tag{65}$$

where $\mathscr{E}_{ps_{max}}$ is the maximum or equilibrium charge level; $\mathscr{E}_{ps}$ is the charge level attained after time $t$; $\beta_i$ is the ion mobility; $n_i$ is the unipolar concentration; $\lambda_i$ is the mean free path of the ions; $\delta$ and $\delta_p$ are the dielectric constants for the fluid and the particle, respectively; and $\mathscr{E}_0$ is the strength of the external field. The term in braces is a function of the particle and gas properties. Neglecting the effect of ion mean free path (which will normally be significant only for particles smaller than 1 micron), this term will vary from 1 for particles that are perfect insulators to 3 for particles that are conductors.

The maximum charge is reached in a time that depends on the ion concentration as indicated by Eq. (64). Ion concentrations of the order of $10^8$ ions/cm$^3$ are readily achieved by means of corona from wires and ion mobilities are of the order of 2 (centimeters/second)/(volts/centimeter). For these conditions $4\varepsilon/(\beta_i ni \mathbf{Q})$ is 0.011 sec. It is apparent from Eq. (64), therefore, that the particle will essentially acquire its maximum charge in less than a second. Since $\mathscr{E}_0$ can be of the order of 3 V/micron (breakdown strength of air), it is also apparent from Eq. (65), that particle charges corresponding to the range of 3–10 V/micron can be achieved by this means.

With ambipolar ions, a net charging can still take place. The charging rate for this case can be approximated by letting the term $(\beta_i n_i)$ in Eq. (64) be equal to the difference between the corresponding terms for ions of each charge. Actually significant ambipolar ion concentrations may be short-lived because of ion recombination.

The following relationship (White, W4) gives the electric field for concentric-cylinder electrodes, as modified by the space-charge effect of the corona ions,

$$\mathscr{E} = [(j_L/2\pi\varepsilon\beta_i) + (\mathscr{E}_c D_c/D)^2]^{1/2} \qquad (66)$$

where $\mathscr{E}$ is the field intensity at diameter $D$; $\mathscr{E}_c$ is the field intensity at the wire of diameter $D_c$ ($\mathscr{E}_c$ is the critical field intensity required for corona); $j_L$ is the corona current per unit of wire length; $\beta_i$ is the ion mobility. For large currents and for the points near the outer cylinder, the last term may become small and the field is closely approximated by

$$\mathscr{E} = [j_L/(2\pi\varepsilon\beta_i)]^{1/2} \qquad (67)$$

Ion density is related to current by the basic relationship

$$j_A = n_i \mathbf{Q} \beta_i \mathscr{E} \qquad (68)$$

where $j_A$ is the current per unit area parallel to the electrodes (i.e., normal to the electrical force field), $n_i$ is the ion concentration, $\beta_i$ is the ionic mobility, and $\mathscr{E}$ is the field intensity.

In order to attain a desired value of field strength, the current must be adjusted to the level indicated by rearranging Eq. (66) as follows:

$$j_L = 2\pi\varepsilon\beta_i\mathscr{E}^2\left[1 - (\mathscr{E}_c D_c/\mathscr{E} D)^2\right] \qquad (69)$$

where the term in brackets may often be neglected. This adjustment of current is actually achieved by regulating the voltage across the cylinders. Thus for $\mathscr{E} = 1$ V/micron, $\beta_i = 2$ (centimeters/second)/(volt/centimeter), neglecting the term in brackets, a current flow of 3.4 mamp/ft of wire length is required.

The corresponding ion density is obtained by combining Eq. (68) with Eq. (69) to yield

$$n_i = \{2\varepsilon\mathscr{E}/(\mathbf{Q}D)\}[1 - (\mathscr{E}_c D_c/\mathscr{E} D)^2] \qquad (70)$$

Thus for $\mathscr{E} = 1$ volt/micron, $D = 2$ in., and neglecting the terms in brackets, $n_i = 2.18 \times 10^9$ ions/cm³.

Loeb (L10) gives a detailed discussion of the factors influencing ion mobility. The electrostatic mobility of a charged particle, $\beta_{pe}$, is defined as the ratio of the terminal migration velocity to the field strength applied. The electrical accelerating force acting on a particle is given by

$$F_e = \mathcal{Q}_p\mathscr{E} \qquad (71)$$

where $\mathcal{Q}_p$ is the charge on the particle and $\mathscr{E}$ is the external field. The force resisting motion is given by Stokes' law of resistance.

$$F_r = 3\pi\mu u D_p/k_C \qquad (72)$$

When $F_e$ and $F_r$ are equal and opposite, the particle has attained its terminal migration velocity, $u_m$,

$$u_m = k_C\mathscr{E}\mathcal{Q}_p/(3\pi\mu D_p) = k_C\varepsilon\delta D_p\mathscr{E}_{ps}\mathscr{E}/3\mu \qquad (73)$$

or

$$\beta_{pe} = u_m/\mathscr{E} = k_C\mathcal{Q}_p/(3\pi\mu D_p) = k_C\varepsilon\delta D_p\mathscr{E}_{ps}/3\mu \qquad (74)[6]$$

Cochet (C3), in measurements with particles in the size range of 0.1–1 micron, found a minimum mobility and migration velocity at a particle diameter of $\sim\frac{1}{4}$ micron. The following are some of his results for the minimum velocity

---

[6] It will be noted by comparison with Eq. (1) that electrostatic mobility, $\beta_{pe}$, is directly proportional to specific particle potential, $E_{ps}$.

and mobility at an ion concentration of $8 \times 10^7$ ions/cm³ with equal charging and migrating field strengths:

| Field strength (V/micron) | Charging time (sec) | Migration velocity (cm/sec) | Electrostatic particle mobility (cm/sec)/(V/micron) |
|---|---|---|---|
| 0.15 | 0.04 | 0.5 | 3 |
| 0.50 | 0.01 | 3 | 6 |

Recently, Foster (F1) reported measurements on corona charging of wood-smoke particles of 0.1–0.3 micron, and obtained charge levels of 5–32 electrons per particle ($\mathscr{E}_{ps} = 1$–5 V/micron). He concludes that these levels agreed well with predictions assuming that field induction and diffusion charging effects are additive. Migration velocities varied from 0.5 to 23 cm/sec depending on the voltage and current applied. Over the limited range covered there was no significant effect of particle size on migration rate.

## D. Diffusion Charging

Whenever ions are present in a gas, they will diffuse and deposit on any surface (bounding wall or suspended particle). As the surface becomes charged subsequent deposition of ions is retarded. The rate of charging will, therefore, drop off very rapidly as charging progresses. There is, however, no real limiting charge for this mechanism since, on a probability basis, there will always be an ion of sufficient energy to penetrate through the electrostatic field set up around the surface by previously deposited particles.

Arendt and Kallman (A3) have derived the following relationship for the rate of charging a particle by diffusion of unipolar ions:

$$\frac{d\mathscr{Q}_p}{dt}\left[1 + \frac{\pi\varepsilon\bar{u}_i k_s D_p^{\ 2}}{4\beta_i \mathscr{Q}_p}\right] = \left[\frac{\pi D_p^{\ 2} k_s \bar{u}_i Q n_i}{4}\right] \exp\left\{-\frac{Q\mathscr{Q}_p}{2\pi\varepsilon D_p kT}\right\} \quad (75)$$

This equation may also be written in the form

$$\frac{d\mathscr{E}_{ps}}{dt}\left[1 + \frac{\bar{u}_i k_s}{4\delta\beta_i \mathscr{E}_{ps}}\right] = \left(\frac{k_s \bar{u}_i n_i Q}{4\varepsilon\delta}\right) \exp\left\{-\frac{\delta Q D_p \mathscr{E}_{ps}}{2kT}\right\} \quad (76)$$

where $\mathscr{Q}_p$ is the charge on the particle after time $t$; $\mathscr{E}_{ps}$, the corresponding specific particle surface gradient; $\bar{u}_i$, the rms ionic velocity [given by $(3kT/m_i)^{1/2}$ or $(3RT/M_i)^{1/2}$]; $k_s$, a sticking coefficient (found to be essentially unity); $\beta_i$, the ionic mobility; and $n_i$, the ionic concentration in the bulk of of the gas phase. White (W4) has given a simpler derivation that yields a result identical with Eq. (75) except that the term in square brackets on the left is omitted. For small particles, this term will normally not have any significant effect compared with that exerted by the exponential term. White's equation also lends itself to ready integration to yield

$$\mathcal{Q}_p = \left[\frac{2\pi\varepsilon D_p kT}{Q}\right] \ln\left[1 + \frac{k_s \bar{u}_i n_i Q^2 D_p t}{8\varepsilon kT}\right] \tag{77}$$

or

$$\mathcal{E}_{ps} = \left(\frac{2kT}{\delta Q D_p}\right) \ln\left[1 + \frac{k_s \bar{u}_i n_i Q^2 D_p t}{8\varepsilon kT}\right] \tag{78}$$

Table IX demonstrates the order of magnitude of the various terms, for $D_p = 1$ micron, $\delta = 1$, $n_i = 10^8$ ions/cm³, $M_i = 32$ gm/gm mole, $\beta_i = 2$ (centimeters/second)/(volt/centimeter), $k_s = 1$.

### TABLE IX

ILLUSTRATIVE VALUES OF TERMS OF EQS. 75–78

|  | T = 298°K | T = 1000°K |
|---|---|---|
| $\bar{u}_i$, m/sec | 482.0 | 883.0 |
| $(\bar{u}_i k_s/4\delta\beta_i)$, V/micron | 0.602 | 1.103 |
| $(2kT/\delta Q D_p)$, V/micron | 0.0513 | 0.1723 |
| $(k_s \bar{u}_i n_i Q/4\varepsilon\delta)$, (V)/(micron)(sec) | 218.0 | 399.0 |
| $(2\pi\varepsilon D_p kT/Q)$, C | $1.430 \times 10^{-18}$ | $4.80 \times 10^{-18}$ |
| $(8\varepsilon kT/k_s \bar{u}_i n_i Q^2 D_p)$, sec | $2.355 \times 10^{-4}$ | $4.32 \times 10^{-4}$ |

The above equations do not allow for kinetics associated with the mean free path discontinuity at the particle surface. A correction for this would increase the charging rate. As a first approximation, this can be allowed for by replacing $D_p$ in Eqs. (75) and (77) with the product $D_p[1 + (2\lambda_i/D_p)]$. This is based on the reasoning that ions migrating to within a mean free path of the particle surface will be deposited and that the ion concentration will drop effectively to zero within a mean free path of the surface.

$$\mathcal{E}_{ps} = \left(\frac{2kT}{\delta Q D_p}\right)\left[1 + \frac{2\lambda_i}{D_p}\right] \ln\left\{1 + \left[\frac{k_s \bar{u}_i n_i Q^2 D_p t}{8\varepsilon kT}\right]\left[1 + \frac{2\lambda_i}{D_p}\right]\right\} \tag{79}$$

Thus, since $\mathcal{E}_{ps}$ will be relatively insensitive to the items inside the logarithmic term, the main effect of this correction is to increase the value of $\mathcal{E}_{ps}$ or $\mathcal{Q}_p$ by the factor $[1 + (2\lambda_i/D_p)]$.

The values in Table X have been calculated from Eqs. (77) and (78) for $n_i = 10^8$ ions/cm³, $\bar{u}_i = 482$ meters/sec (corresponding to $O_2$), $T = 25°C$, $\delta = 1$, neglecting the mean free path correction factor.

TABLE X

ILLUSTRATIVE VALUES OF DIFFUSIONAL CHARGING OF PARTICLES

| Particle diameter (microns) | Charging time (sec) | Particle charge $\mathcal{Q}_p$ (electrons) | $\mathscr{E}_{ps}$ (V/micron) |
|---|---|---|---|
| 0.1 | 0.01 | 1.5 | 0.85 |
|  | 0.1 | 3.4 | 1.94 |
|  | 1 | 5.4 | 3.11 |
|  | 10 | 7.5 | 4.29 |
| 1.0 | 0.01 | 33.6 | 0.19 |
|  | 0.1 | 54.0 | 0.31 |
|  | 1 | 74.6 | 0.43 |
|  | 10 | 95.1 | 0.55 |
| 10.0 | 0.01 | 540 | 0.031 |
|  | 0.1 | 746 | 0.043 |
|  | 1 | 951 | 0.055 |
|  | 10 | 1157 | 0.067 |

It is apparent from the values of $\mathscr{E}_{ps}$ that diffusional charging is primarily effective with very small particles, as might be expected. With particles as large as 10 microns the charge level produced in any reasonable time is small compared with what can be achieved with corona charging. For the 0.1-micron particle, however, the charge level attained by diffusion is of the same order and may be larger if the mean free path correction is allowed for.

Whitby (W3) has developed an ac ion generator specifically for the purpose of producing electrically neutral particles. This neutralization is accomplished by adding a gas stream loaded with gaseous ions (both positive and negative) to the aerosol to be neutralized. Although exact neutralization cannot be assured by this means, actual neutralization to within a few electrons has been achieved. This is probably due to the fact that diffusion charging (discharging in this case) is not sensitive to ion concentrations that appear only in the logarithmic terms of Eq. (77).

E. THERMIONIC EMISSION CHARGING

All materials emit electrons to an extent dependent on their temperature. Such emission of electrons will cause a body to become positively charged. This effect does not usually become significant until the temperature levels are above the normal incandescent levels, which are to some extent related to this phenomenon. Einbinder (E3) reports that the ionization levels are given by

(a) for $n_e/n_p$ very small:

$$\frac{n_e}{n_p} = \left(\frac{2}{n_p}\right)\left(\frac{2\pi m_e kT}{h^2}\right)^{3/2} \exp\left(-\frac{\varphi}{kT}\right) \tag{80}$$

(b) for $n_e/n_p \geq 3$:

$$\frac{n_e}{n_p} + \left(\frac{2\pi\varepsilon D_p kT}{Q^2}\right) \ln\left(\frac{n_e}{n_p}\right) =$$

$$\frac{1}{2} + \left(\frac{2\pi\varepsilon D_p kT}{Q^2}\right) \ln\left[\left(\frac{2}{n_p}\right)\left(\frac{2\pi m_e kT}{h^2}\right)^{3/2} \exp\left(\frac{-\varphi}{kT}\right)\right] \tag{81}$$

where $n_e$ is the concentration of electrons in space; $n_p$, the concentration of particles in space; and $\varphi$, the work function for the particle surface. These equations assume that equilibrium is reached between the spatial concentration of electrons and particles. Thus the number of electrons emitted must be equal to the number of positive charges left on the particles. Since $\mathscr{E}_{ps} = (n_e/n_p)(Q/\pi\varepsilon\delta D_p^2)$, the above can be written in terms of the average value of $\mathscr{E}_{ps}$ to give

(a) for $n_e/n_p$ very small:

$$\mathscr{E}_{ps} = \left(\frac{2}{n_p}\right)\left(\frac{Q}{\pi\varepsilon\,\delta D_p^2}\right)\left(\frac{2\pi m_e kT}{h^2}\right)^{3/2} \exp\left(\frac{-\varphi}{kT}\right) \tag{82}$$

(b) for $n_e/n_p \geq 3$:

$$\mathscr{E}_{ps} + \frac{2kT}{\delta Q D_p} \ln\left(\frac{\pi\varepsilon\,\delta D_p^2 \mathscr{E}_{ps}}{Q}\right) =$$

$$\frac{Q}{2\pi\varepsilon\,\delta D_p^2} + \frac{2kT}{\delta Q D_p} \ln\left[\left(\frac{2}{n_p}\right)\left(\frac{2\pi m_e kT}{h^2}\right)^{3/2} \exp\left(\frac{-\varphi}{kT}\right)\right] \tag{83}$$

Table XI demonstrates the order of magnitude of the various terms, for $D_p = 1$ micron, $\delta = 1$, $\varphi = 5$ eV.

TABLE XI

Illustrative Values of Terms of Eqs. 80–83

|  | $T = 1000°K$ | $T = 2000°K$ | $T = 3000°K$ |
|---|---|---|---|
| $(2\pi m_e kT/h^2)^{3/2}$, particles/cm$^3$ | $0.760 \times 10^{20}$ | $2.15 \times 10^{20}$ | $3.95 \times 10^{20}$ |
| $2\pi\varepsilon D_p kT/Q^2$, dimensionless | 30.0 | 60.0 | 90.0 |
| $\varphi/kT$, dimensionless | 58.0 | 29.0 | 19.32 |
| $e^{-\varphi/kT}$, dimensionless | $6.47 \times 10^{-26}$ | $2.54 \times 10^{-13}$ | $4.03 \times 10^{-9}$ |
| $Q/\pi\varepsilon\delta D_p^2$, V/micron | 0.00576 | 0.00576 | 0.00576 |
| $2kT/\delta Q D_p$, V/micron | 0.1723 | 0.346 | 0.518 |

For $n_p$ in the range of $10^6$–$10^{10}$ particles/cm$^3$, it is evident from these values that significant charging does not occur by thermal emission until temperatures are of the order of 2000°K or over.

The above indicates the charge attained in the presence of the electrons, the cloud being electrically neutral. As the cloud is cooled or diluted, the cloud will remain electrically neutral but some recombination of electrons with the charged particle may occur to reduce the net level of particle charge. In the above, ionization of the gas phase has been neglected. Such gas ionization would also occur either at high temperature or at high particle charge level and could exert a significant interactive effect on the process of particle charging by thermal means.

At elevated temperatures emission of ions can also occur but this normally requires an even higher temperature level than does electron emission.

## F. Charging by Interface Alteration

When materials are brought into contact and then separated or when a bulk material is separated by formation of a new surface, the two separated bodies usually acquire an equal and opposite charge. The sign and magnitude of charge on each body depend on a large number of factors, which are (1) the physical and chemical composition of each body; (2) the nature of contacting (gentle contact, impact, or rubbing); (3) the method and speed of separation after contact; (4) the presence of contaminants or ions on the surface; (5) the nature of the atmosphere surrounding the bodies; and (6) the presence of any external electrostatic fields. The final charge observed is the residual charge left after separation. The charge present just before separation takes place could be higher because of the back discharge that may occur during the separation process. The final observed charge separation is, therefore, the net result of both a charging process and a discharging process.

All available investigations in this area are essentially reviewed in recent survey articles by Harper (H3, H4), Loeb (L9, L11, L12), and Montgomery (M7). This area has been the subject of investigations since ancient times. Despite the extensive investigations made in this area, considerable uncertainty and confusion remain as to the details of the charging processes involved. Table XII gives a summary of the types of systems that are subject to charging as the result of interface alterations and the possible mechanisms that may be responsible for the charging process. The subsequent sections discuss the magnitude and mechanism of charging in each of these cases in more detail.

### 1. *Liquid Breakup*

The charge separation resulting from the alteration of a liquid–gas interface has been variously termed spray electrification and balloelectricity and

TABLE XII: Processes and Mechanisms Involved in Charging by Interface Alteration[a]

| General process | Special case or conditions | | | Possible charging mechanisms |
|---|---|---|---|---|
| Liquid breakup | High resistivity liquids ($\mathscr{R}_0 > 10^{10}$ ohm cm) | | | Statistical distribution due to ion concentration fluctuations (most probable charge = 0) Double layer disruption Unequal ion mobility or migration rates Interface contaminants (incl. surfactants) |
| | Ionic liquids ($\mathscr{R}_0 < 10^{10}$ ohm cm) | | | Statistical distribution due to ion concentration fluctuations Double layer (zeta potential) disruption Volta potential (for electron conducting materials) Electrolytic (galvanic) potential (for ionic systems) |
| Liquid–solid contact Liquid–liquid contact | | | | Volta potential (equalization of Fermi levels) Electrolytic potential (where adsorbed water films may be present) |
| Solid–solid contact (inc. solid breakup) | Metal to metal Metal to semiconductor Semiconductor to semiconductor | | | |
| | Metal to insulator Semiconductor to insulator, or Insulator to insulator | Light contact (touching) | | Ion migration (due to inherent or unavoidable ion contamination; electron traps (a) by random adhesion of ions for contact of dissimilar materials (b) by diffusion due to differences in ion concentration or mobilities (c) by image attraction Anomalous (due to avoidable surface contamination) |
| | | Intense contact (compression or rubbing) | | Same as for light contact plus the following Preferential local heating (to cause ion migration or electron transfer) Rupture of chemical bonds (Asymmetry of either surface structure or nature of contact or both is necessary to achieve charging) |
| Solid–gas contact Liquid–gas contact | | | | No known charging provided solid or liquid does not break up and gas contains no free ions or suspended particles |

[a] Covers charging due to relative motion or separation of two contacting phases in the absence of external electrostatic fields, radiation,

involves the formation of charged droplets as the result of bubbling, splashing, spraying, and aerodynamic shattering. Some of these operations, such as splashing, could involve a liquid–solid interface as well. This type of charging has been of historical interest in connection with apparent atmospheric electricity around waterfalls and currently in connection with establishing the method of generation of thunderstorms.

Charge levels produced by this mechanism on individual droplets are relatively small, normally corresponding to $\mathscr{E}_{ps}$ values of less than 0.01 V/micron, when there is no concentration of charge due to evaporation of liquid in the drop. Dodd (D5), in generating organic liquid drops of from 1 to 30 microns in diameter with a nebulizer, found that the charge followed a Gaussian distribution with zero charge being the most probable (i.e., equal quantities of positively and negatively charged drops). The average absolute charge was proportional to the $\frac{3}{2}$ power of particle diameter (or $\mathscr{E}_{ps}$ proportional to the inverse square root of particle diameter). His results for 1-micron drops would correspond to a range in $\mathscr{E}_{ps}$ values of 0.0018 V/micron for paraffin oil to 0.024 V/micron for dibutylphthalate. For larger droplets the $\mathscr{E}_{ps}$ values would be lower. Natanson (N1), using a compressed air nebulizer to produce 1- to 4-micron transformer oil drops, obtained similar results. The larger drops were almost all charged but only 10–75% of the smaller drops were charged. In general he obtained charges of 0–30 electrons per particle with an average charge corresponding to a value of $\mathscr{E}_{ps}$ covering the range of 0.002–0.006 V/micron. These levels agreed with magnitudes that would be calculated on the basis of statistical fluctuations in the local ion concentration levels. Harper (H1) reports that for liquids with resistivities greater than $10^{13}$ $\Omega \cdot$ cm, no charging on bubbling was observed.

With the more conductive liquids, the ion concentration becomes so great that ion concentration fluctuations on a statistical basis are likely to be small. However, charging can take place by three other mechanisms: (1) mechanical disruption of any double layer of ions that may exist at the surface in times that are short compared with the relaxation time, with a predominance of the surface ions going to the portion of fluid coming from the surface; (2) unequal ion mobility with the larger ions unable to return to the bulk of liquid as readily as the smaller and more mobile ones; and (3) contaminating materials, such as dust or surfactants at the interfaces serving as ion carriers into one portion or the other of the ruptured liquid.

The nature and magnitude of charge resulting from liquid breakup are illustrated by the following representative investigations. Blanchard (as quoted by Loeb, L9) reports charges corresponding to $\mathscr{E}_{ps}$ values of 0.02–0.05 V/micron when droplets of 4–40 microns are formed as the result of the breaking of air bubbles in salt water. When exposed to external electrostatic fields of 0.005–0.030 V/micron, the charge levels were increased by an order of magnitude. Because of droplet evaporation, it is possible that the actual initial droplet sizes were larger than reported by Blanchard. This would make the actual corresponding $\mathscr{E}_{ps}$ values smaller. Loeb (L9) reports that when bubbles are allowed to rise through many centimeters of sea water (as in Blanchard's case), the resultant droplets are positive. However, when allowed to rise through about 1 cm of water, the drops are negative. Matthews and Mason (M3) report negligible charges as the result of aerodynamic breakup of 15-mm-diam water drops in the absence of an electrical field. Charges of the order of $3 \times 10^{-9}$ C/kg (corresponding to an $\mathscr{E}_{ps}$ value of 0.001 V/micron for a 15-mm-diam drop) were found. With external fields ranging from 0.003 to 0.15 V/micron, charges of $10^{-7}$–$2 \times 10^{-6}$ C/kg (or $\mathscr{E}_{ps}$ for 15-mm drop of 0.03–0.5 V/micron) were obtained. A reversal of the field also reversed the sign of the charge on the fine drops formed.

2. *Liquid–Solid Contact*

When liquids come into contact with a solid surface a transverse potential difference may be set up at the interface because of a preferential adsorption of some ions on the surface with a resultant electrical double layer. This is commonly called a *zeta potential*. If a relative motion is set up between the liquid and the solid surface, a charge displacement will take place. This will result in a buildup of potential difference in the direction of motion until the resultant back migration of the unbound ions just balances the rates at which they are transported forward by the relative fluid motion. This potential difference is called a *streaming potential*. Conversely, if an external potential gradient is imposed along the interface or across a porous boundary, the resultant charge displacement will set up a relative force between the liquid and the wall. If the liquid is not otherwise constrained, it will move relative to the solid. This is called *electrical endosmosis* If the liquid is constrained and the solid is free to move, the resultant motion of the solid body through the liquid is called *electrophoresis* (or cataphoresis in older literature). Endosmosis rates are established by the equilibrium between the driving force and any frictional resistance to motion. Endosmosis will cease if the hydrostatic head of transported liquid is allowed to build up to a point where the driving force is balanced. Similarly electrophoresis rates will be established by the equilibrium between the driving force and frictional forces on the moving solids.

When a low conductivity liquid containing some ions (and there are almost always some present in all liquids) is forced through a conduit under a high pressure gradient, enormous voltages can be generated. This condition is treated mathematically by Cooper (C6) who cites a critical resistivity of $10^{11}$ $\Omega \cdot$ cm above which voltages high enough to result in sparking to ground can be generated, with resultant fire and explosion hazards, during industrial transfer of such liquids from one vessel to another.

It should be noted that the above phenomena all involve relative motion or charge segregation but no static electrification or true charge separation. The latter can only occur when the liquid is separated from the solid or is broken up. The above phenomena consequently can only create conditions that will permit charge separation. Any net charging process itself will, therefore, be very similar to that involved in liquid breakup except that the initial charge segregation will be influenced by double layers at the solid surface as well as by those at a liquid-gas interface.

The net charge resulting from a simple liquid-solid contact is likely to be small, as for liquid breakup. However, if the liquid-solid contact is extensive, as in the flow through a long capillary, the potential difference buildup could cause extensive charging of any liquid drops formed at the end of the pipe. This charging, however, is a result of the electrostatic field between the conduit and surroundings. Thus, although this field is a direct result of the liquid-solid contact, the droplet charging is only an indirect result since it could have been obtained by direct application of an external electrostatic field to the discharge end of the conduit.

With higher ion concentrations or electron-conductive liquids, the effect of the zeta potential is essentially superseded by electrolytic (galvanic) or volta potentials, respectively. Although these potentials are of the order of 1 V while zeta potentials are one or two orders of magnitude lower, the high conductivities of the more conductive liquids make it normally impossible to build up the high streaming potentials that are possible with low conductivity liquids by virtue of the zeta potential.

Vonnegut et al. (V3) impinged deionized water ($\mathscr{R}_0 = 10^6 \, \Omega \cdot \text{cm}$) against an aluminum plate. The resulting 2-mm-diam drops had a positive charge of $4 \times 10^{-14}$ C/drop (corresponding to $R_c = 1 \times 10^{-8}$ C/kg or $\mathscr{E}_{ps} = 0.0004$ V/micron).

Mercury is a special type of liquid and behaves essentially as a metal for which charging mechanisms are discussed in Section V, F, 3. Dodd (D5) determined the charge acquired by mercury drops in contact with glass and various insulators. The mercury was always positively charged with $\mathscr{E}_{ps}$ values of the order of 0.01 V/micron.

Medley (M4) working in vacuum also found that mercury became positively charged relative to nylon, mica, and Alkathene[6a] film. He obtained the contact by bringing a column of mercury up against a film of the other material, the back of which was covered with a stationary film of mercury connected to an electrometer, which measured the charge after the contact

---

[6a] Alkathene is a trademark of I.C.I. and is a form of polyethylene.

was broken. By varying the thickness of the film, he measured a maximum charge of the order of 500 esu/cm² (equivalent to a value of $\mathscr{E}_{ps}$, as defined by Eq. (1) for a value of $\mathscr{Q}_{ps}$ of 500 esu/cm², of 180 V/micron).[7] When small amounts of sodium, zinc, lead, or tin were added to the mercury, the sign of charge was reversed but the maximum charge level observed was the same. This suggested that actual surface charges were initially much higher than those measured, the latter being limited by considerations of back discharge during separation.

Liquid–liquid contact will be basically similar to liquid–solid contact. In this case, both phases would be mobile, however, and some of the extreme charge separations possible in liquid–solid contacts would not be possible.

## 3. Solid–Solid Contact

The terms contact electrification and triboelectrification[8] have been applied to the process where charge is acquired when two solids are contacted and separated. It is in this area of static electrification that the greatest confusion exists. This confusion comes about because of the complications produced by: (1) the variety of mechanisms that can operate (electrolytic films, ion transfer, electron transfer); (2) the sensitivity to impurities or surface contamination; (3) the manner in which contacting is achieved (normal, sliding, rubbing, or rolling contact); (4) the manner of making measurements; (5) the presence of external or extraneous electrical fields; (6) variations in ambient atmosphere; and (7) back-discharge phenomena during separation. In order to understand the basic mechanisms involved, it is informative to consider first the current theories of solid state structure.

*a. Theory of Solid State Structure.* The modern concept of structure of solids visualizes a discrete atom as composed of a nucleus surrounded by electrons in distinct energy levels, the number of electrons and their energy levels being different for different materials. When atoms are brought into proximity to form molecules or crystals, the corresponding electrons in each original energy level will be forced to occupy a band of energy levels instead of the original single energy level because of the Pauli exclusion principle (which states that no two electrons can occupy the same quantum energy state at the same time). The outermost or highest energy electrons are the valence electrons. Thus, for a proximate assemblage of atoms, this highest

---

[7] The value of $\mathscr{E}_{ps}$ so calculated would be equivalent to the field existing outside of a particle the entire surface of which has the specified charge surface density. Because of the stationary mercury film behind the material film, the actual field outside of the material film will be considerably less than $\mathscr{E}_{ps}$.

[8] From the Green word *tribein* meaning "to rub."

energy level reverts to a valence band rather than being a discrete level, the spread in the bandwidth becoming larger the closer the spacing of atoms.

In order for an electron to travel from atom to atom (i.e., for electron conduction to take place), the electron must first be raised to an unoccupied, allowable higher energy level (or band). For assemblages of atoms this is called the conduction band.

With conductors either the valence band is only partially filled (as with metals) or it overlaps with the next higher allowable empty band. Hence only small amounts of energy are required to raise a valence electron into a higher-level empty band where it is free to move from atom to atom.

With insulators or semiconductors, the valence band is completely filled and is separated from the next highest allowable energy band (conduction band) by a gap (or forbidden band). In order for valence electrons to be raised into the conduction band, they must be given a sizable amount of energy. With insulators this gap corresponds to several electron volts. For semiconductors the gap is sufficiently small so that a significant number of electrons can acquire the necessary additional energy by thermal means at room temperature.

Electrons in thermal equilibrium in the solid state obey Fermi–Dirac statistics as given by

$$f_p = 1/(1 + e^{(\varphi_T - \varphi_F)/kT}) \tag{84}$$

where $f_p$ is the probability that an electron will occupy an allowed quantum energy level $\varphi_T$ at temperature $T$. The term $\varphi_F$ is known as the Fermi level and represents the energy of the highest occupied level at $0°K$; $\varphi_F$ is usually very large compared to $kT$. Thus at $0°K$, $f_p$ drops abruptly to zero at $\varphi_T = \varphi_F$. For higher temperatures, the Fermi level corresponds to that energy level $\varphi_T$ at which the probability of occupancy is $\frac{1}{2}$. With metals, there actually exists an occupied energy level corresponding to the Fermi level. With insulators, it is a hypothetical energy level above which electron traps due to imperfections will probably be empty and below which they will probably be full. This results from the fact that the Fermi level lies in the forbidden band for insulators.

*b. Mechanisms of Charge Transfer.* When two materials are brought together, the Fermi levels for each must equalize to give a common Fermi level for the combined system. Thus electrons will flow from the material with the higher initial Fermi level to that with the lower. The material with the higher initial Fermi level, therefore, becomes positively charged. For metals, the difference in Fermi levels is equal to but opposite in sign to the differences in the work functions (the energy required to remove an electron from the surface of the metal). Thus, when two metals are contacted, the one with the higher work function becomes negative. This type of contact charging, however, is possible only if there is an allowable (i.e., not forbidden) unoccupied energy level in the material with the lower Fermi level to receive the electrons from the material with the higher Fermi level. This will always be the case with metals and some semiconductors. However, as pointed out by Vick (V2), the Fermi level of most insulators is below that of metals but the conduction band is above that of the Fermi level for metals (i.e., the Fermi level of the metal lies in the forbidden band for the insulator). Thus no electrons can flow from the metal to the insulator since there is no allowable energy band open and no electrons can flow from the insulator to the metal because the insulator electron energy levels are too low. In unusual cases, the conduction band for the insulator is below the Fermi level for the metal so that electron flow from the metal to the insulator can occur. With two insulators in contact, electron transfer could only occur if the conduction band for one were below the valence electron level (maximum filled energy level) for the other, an unlikely condition. Strictly speaking the above statements are valid only for 0°K where no electrons exist in the conduction band. At any finite temperature some electrons will occupy the conduction band to an extent dictated by Fermi–Dirac statistics. Thus there can always be flow between the conduction bands although the rate may be insignificant until temperatures become very high.

If the temperature of one insulator is raised (as by rubbing), electrons may be transferred to the conduction band or the band levels may be altered to an extent that would permit appropriate electron flow. The presence of surface states may also alter the general picture. Such states, acting as additional levels within the forbidden band for trapping electrons, may originate in various ways, including imperfections of the lattice structure at the surface and the presence of other adsorbed atoms.

Thus the charging resulting from metal–metal contact is readily explained in terms of modern concepts of structure and would follow the principles described above in Section V, B. For perfect and pure insulators the same theory would predict that charging by simple contact may be very slow. However, the alteration of surface states, either by the presence of imperfections or contaminants or by intense contact (pressure or rubbing), could explain significant charging by electron transfer. Harper (H3), for example, has shown that simple contact between amber and a metal, silica, or polystyrene, will yield no significant charge whereas rubbing is known to give a very large charge.

With metal–metal contact, it is also possible to obtain an electrolytic type contact potential if adsorbed water films are present. Deaglio (as quoted by Harper, H3), in investigating the contact potential between silver and nickel balls, observed that when he carefully controlled the gap between the surfaces the sign of charge reversed at a gap width of the order of 100 Å. Thi he explained by the formation of an electrolytic cell as the result of water adsorption.

Charging of contacting bodies could occur by transfer of ions between the surfaces rather than electrons. This would require, however, that free ions be present in the surface. If one in every 100,000 surface molecules were an ion, the charge level would correspond to the order of $\mathscr{E}_{ps} = 1$ V/micron. The presence of such small concentrations of surface ions is not unreasonable to account for the charge levels that have been observed. Harper (H3) prefers such ionic transfer as the probable explanation of the charging of insulators on light contact. The early experiments by Knoblauch (K5) suggest an ionic transfer mechanism. He determined the charge acquired by a plate of platinum, glass, or sulfur when various substances were allowed to slide off of it. In general, the plates were charged positively by acidic substances and negatively by basic substances.

From the above, it would be expected that, except for metal–metal contacts, charging can be very sensitive to the method of contacting and to the surface condition (or state of contamination) of the material. This is, in general, borne out by the seeming contradiction of available data, which are based on widely different methods of contacting and degrees of cleanliness.

Van Ostenburg and Montgomery (V1) have reported calculations of the surface charge density that might be expected when contacting specific types of materials. Their calculations are based on thermodynamic equilibrium in which the Fermi level for the entire system becomes constant and ignore consideration of the time that might be involved in achieving equilibrium when depending on chance transfer of electrons through the conduction band by virtue of thermal energy. In many of their cases the calculated charge levels are extremely high ($>0.01$ C/meter$^2$ or $\mathscr{E}_{ps} > 1000$ V/micron). This might imply that final charge levels are determined largely by considerations of back discharge upon separation of the surfaces. Montgomery (M7) and Harper (H3) believe that back discharge is mainly due to electron tunneling[8a] rather than air breakdown.

The general concept that work functions were somehow involved has led a number of investigators to set up a triboelectric series in which any substance in the series will acquire a positive charge if contacted or rubbed by a material lower in the series and a negative charge if rubbed by one higher in the series. Shaw and Jex (S8) claim it is impossible to devise a simple series. To illustrate this they found that they could obtain a series (most positive given first)—filter paper, cotton, glass, zinc, silk, filter paper—that is actually

---

[8a] Electron tunneling is the process whereby electrons are emitted from a body as the result of the probability of an electron penetrating a potential barrier according to quantum mechanics considerations.

## TABLE XIII
### Typical Triboelectric Series (Most Positive at Top of List)

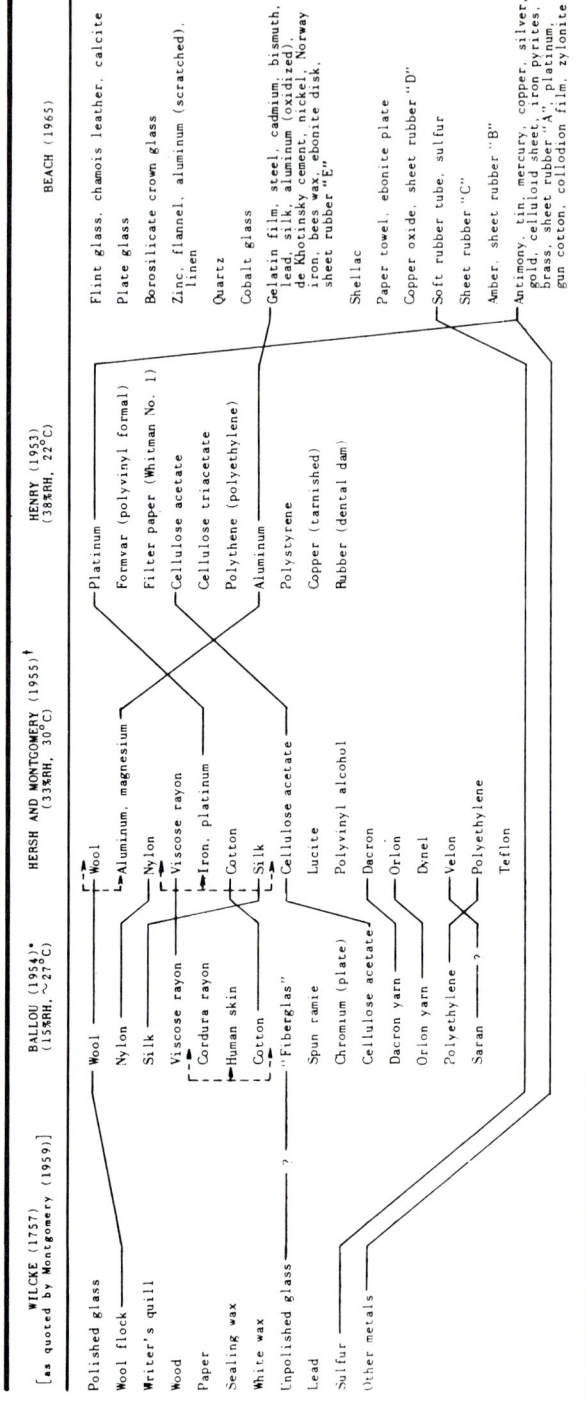

| WILCKE (1757) [as quoted by Montgomery (1959)] | BALLOU (1954)* (15%RH, ~27°C) | HERSH AND MONTGOMERY (1955)† (33%RH, 30°C) | HENRY (1953) (38%RH, 22°C) | BEACH (1965) |
|---|---|---|---|---|
| Polished glass | Wool | Wool | Platinum | Flint glass, chamois leather, calcite |
| Wool flock | Nylon | Aluminum, magnesium | Formvar (polyvinyl formal) | Plate glass |
| Writer's quill | Silk | Nylon | Filter paper (Whitman No. 1) | Borosilicate crown glass |
| Wood | Viscose rayon | Viscose rayon | Cellulose acetate | Zinc, flannel, aluminum (scratched), linen |
| Paper | Cordura rayon | Iron, platinum | Cellulose triacetate | Quartz |
| Sealing wax | Human skin | Cotton | Polythene (polyethylene) | Cobalt glass |
| White wax | Cotton | Silk | Aluminum | Gelatin film, steel, cadmium, bismuth, lead, silk, aluminum (oxidized), de Khotinsky cement, nickel, Norway iron, bees wax, ebonite disk, sheet rubber "E" |
| Unpolished glass | "Fiberglas" | Cellulose acetate | Polystyrene | |
| Lead | Spun ramie | Lucite | Copper (tarnished) | Shellac |
| Sulfur | Chromium (plate) | Polyvinyl alcohol | Rubber (dental dam) | Paper towel, ebonite plate |
| Other metals | Cellulose acetate | Dacron | | Copper oxide, sheet rubber "D" |
| | Dacron yarn | Orlon | | Soft rubber tube, sulfur |
| | Orlon yarn | Dynel | | Sheet rubber "C" |
| | Polyethylene | Velon | | Amber, sheet rubber "B" |
| | Saran ? | Polyethylene | | Antimony, tin, mercury, copper, silver, gold, celluloid sheet, iron pyrites, brass, sheet rubber "A", platinum, gun cotton, collodion film, zylonite |
| | | Teflon | | |

| SILSBEE (1942) | GANOT (1917) [as quoted by Guest (1933)] | GRUNER (1953) [as quoted by Shashoua (1958)] | ROSE AND WARD (1957) | FUKADA AND FOWLER (1958) |
|---|---|---|---|---|
| Asbestos | Catskin | Wool +42 | Ethyl Cellulose | Mica |
| Glass | Flannel | Perlon II +20 | Casein | PMMA[§] (unplasticized) |
| Mica | Ivory | Dacron +14 | Perspex | PMMA (red) |
| Wool | Rock crystal | Paper +12 | Tufnol | PMMA (plasticized) |
| Cat's fur | Glass | Glass, steel +10 | Ebonite | Amber (natural) |
| Lead | Cotton | Nylon +7 | Cellulose acetate | Amber (molded) |
| Silk | Silk | Cotton +5 | Glass | Polystyrene (British) |
| Aluminum | Human hand | Brass 0 | All metals | Polystyrene (U.S.A.) |
| Paper | Wood | Orlon −4 | Polystyrene | Polyethylene |
| Cotton | Metals | Hard rubber −14 | Polyethylene | Polytetrafluorethylene |
| Wood, iron | India rubber | Rubber −20 | Teflon (polytetrafluorethylene) | |
| Sealing wax | Sealing wax | (numbers above are contact potentials against brass) | Cellulose nitrate | |
| Ebonite | Resin | | | |
| Nickel, copper, silver, brass | Sulfur | | | |
| Sulfur | Gutta-percha | | | |
| Platinum, mercury | Gun cotton | | | |
| India rubber | | | | |

* Ballou points out that there appears to be a relation between structure of insulators and their position in the series. Amide groups (wool, nylon, silk) are at the positive end; hydroxyl-rich polymers are in the middle; hydrocarbons and halogenated hydrocarbons (polyethylene, saran) are at the negative end. At 60% relative humidity the positions in the series were the same except that "Fiberglas" moved to a position between wool and nylon. Also a fire-polished glass rod rubbed with silk is charged positively. When the surface was roughened with emery paper, the rod took on a negative charge when rubbed with either silk or orlon.

† Arrows indicate that metal shown is somewhere in range indicated by extremes of dashed line. These data for metals are based on scraped surfaces. The data for unscraped (presumably oxidized) metals were somewhat lower (more negative) in the series.

§ PMMA is polymethylmethacrylate.

cyclic. Such results could reflect different predominating mechanisms (as, for example, contact potentials vs electrolytic effects). Table XIII lists a number of such series. Because of the widely different substances covered by each investigator, it is difficult to compare the results directly. Some of the substances common to several of the series have been connected by lines to give some indication of the level of agreement. It should be noted that the inclination of these connecting lines does not indicate disagreement. It is only when these lines cross between adjacent series that a discrepancy is implied. Although there are a number of discrepancies, the general agreement is good, especially considering the variety of ways in which these data were obtained and considering the problems discussed above. One of the worst discrepancies is that for platinum reported by Henry. It has also been rather generally agreed that charge levels are not additive even if a valid triboelectric series should exist. It is possible, however, that this conclusion is the result of back-discharge processes rather than initial charging processes.

Based on electroendosmotic experiments, Coehn and Raydt (C4) came to the conclusion (later known as Coehn's rule) that the material of higher dielectric constant always assumes a positive charge and that the charge is proportional to the difference in dielectric constants. This can be written in the form

$$\mathcal{Q}_s = K_\delta (\delta_1 - \delta_2) \tag{85}$$

where $\delta_1$ and $\delta_2$ are the dielectric constants of the two materials rubbed together and $K_\delta$ is a constant with a value of $1.48 \times 10^{-5}$ C/meter$^2$, based on data reported by Richards (R4).

If an entire particle is assumed to have a surface charge density equal to $\mathcal{Q}_s$ (i.e., $\mathcal{Q}_{ps} = \mathcal{Q}_s$), this may be written in terms of $\mathscr{E}_{ps}$ as

$$\mathscr{E}_{ps} = K_{\delta\varepsilon}(\delta_1 - \delta_2) \tag{86}$$

where $K_{\delta\varepsilon}$ is a constant equal to $1.67 \times 10^6$ V/meter (1.67 V/micron) based on Richard's data.

Actually Coehn's rule is not generally applicable. In order to obtain the seeming correlation above, Richards had to resort to the artifice of assigning a value of 3.1 to the dielectric constant of steel. Nevertheless, in the absence of any other basis, Coehn's rule can be used as a guide to the sign of charge that may be expected when contacting various *insulators*. It is also interesting to note that positive ion transfer by virtue of image forces resulting from dielectric constant differences between two contacting surfaces would account for Coehn's rule that high dielectric constant materials become positively charged in contact with low dielectric materials. An exception could be expected when the substance of high dielectric constant normally contains a very high concentration of positive surface ions.

*c. Typical Data.* The literature on solid–solid charging is immense and covers all conceivable ways in which solids may be contacted. The following are summaries of some representative results in various areas to illustrate both the nature and the magnitude of charges attained.

(1) *Simple Contact*: Harper (H4) indicates that typical charge levels obtained in contacting surfaces range between $3 \times 10^{-6}$ and $3 \times 10^{-5}$ C/meter$^2$ (equivalent to $\mathscr{E}_{ps} = 0.4$–4 V/micron)[9]; with powders, charge densities may be an order of magnitude lower.

In experiments involving the contacting of quartz spheres having specific orientations of their crystal faces, Harper (H2) attained local charges that were estimated to be as high as 0.029 C/meter$^2$. If this charge density had existed over the entire sphere, the corresponding value of $\mathscr{E}_{ps}$ would be 3300 V/micron.[9] Actually, for this localized charge, the surface field intensity would be only some 600 V/micron.[9] He attributes the failure for this charge to leak off by air breakdown to the extremely small volume of air exposed to the charge and the corresponding low probability of finding a free electron in this volume to initiate gas breakdown.

Harper (H3) gives an analysis which he claims shows that the final charge, when contacting two metals, is not significantly affected by the speed with which the surfaces are separated, although he presents too few details of his analysis to permit a critical evaluation of it. In his analysis he assumed that transfer due to electron tunneling could occur during the separation step. Bredov and Kshemyanskaya (B13) report that the charge that flows away from the contacted surfaces is reduced by a factor of 4 as the speed of breaking the contact is increased by a factor of 2. They claim that a theoretical calculation, assuming the presence of electron tunneling, agrees with their experimental results, but present no details of either this experiment or their calculation. It is implied that the contact was between two metals. Their data also show essentially a direct relationship between surface charge level and the contact potential difference for metal-to-metal and metal-to-semiconductor contacts.

(2) *Compressive Contact*: Rose and Ward (R8) made measurements of charges resulting from intensively pressing $\frac{1}{2}$- to 1-in.-diam dielectric spheres against flat metal or dielectric plates. They concluded that the amount of charge was directly proportional to the actual contact area, regardless of the time or intensity of the contact, the latter simply determining the area of contact. This could be construed to indicate that ion rather than electron

---

[9] The $\mathscr{E}_{ps}$ values are calculated assuming that the entire particle has a surface charge density corresponding to that over the contact area (i.e., assuming $\mathscr{Q}_{ps} = \mathscr{Q}_s$). Where the charge exists only over a limited surface area, the actual field intensity just outside of this area will be less than these values of $\mathscr{E}_{ps}$ by a factor $[\delta/(\delta + \delta_p)]$. This local field intensity will drop even lower as the charge has time to be conducted to other regions of the particle.

transfer is involved or that the final charge levels are determined primarily by considerations of back discharge. The area of contact was separately measured by coating the surface with a thin layer of soot and found to agree with predictions from Hertz's (H9) theory of sphere deformation on compression. Maximum areas of contact were 0.03 cm$^2$ (or less than 1% of the total sphere surface). Although they obtained a consistent triboelectric series as shown in Table XIII, they conclude that charge levels in the series were not additive. The charge attained when dielectric spheres were pressed against metal was independent of the nature of the metal and ranged from $4 \times 10^{-5}$ to $3 \times 10^{-4}$ C/meter$^2$ of contact area (corresponding to $\mathscr{E}_{ps}$ values of 4.5–30 V/micron).[9] When dielectric surfaces were pressed together, charge levels of $7 \times 10^{-5}$ to $2 \times 10^{-4}$ C/meter$^2$ were obtained ($\mathscr{E}_{ps} = 7.5$ to 20 V/micron). By way of comparison, Richard's (R4) values ranged from $5 \times 10^{-6}$ to $7 \times 10^{-5}$ C/meter$^2$ (corresponding to $\mathscr{E}_{ps}$ values of 0.6–8 V/micron).[9] Rose and Ward also found no significant charging when like dielectric surfaces were pressed together. Although Rose and Ward claim some correlation of charge level with the lowest dielectric constant of the surfaces pressed together, their plots of data are not convincing on this score. With dielectric-metal contacts, they did find the highest charge levels when the dielectric resistivity was greater than $10^{11}$ $\Omega \cdot$ cm.

(3) *Pneumatic Conveying Contact*: Kunkel (K11) pneumatically conveyed silica, sulfur, starch, talc, and nickel powders through quartz, Pyrex, sulfur, nickel, and platinum tubes. The highest charge level reported was 2700 electrons for a particle 6.6 microns in diameter blown out of a platinum tube (corresponding to $\mathscr{E}_{ps} = 0.4$ V/micron). Whenever contact between unlike substances was avoided (e.g., blowing quartz powder out of a quartz tube), there were an equal number of positive and negative charges. With unlike materials, the charge distribution may become asymmetric. In the most asymmetric case the ratio of charges of opposite sign was 4:1. When insulators were impacted on metals, the charge was always asymmetric, with the insulator particles having net negative charges. Changing relative humidity up to a level of 90% had no appreciable effect. The total charge acquired by particles was proportional to between the first and second power of particle diameter although there was a very wide scatter in the data.

Berg and Flood (B8) and Berg, Fernish, and Flood (B5, B6) made measurements on the charging of Carbowax 6000[9a] and saccharin (with and without 1% Cab-O-Sil additive), Cab-O-Sil,[9a] MgO, and NH$_4$Cl when passed pneumatically through a 0.8-mm-diam stainless steel tube (see Section VI for

---

[9a] Carbowax and Cab-O-Sil are trademarks of commercial materials. Carbowax is a series of polyethylene glycols manufactured by Union Carbide. Cab-O-Sil is a superfine silica from Cabot Corp.

details of equipment), at velocities of 3–15 meters/sec. Cab-O-Sil was predominantly negatively charged (ratio between negative and positive charge as much as 12:1). The charge for the others was about evenly split (within a factor of 2) between positive and negative. The charge level for the Carbowax and saccharin was in the range of 0.001–0.01 C/kg, while for Cab-O-Sil it was between 0.01 and 0.3 C/kg. A 1% addition of Cab-O-Sil to Carbowax or saccharin had only a small effect on charge. The charge level increased as the first to second power of gas flow rate but was independent of any initial charge on the particles fed into the tube. Mass median diameters were given as follows: Carbowax 6000, 14 microns; Cab-O-Sil, 0.02 microns; saccharin, 7 microns. Based on these diameters, the charge levels would correspond to $\mathscr{E}_{ps}$ values of 0.3–3 V/micron for Carbowax and saccharin and 0.004–0.1 V/micron for Cab-O-Sil. The low values of $\mathscr{E}_{ps}$ for Cab-O-Sil despite the high charge-to-mass ratios are undoubtedly due to failure to disperse the Cab-O-Sil to anything approaching its ultimate particle size. With MgO and $NH_4Cl$, charge levels were much higher, 0.1–1 C/kg with one value of 50 C/kg reported. Assuming a particle diameter of 1 micron (the nominal discrete particle diameter reported), these charge levels would correspond to $\mathscr{E}_{ps}$ values of 2–20 V/micron with the one extreme charge corresponding to 1000 V/micron. Any failure to disperse the particles would mean that the effective value of $\mathscr{E}_{ps}$ would be even larger. It seems unlikely that the extremely high value reported could be real.

Taneya (T1) measured the voltage within the cloud formed when skim milk was hydraulically atomized. This voltage varied from positive to negative as the distance from the nozzle increased. From this, he concluded that the small particles are charged positively, and the large ones negatively. It should be noted, however, that his probe (which was measuring voltages of the order of 0–300 V) need not necessarily reflect only the space charge. In pneumatic conveying of dried skim-milk powder, he found that the powder became negatively charged and that the duct potential (0–2000 V) increased with both concentration of particles in the air stream and with air velocity in the duct.

(4) *Sliding Contact*: Taneya (T1) also allowed five kinds of powder (dried skim milk, dried milk, dried buttermilk, flour, and α-lactose hydrate) to slide down a grounded V-shaped chute and measured the charge of individual particles in a modified type of Millikan cell. The average particle size ranged from 30 to 48 microns, and the average absolute charge from 1 to $4 \times 10^{-14}$ C/particle, with corresponding $\mathscr{E}_{ps}$ values of from 0.4 to 0.7 V/micron. From one-third to one-half of the particles were uncharged, and the rest were approximately evenly divided between positive and negative particles. The net charge of the powder was positive in all cases and corresponded to charge-to-mass ratios of $3 \times 10^{-6}$ to $4 \times 10^{-5}$ C/kg. Particles

larger than 40 microns did not appear to be charged. This was believed to be due to flocculation with smaller particles to achieve charge neutralization. He also found that the net charge-to-mass ratio appeared to increase with fat content of the material.

(5) *Rolling Contact*: A number of investigators have studied charges developed during rolling contact. This process is useful for assessing certain charging mechanisms. Peterson (P4, P5) rolled 2-mm borosilicate glass spheres and 6-mm quartz spheres over nickel at a variety of ambient air pressures and rolling speeds. Both types of spheres acquired a negative charge, which was at a minimum for an air pressure of about 1 Torr absolute. This was interpreted as indicating that at significant air pressures (of the order of 1 Torr and over) the maximum charge that the spheres could acquire is limited by discharge due to gas breakdown. Wagner (W1) conducted a similar series of experiments using a wide variety of sphere materials rolled on nickel at very high vacuum ($10^{-5}$ Torr) in an attempt to measure maximum charge levels without being limited by the question of discharge by gas breakdown. When rolled on nickel, aluminum oxide became negatively charged while the alkali halides became positively charged. Clean magnesium oxide (periclase) was positively charged but became negative after heating to 1200°C. This was attributed to loss of excess oxygen to produce the stoichiometric compound on heating, since it regained its original positive charge after successive polishing. Both Peterson and Wagner found a maximum rate of charging at rolling speeds of some 10–15 cm/sec. The increase in rate of charging with rolling speed is expected because of the equilibrium that is established between the contact charging and the discharging due to surface or bulk conductivity of the spheres. The reduced net charging rate at the high speeds was attributed either to a change in surface state as a result of the more intensive contacting or to a local heating effect with an attendant increased surface conductivity. From a detailed study of the maximum charges acquired at high vacuum, it was concluded that there was one case (silica) where there was electron transfer of a type related to the contact potential of metals. In other cases, there were indications of both ion transfer and electron transfer not related to work function charging.

Berg (B4) and Berg and Gaukler (B9) measured the charge attained by 5-mm-diam borosilicate glass beads when they were rolled down a 24-in-long, 10-mm-i.d. Pyrex tube at various frequencies and at various humidities. In all cases, the bead charge increased with the number of times it was rolled, usually reaching a maximum charge after some 400 cycles or rollings. At relative humidities below 1.5%, the initial bead charge was random (half of the beads positive, half negative), but the beads all eventually acquired a negative charge. At humidities of 1.5% and over, almost all the beads were charged negatively at the start, but became positive by the time the maximum average charge was reached. At humidities of under 1% and over 35%, the average maximum absolute charge attained was approximately $2 \times 10^{-9}$ C per bead, while at intermediate humidites, it was $1.7 \times 10^{-8}$ C (corresponding to values of $\mathscr{E}_{ps}$ of 3 V/micron and 25 V/micron respectively). It would appear from these results that the beads accumulated charge up to a maximum level dictated by the point where charge is lost owing to either air ionization at the high charge levels or surface conductivity at the high humidities. The charge level of 25 V/micron is difficult to explain, however, since it is almost an order of magnitude larger than the field intensity needed for air breakdown from a 5-mm bead.

(6) *Miscellaneous Data*: Shaffer (S5) reported net charges for BZ,[9b] Carbopol,[9c] and saccharin of $10^{-5}$ C/kg after working them with a spatula. BZ was positive, the others negative, and only about half the charge disappeared after 6 days storage. He also measured the average charge of saccharin power (1- to 6-micron particles) after spraying from a DeVilbiss atomizer to be of the order of 0.001 C/kg (corresponding to $\mathscr{E}_{ps} \approx 0.05$ volt/micron). The sign distribution was 25% positive; 64% negative; 11% neutral. The resistivity of saccharin was measured at $10^{15}$ $\Omega \cdot$ cm.

Fukada and Fowler (F5) established a triboelectric series by lightly rubbing specimens of insulators together (see Table XIII for their results). They made no attempt to obtain quantitative measurements of charge because of the different geometries of the bodies used, but did attempt to correlate the series with corresponding measurements of the number and distribution of electron traps for each of the substances. While it was concluded that those substances with the greatest number of traps and the steepest distribution would become negatively charged, the data reported are too erratic to be convincing.

The importance of surface states is demonstrated by the data of Thomas (T2) who found that the charge acquired by coal when contacted with iron was a function of its degree of oxidation. Raw coal had a positive charge. As the coal became oxidized (by exposure to air at 350°C for various lengths of time), the coal acquired less of a positive charge, finally becoming strongly negative with a highly oxidized coal. This phenomenon was proposed as the basis for a method of process control during coal carbonization. Brasefield (B12) found similar results for carbon blacks when allowed to slide down a grounded nickel surface. Blacks with a low oxygen content acquired a large positive charge, while blacks with a high oxygen content became negative.

---

[9b] BZ is a chemical incapacitating agent. Its composition is security classified.
[9c] Carbopol is a thickening agent from B. F. Goodrich Chemical Company.

The charge was zero at a 4% oxygen content. Similar phenomena in the inorganic field have previously been discussed in connection with Wagner's (W1) results on magnesium oxide.

### 4. *General Comments*

From the above discussion it is apparant that, in the absence of external electrostatic fields, charging of liquids by means of contacting or interface alteration mechanisms are capable of yielding only low charge levels, corresponding to $\mathscr{E}_{ps}$ values of less than 0.1 V/micron, probably less than 0.01 V/micron. With solids however, relatively high charge levels ($\mathscr{E}_{ps} = 1 - 10$ V/micron) can be attained. Although some values have been reported that are radically higher than this, there is reason to question their validity.

Although true homopolar charging is basically possible in contact charging of solids, and has been obtained with single bodies, it has never been achieved with powders, which have always yielded a significant amount of ambipolar charges. This is because of the contact that invariably occurs between like particles, either in suspension or as the result of some particles adhering to and coating pipe walls. From a practical standpoint it seems doubtful that homopolar charging of powders can be realized by these mechanisms. The closest approach to homopolar charging is probably an extreme asymmetric charging.

It should also be recognized that conducting particles are charged on the first contact; repeated contacts will not yield added charging (assuming that the particle shape is not altered in the process). With insulators however, successive contacts will permit additional charging until the maximum possible stable charge exists over the entire particle surface. With conducting particles it is possible that flattened shapes will permit high charge levels to be attained. Distortion during contacting will not be useful in this respect unless the elastic limit is exceeded so that the distortion is permanent.

It seems fortuitous that the same order of maximum charge levels has been reported for metal–metal contacts as for insulator–insulator contacts. This might indicate that back discharge could be a controlling factor in establishing final charge levels.

In a case where charge is acquired by rolling, the surface conductivity of the particle may be important in establishing the final level of charge acquired. Thus an insulator of high conductivity (say $\mathscr{R}_0 < 10^{10}$ Ω · cm) may not only be incapable of acquiring the high local charge possible with a metal–metal contact, but, because of its high conductivity, may not benefit from successive contacts along other parts of its surface. On the other hand, the fact that its conductivity is high would imply a high concentration of surface ions. Thus, if charging is due to ion transfer, a higher local charging may be possible. Current data are not yet adequate to resolve questions of this type.

## G. CHARGING BY PHASE CHANGE

### 1. *Solidification or Freezing*

A number of investigators have reported a net electrification of ice crystals formed as the result of the freezing of water droplets. Mason and Maybank (M1), Latham and Mason (L6), Latham (L4), and Evans and Hutchinson (E4), have reported charge levels corresponding to $\mathscr{E}_{ps}$ values in the range of 0.004–0.02 V/micron. Kachurin and Bekryaev (K1) report values of the order of 0.1 V/micron. In all cases, the fine ice crystals are positively charged. The most likely explanation of this phenomenon is that proposed by Latham and Mason (L5). They suggest that charging is not a direct result of the freezing process but that charge displacement occurs because of the combined effect of different ion mobilities and temperature gradients. If a temperature gradient exists within a drop, there will also be a gradient in degree of water ionization, the concentration of both $H^+$ and $OH^-$ ions being greatest in the region of highest temperature. Because of the much higher mobility of the $H^+$ ion, it will diffuse more rapidly than the $OH^-$ ions toward the colder regions of lower ionic concentration. This displacement will set up a potential difference, which will then induce a backflow of $H^+$ ions. An equilibrium potential difference or charge segregation will be reached when the back-flow rate equals the forward diffusion rate. A net charge separation, however, cannot be realized until the drop is shattered. While this phenomenon should occur with either a liquid or solid across which a temperature gradient exists, the difference in mobilities between $OH^-$ and $H^+$ ion is greater in ice than in water.[10]

Since the outside of the drop will be the coldest and the first to freeze, it will tend to contain an excess of $H^+$ ions with the compensating $OH^-$ being concentrated at the center of the drop. When the drop shatters as the result of stressess set up by freezing, it will send out fine crystals from the surface and these will be positively charged. Latham and Mason (L5) have

---

[10] Based on data reported by Eigen and De Maeyer (E2), the mobilities at 0°C in units of (cm/sec)/(V/cm) are

| Type ion | In water | In ice |
|---|---|---|
| $H^+$ | 0.0024 | 0.1–0.5 |
| $OH^-$ | 0.0011 | ≤0.05 |

predicted that relatively high charge levels can be attained with sufficiently high temperature gradients.

## 2. *Condensation*

Very few data are available on possible charges arising as the result of condensation. Dubois (D8) made measurements on the evaporation of and condensation on 20- to 60-micron droplets consisting of water solutions of various salts. He reports the appearance of a positive charge on the droplet as the result of condensation and negative charge as the result of evaporation. The charge level corresponded to $\mathscr{E}_{ps}$ values of the order of 0.01 V/micron or less. Dubois attempts to explain this on the basis of the formation of ions by rupturing of molecular bonds during evaporation, with the electrons remaining in the liquid phase. He gives no reason, however, why the positive ions should go preferentially into the vapor phase.

### H. OTHER METHODS OF CHARGING

There are two other methods by which particles can become charged. These both involve emission of electrons or ions: photoemission and field emission. Photoemission results from the bombardment of the particle surface by electromagnetic radiation. Field emission is the result of subjecting the particle surface to a high electric stress (field intensity).

Visible light or other electromagnetic radiation incident on a solid, liquid, or gas can liberate electric charges. This is called photoelectricity. Ejection of electrons from the surface is usually called photoemission. Electrons or positive ions formed in a gas as the result of such radiation is called photo-ionization. Such a process, however, cannot charge a particle directly. The charging process in that case is a direct result of subsequent diffusion.

When a material is subjected to an electric field intensity greater than $10^7$ V/cm (1000 V/micron), electrons may be emitted. At intensities greater than $10^8$ V/cm ($10^4$ V/micron), ions may be emitted as the result of electron bombardment while at $5 \times 10^8$ V/cm direct field desorption may occur.

Photoemission is unlikely to be a significant contributor to the charging of particles. Field emission, however, could be involved in some of the intensive electrostatic atomization processes where very fine highly charged particles are produced. Because of the dubious current applicability, no effort has been made to review the literature in these cases.

Several other mechanisms could be construed as potential means for charging particles, but they are all secondary factors that depend on some other mechanisms for achieving a charge on the particle. Nuclear fragments, for example, may produce gaseous electrons, ions, or photons. It is then these that actually result in particle charging by means of diffusion or emission. Photoionization, previously mentioned, is in this category. The use of radioactive substances for alleviating problems with static electricity in industry are also in this category. The radioactivity ionizes the surrounding gas, which then acts to neutralize any free charges by diffusion and electrostatic attraction. Shock waves can produce some form of radiation or temperature rise, which then permits charging by one of the previously described mechanisms. Chemical reactions may generate ions (or free radicals), which are then available for diffusion charging of particles or direct charging if the reaction is allowed to take place on the particle surface.

## VI. Charge Measurement Techniques

Particle charge can be determined directly by measuring either (1) the current flowing from (or to), or the change in potential of, a surface from which the particles are disseminated or upon which they are deposited, or (2) the space charge due to the presence of the particles. Particle charge can also be determined indirectly by measuring the behavior of the particles in an electric or electromagnetic field. The direct measurements will normally measure particle charge by itself, whereas the indirect measurements will measure some combination of particle charge with some other particle property (such as diameter or mass). Thus, to obtain charge alone by the indirect methods requires another independent measurement of the other particle property included. Some of the indirect measurement methods are designed to do this in the same apparatus; others are not.

The charge on particles may be measured individually or collectively. The individual measurements will give a complete charge distribution if a large number of particles are measured. Collective measurement may measure the net charge (i.e., algebraic sum of the individual charges), the total charge (i.e., sum of the individual charges regardless of sign), or the charge distribution (i.e., relative quantity of positive, negative, and neutral). Such measurements may be made to yield an average value of charge for the entire aerosol or to yield a complete charge distribution.

Various types of measurements are outlined in Table XIV. Table XV gives a summary of many of the specific techniques that have been reported in the literature, indicating the general nature of each technique and what it measures. The following will describe each technique in more detail, approximately in the order that it is mentioned in Table XV.

Hendricks (H5) and Cho (C2, Method No. 1) both measured particle charge by measuring the voltage pulse on an oscilloscope due to the passage of a charged particle through a drift tube detector. The particle mass was

## TABLE XIV

### Types of Measurements for Determining Particle Charge

I. Direct
   A. Measurement of Current
      1. Current flowing from the surface (or electrode) due to continuously depositing particles
         a. permanently deposited
         b. momentarily deposited
      2. Current flowing from a surface (or electrode) from which particles are continuously disseminated
   B. Measurement of Voltage
      1. Due to cumulative charge
         a. from deposition of particles on a surface (or electrode)
         b. from dissemination of particles from a surface (or electrode)
      2. Due to instantaneous charge
         a. from an enclosed space charge
         b. from momentary deposition of particles
II. Indirect
   A. Measurement of Particle Path
      1. In quiescent fluid
         a. in dc field
         b. in ac field
      2. In moving fluid
   B. Measurement of Balancing Forces on Particles
      1. Static (Millikan cell)
      2. Dynamic (quadrupole mass spectrometer)
   C. Measurement of Particle Deposition

determined by measuring the velocity attained by the particle when it was accelerated through a known distance by a known potential gradient. A high vacuum was used in this apparatus.

Hopper and Laby (H14) used a modified version of the Millikan cell. A charged particle falling between two parallel vertical plates is photographed both in the absence and in the presence of an electric field. In the absence of the field a vertical track is obtained (provided no convective air currents are present). By using a timed, intermittent illumination, the particle settling velocity can be calculated from this track. In the presence of the field, the particle will have a path that is inclined to the vertical. The angle of the path with the vertical is a direct measure of the charge-to-mass ratio. The direction of the inclination gives the sign of the charge. Particle mobility is obtained from the combination of the charge-to-mass ratio and settling velocity. It could, however, have been obtained directly by measuring the horizontal drift rate in the presence of the field. Kunkel and Hansen (K12) and Dodd (D5) used the same basic technique. Kunkel and Hansen improved the optical system to permit simultaneous determinations for a large number of particles. They also state that the entire apparatus must be maintained isothermal within $\pm 0.01°C$ to avoid excessive convection currents.

Wells and Gerke (W2) used a method similar to that of Hopper and Laby (H14) except that the field was reversed at a fixed low frequency ($\sim 1$ cps), giving the equivalent of a low-frequency square-wave alternating current. This caused the particle to pursue a zigzag path, the amplitude of which is a direct measure of the particle mobility. Any convection currents can be detected by a drift in the zigzag path. Natanson (N1) and Solov'yev (S10) used this same basic method. This method does not permit the sign of charge to be established unless a prolonged period of dc operation is superimposed.

Taneya (T1) also photographed the fall of particles between two vertical parallel electrodes using a 50-cycle alternating current. The mobility of the particles in the electrostatic field is obtained from the horizontal amplitude of the particle path and the current frequency. The gravitational settling velocity is obtained from the wavelength of the vertical path traversed and the frequency. Sign of charge was determined by a momentary application of a dc potential. This method minimizes sensitivity to convection currents.

Goyer, Gruen, and LaMer (G7), Shaffer (S5), and Doyle, Moffett, and Vonnegut (D6), measured particle charge-to-mass ratio by balancing the particle, with an electric field counteracting the effect of gravity. They all used either the classical Millikan cell procedure, or some modification thereof. A separate measurement in the absence of the electric field is needed to obtain particle size. In the Millikan cell, this is obtained by measuring the gravity settling velocity of the identical particle in the absence of the electric field. Thus, in effect the Millikan cell measures charge-to-mass ratio and mobility. Doyle *et al.*, (D6) measured drop size by a calibrated filter-paper absorption technique of other particles assumed to be of the same size.

Vonnegut *et al.* (V3), in addition to reporting the use of the same method described by Doyle *et al.* (D6), suggested a modification in the same category that would permit measurement of both charge and mass. Because of the ionic image effect, the electric field required to balance a charged particle against the action of gravity will become less as the particle approaches the top electrode. They show that, by measuring the necessary balancing force at two clearances between the particle and the electrode, both particle charge and particle mass are established, provided the clearance is also measured in each case. There is, however, no report of any actual application of this method.

Shapiro and Watson (S6) and Frickel (F2) considered a device which in principle is similar to the quadrupole mass spectrometer to be discussed later. In this device individual particles are suspended and balanced in an alternating field in such a way that they cannot escape. The frequency of

## TABLE XV
## SUMMARY OF CHARGE MEASUREMENT TECHNIQUES REPORTED IN LITERATURE

| Type of Analysis | | Type of Charge Measurement[†] | Scope of Charge Analysis | | | Quantity Actually Measured[**] | | | | Pressure (atm) | Author(s) | Year | Ref. |
|---|---|---|---|---|---|---|---|---|---|---|---|---|---|
| | | | Gives Single Average Values Only | Gives Distribution Data | | Charge ($Q_p$) | Charge-to-Mass Ratio ($Q_p/m_p$) | Charge and Mass ($Q_p$ & $m_p$) | Electro-Static Mobility[***] ($Q_p/D_p$) | | | | |
| Particles are measured individually | | I B 2 a | | * | | | | X | | $\sim 10^{-8}$ | Hendricks | 1959 | H5 |
| | | I B 2 a | | * | | | | X | | $\sim 10^{-8}$ | Cho (Method 1) | 1964 | C2 |
| | | II A 1 a | * | | | | X | | | 1 | Hopper,Laby | 1941 | H14 |
| | | II A 1 a | | * | | | X | | X | 1 | Kunkel,Hansen | 1950 | K12 |
| | | II A 1 a | | * | | | X | | X | 1 | Dodd | 1953 | D5 |
| | | II A 1 b | | * | | | X | | X | 1 | Wells,Gerke | 1919 | W2 |
| | | II A 1 b | | * | | | X | | X | 1 | Natanson | 1949 | N1 |
| | | II A 1 b | | * | | | X | | X | 1 | Solov'yev | 1957 | S10 |
| | | II A 1 b | | * | | | X | | X | 1 | Taneya | 1963 | T1 |
| | | II B 1 | | * | | | X | | X | 1 | Goyer,Gruen,LaMer | 1954 | G7 |
| | | II B 1 | | * | | | X | | X | 1 | Shaffer (Method 1) | 1962 | S5 |
| | | II B 1 | | * | | | X | | | 1 | Doyle,Moffett,Vonnegut | 1964 | D6 |
| | | II B 2 | | * | | | X | | | $\sim 10^{-8}$ | Shapiro,Watson | 1963 | S6 |
| | | II B 2 | | * | | | X | | | 1 | Frickel | 1964 | F2 |
| Particles are measured collectively | (a) Net charge only | I A 1 a | X | | | | | X | | 1 | Penney,Lynch (Method 1) | 1957 | P2 |
| | | I A 1 a | X | | | | | X | | 1 | Hewitt (Method 1) | 1957 | H10 |
| | | I A 1 b | X | | | X | | | | 1 | Soo et al. | 1964 | S11 |
| | | I A 2 | X | | | | | | | 1 | Graf | 1962 | G8 |
| | | I B 1 a | X | | | | | X | | 1 | Shaffer (Method 2) | 1962 | S5 |
| | | I B 1 a | X | | | | | X | | 1 | Masters | 1953 | M2 |
| | | I B 1 b | X | | | X | | | | 1 | Whitman | 1926 | W5 |
| | | II B 2 | | | X | | X | | | $\sim 10^{-8}$ | Schultz, Branson | 1959 | S2 |
| | | II B 2 | | | X | | X | | | $\sim 10^{-8}$ | Schultz, Wiech | 1960 | S3 |
| | | II B 2 | | | X | | X | | | $\sim 10^{-8}$ | Hogan | 1963 | H12,H13 |
| | | II B 2 | | | X | | X | | | $\sim 10^{-8}$ | Cho (Method 2) | 1964 | C2 |
| | | II C | | | X | | | | X | 1 | Cohen | 1964 | C5 |
| | (b) Positive, negative, and neutral fractions | I A 1 a | X | | X | | | X | | 1 | Penney,Lynch (Method 2) | 1957 | P2 |
| | | I A 1 a | X | | X | | | | | 1 | Berg,Fernish,Flood | 1964 | B6 |
| | | II A 2 | | X | | | | X | | 1 | Daniel,Brackett | 1951 | D1 |
| | | II A 2 | | X | | | | X | | 1 | Gillespie,Langstroth | 1952 | G4 |
| | | II A 2 | | X | | | | X | | 1 | Langer,Radnik | 1961 | L3 |
| | | II A 2 | X | | | | | X | | 1 | Sergiyeva | 1958 | S4 |
| | | II A 2 | X | | | | | X | | 1 | Hewitt (Method 2) | 1957 | H10 |
| | | II A 2 | X | | | | | X | | 1 | Kraemer,Ranz | 1952 | K8 |
| | | II A 2 | X | | | | | X | | 1 | Suzuki,Tomura | 1962 | S13 |
| | | II A 2 | X | | | | | X | | 1 | Hinkle,Orr,DallaValle | 1954 | H11 |

oscillation of these particles is a measure of their charge-to-mass ratio. Shapiro and Watson gave only a theoretical analysis.

Penney and Lynch (P2, Method No. 1) and Hewitt (H10, Method No. 1) collected aerosol in a fibrous filter. They measured the current flow to ground from this filter during the deposition and the total mass of aerosol collected during the run.

Soo *et al.* (S11) allowed the particles, moving in a duct at high velocity, to impinge on a $\frac{1}{8}$-in.-diam stainless steel ball. The discharge current to ground from the $\frac{1}{8}$-in ball was measured. In interpreting the results it was assumed (without being proven) that each particle is discharged completely before being reentrained from the ball and that no recharging of the particles occurs because of contact potential effects. Several other assumptions were also made but were less basic in nature and are subject to calibration or correction.

Graf (G8) measured the total particle charge by observing the current flow to or from the metallic needle from which liquid was being atomized electrostatically, assuming that all this charge is transmitted to the liquid drops. The mass of atomized liquid was determined by weighing the amount subsequently collected on a grounded plate.

Shaffer (S5) collected the particles in a Faraday cage and measured the change in potential and the change in mass of the cage to establish particle charge and mass. Masters (M2) used a similar procedure using a filter paper inside of a duct, letting the duct act as a Faraday cage. Both methods are similar to those used by Penney and Lynch and by Hewitt except that a total voltage change was measured rather than an integrated current flow in order to obtain total charge.

In Whitman's (W5) experiments. particles became charged by passing the aerosol through a metallic tube. The charge acquired was measured by noting the rise in potential of the tube by an electroscope and assuming the corresponding charge was all transferred to the particle. This is in effect similar to Graf's (G8) technique except that Graf measured current rather than voltage.

Schultz and Branson (S2), Schultz and Wiech (S3), Hogan (H12, H13), Cho (C2, Method No. 2), and Cohen (C5) all used the quadrupole mass spectrometer. This involves passage of accelerated aerosol particles between four longitudinal electrodes in high vacuum. These electrodes have imposed

---

Footnotes to Table XV:

\* Individual particles are measured; hence distribution data are obtained by measuring a number of particles.

\*\* Does not include measurements which authors may have made by a separate apparatus or technique.

\*\*\* Actually measures $k_C \mathcal{Q}_p/\mu D_p$, thus including some fluid properties as well.

† See Table XIV for details.

on them an oscillating electric field adjusted to allow only particles of a given mass-to-charge ratio to penetrate longitudinally. A high vacuum ($\sim 10^{-8}$ atm) is essential to the operation of this device since any dampening factor (such as fluid friction) interferes with the discriminating ability of the instrument.

Penney and Lynch (P2, Method No. 2) passed an aerosol between two parallel plates with an adjustable potential across them. The collection efficiency achieved is a measure of the average electrostatic mobility of the particles.

Berg, Fernish and Flood (B6) passed an aerosol between two shielded parallel electrodes with a potential of $+1000$ and $-1000$ V, respectively. The clearance between the electrodes was 4 mm. The positively charged particles are caused to migrate toward the negative electrode, the negatively charged ones toward the positive electrode, and the neutral particles are collected in a filter. The magnitude of the charge is determined by measuring and integrating the discharge current from each electrode. The total mass is determined by weighing the deposit on each electrode and on the filter. Air flow rates were 50–500 cm$^3$/min and the total mass collected was 0.1–2 mg during a run. The air was introduced through a 0.8-mm-i.d. tube at velocities of 1.5–15 m/sec. Since runs were of the order of 1 min, aerosol concentrations were of the order of 1,000–10,000 μg/liter (corresponding to $10^6$–$10^7$ particles/cm$^3$ for 1-micron-diam particles). No significant quantity of uncharged particles was ever found.

The apparatus of Daniel and Brackett (D1) was in principle similar to that of Berg, Fernish, and Flood (B6) except that they passed the aerosol through two sets of parallel plates in series, $\frac{3}{16}$ in. apart in each set. Potentials across the plates were varied over the range $\pm 800$ V. The second series of parallel plates is essentially a basis for checking results from the first set. The current from each set of plates was measured as a function of the applied voltage. By differentiating the curves obtained in this fashion, it is possible to obtain a distribution in terms of electrostatic particle mobility. This technique is essentially the electrostatic counterpart of the continuous gravity settling chamber or of the Oden sedimentation balance.

Gillespie and Langstroth (G4) allowed a flat aerosol stream, surrounded by a shielding gas stream of like velocity, to flow between two parallel electrodes. The particles were collected on the electrodes and examined microscopically. The position on the electrode is a measure of mobility. The additional independent measurement of size gives a means for calculating the corresponding charge. A distribution is obtained by counting particles along the entire electrode length. Langer and Radnik (L3) used a similar device.

Sergiyeva (S4) employed a modification of Gillespie and Langstroth's method in which the aerosol is allowed to flow across the entire width of the apparatus. However, the aerosol first flows between a series of charged parallel plates where aerosol particles are precipitated from all but a thin stream. He prefers this procedure to the use of a shielding stream to minimize evaporation problems with liquid drops.

Hewitt's (H10, Method No. 2) method is similar to that of Gillespie and Langstroth except that, instead of collecting the particles, the actual concentration of aerosol at various points along the electrode was measured by passing it through a slit that could be moved.

Kraemer and Ranz (K8) also used equipment similar to that of Gillespie and Langstroth. However, since they were dealing with essentially homopolar aerosols, the aerosol jet was simply deflected from the axis by a fixed angle. They photographed the jet deflection and determined mobility from the angle so observed. Suzuki and Tomura (S13) give a theoretical analysis of the Kraemer and Ranz spectrometer.

Hinkle, Orr, and DallaValle (H11) passed a flat aerosol jet between two cylindrical electrodes, with the electrodes parallel to each other and to the plane of the aerosol jet. The jet was caused to spread toward the electrodes and was photographed. From the lateral displacement and optical intensity, they were able to obtain a distribution between positive, negative, and neutral particles and an approximate value of average charge.

It is interesting to note from Table XV that all methods that have been used to obtain charge distribution data by means of measurements on collections of particles at atmospheric pressure involve a measure of electrostatic particle mobility. Other aspects of charging have only been obtained with those methods in which individual particles are evaluated.

The particle deflection due to an electrostatic field alone is a direct measure of electrostatic particle mobility. This assumes that periods of particle acceleration are negligible, which is usually the case for particles of interest in aerosol dissemination. If this deflection is compared with that due to the gravity field, a direct measure of charge-to-mass ratio is obtained. Such a comparison is possible in all those techniques involving particle motions in a quiescent fluid. With a flowing fluid, however, the fluid velocity is usually so large compared with the gravitational settling velocity that the mobility due to the electrostatic field is the only item that can be measured directly.

It should also be noted that in those methods in which charged particles are deposited on an electrode under the influence of an electrostatic field, there exists the possibility of reverse contact charging of the particle if the particle is sufficiently conductive. Such contact charging can cause the particle to oscillate between the electrodes, carrying current from one to the other. This phenomenon has often been observed in connection with electrostatic printing and has been specifically reported by Cho (C2). Such a transfer of particles would result in a transfer of charge to the electrodes far in excess of that originally carried by the deposited particles and could introduce a serious error in all methods that measure charge transfer to charged, electrodes (Methods IA-1a or IA-1b in Table XIV). Daniel and Brackett (D1), Penney and Lynch (P2, Method No. 1), and Hewitt (H10, Method No. 1) used relatively low potential gradients so that this effect may not have been serious. Berg, Fernish, and Flood (B6) used very high potential gradients and this effect may have been large. From the published data, it is not possible to reach any conclusions on this score. It is noteworthy, however, that they report much higher charge-to-mass ratios than other investigators working with similar systems, a result which could be explained by such extraneous oscillatory transfer. Ruling against this explanation, on the other hand, is the illustrative current-time curve that they present. This curve would indicate that the current flow ceases suddenly once the aerosol flow ceases. If oscillatory transfer were involved, it would be expected that the current would drop gradually after the aerosol flow ceases.

An electronic probe has been evaluated by Guyton (G10) and by Geist, York, and Brown (G2) for measuring particle size. In both devices an aerosol stream impinged on a pickup wire, which was connected to electronic equipment wherein the pulses were amplified, classified by magnitude, and counted. Guyton used primarily a grounded wire and found that all materials gave a positive pulse proportional to the square of particle diameter. Nonconductors (such as talc, dry potassium chloride, and barium carbonate) gave the highest pulse of almost the same magnitude. Conducting particles, such as iron, gave very weak pulses. By charging the pickup to 400 V, either negatively or positively, neither the sign nor magnitude of the pulse was altered when dealing with solid aerosol particles. With water, however, which normally also gave a positive pulse, a positive wire potential tended to neutralize the pulse, completely neutralizing it at $+22$ V, and gave an increasingly large negative pulse at higher positive potentials on the wire. For negative potentials the pulse became increasingly positive. Geist applied a $+430$ V potential on his pickup. He made measurements by allowing previously grounded metal spheres to strike the pickup wire (a 4-in piece of 18-gauge copper wire) and by allowing drops to fall on the wire. Only negative pulses were observed. No measurable pulses were obtained with nonconducting liquids such as carbon tetrachloride, benzene, toluene, or kerosene or by striking the pickup gently with a polystyrene rod. Water, acetone, and alcohol did give a pulse, the latter two giving the smaller pulse. Small pulses (0.1 V) were obtained when the wire was struck a sharp blow with the polystyrene rod. Geist found the magnitude of the pulse with the conductors to be proportional to the 1.6 power of sphere diameter.

Guyton concludes that with nonconductors the charging process is due to particle electrification either upon impact or during passage through the nozzle prior to impact. Geist explains his charges with conducting particles on the basis of the capacity increase when the spheres or drops come in contact with the wire pickup. Guyton arrives at this same explanation for his results with water. It should also be noted that the apparent sensitivity of Geist's measurements was 10,000 $\mu$V at the pickup wire while Guyton's sensitivity at that point was much better, 60 $\mu$V. The impaction velocities in Guyton's tests were also quite high, approximately sonic velocity (350 m/sec). The lower size limit of applicability of Guyton's instrument was 2-3 microns.

Keily (K2) also reports on the development of a similar system for measuring ground fogs in the size range of 5 microns and larger. He reports a pulse signal proportional to particle surface. However, the results indicated either no dependency or a less-than linear dependency of signal on the magnitude of any potential applied to the pickup, for positive potentials up to 450 V.

While these devices were intended as a means for size analysis, the principle should be applicable to charge analysis. There remains, however, the problem of distinguishing between original charge on a particle and charge generated upon interception by the probe and reconciling some of the seeming differences between the results of Guyton and Geist. [Appendix begins on p. 84.]

Appendix. Conversion Factors for Commonly Used Units; Universal Constants and Defined Values; and Properties of Standard Air

Conversion Factors for Commonly Used Units

| Symbol | Quantity Description | To convert to the MKSA units listed below | From the units listed below[a] | Multiply by | Or divide by |
|---|---|---|---|---|---|
| $A$ | area, surface | meter$^2$ | cm$^2$ | $10^{-4}$ | $10^4$ |
|  |  |  | ft$^2$ | 0.09200 | 10.764 |
| $c$ | concentration | kg/meter$^3$ | µg/liter | $10^{-6}$ | $10^6$ |
|  |  |  | mg/meter$^3$ | $10^{-6}$ | $10^6$ |
|  |  |  | mg/ft$^3$ | $3.531 \times 10^{-5}$ | 28,320 |
|  |  |  | grains/ft$^3$ | 0.002288 | 437.0 |
| $\mathscr{C}$ | capacitance | farads or C/V or (C)$^2$/J | statfarads | $(10^5/\mathbf{c}^2) = 1.11265 \times 10^{-12}$ | $8.98755 \times 10^{11}$ |
| $D, L, R$ | diameter, distance, length, radius | meter | microns | $10^{-6}$ | $10^6$ |
|  |  |  | cm | $10^{-2}$ | $10^2$ |
|  |  |  | in. | 0.025400 | 39.370 |
|  |  |  | ft. | 0.3048 | 3.2808 |
|  |  |  | mile | 1609.35 | $6.2137 \times 10^{-4}$ |
| $\mathscr{D}$ | diffusivity | meter$^2$/sec | cm$^2$/sec | $10^{-4}$ | $10^4$ |
|  |  |  | ft$^2$/sec | 0.09290 | 10.764 |
| $E$ | potential | V or J/C | statvolts | $(10^{-6}\mathbf{c}) = 299.79$ | 0.00333564 |
|  |  |  | ergs/electron | $6.24181 \times 10^{11}$ | $1.60210 \times 10^{-12}$ |

| | | | | |
|---|---|---|---|---|
| $\mathscr{E}$ | potential gradient, field intensity | V/meter | V/cm<br>statvolts/cm<br>V/micron | $10^{-2}$<br>$3.33564 \times 10^{-5}$<br>$10^{-6}$ |
| $F$ | force | N or (kg)(meter)/(sec)$^2$ | dyn<br>lb force | $10^{-5}$<br>4.4482<br>$10^5$<br>0.22481 |
| $G$ | mass velocity | (kg)/(sec)(meter$^2$) | (lb)/(sec)(ft$^2$)<br>(gm)/(sec)(cm$^2$) | 4.88240<br>10 | 0.20482<br>$10^{-1}$ |
| $m$ | mass | kg | grain<br>gm<br>lb | $6.480 \times 10^{-5}$<br>$10^{-3}$<br>0.45359 | 15,432<br>$10^3$<br>2.2046 |
| $p$ | pressure | N/meter$^2$ or (kg)/(sec)$^2$(meter) | dyn/cm$^2$<br>lb force/ft$^2$<br>mm Hg(0°C)<br>in. water (15°C)<br>lb force/in.$^2$<br>atm | $10^{-1}$<br>47.882<br>133.322<br>248.87<br>6895.<br>$1.01325 \times 10^5$ | 10<br>0.020885<br>0.0075006<br>0.004018<br>$1.4503 \times 10^{-4}$<br>$9.869 \times 10^{-6}$ |
| $\varphi$ | energy, work | $J$, (C)(V), or (kg)(meter)$^2$/(sec)$^2$ | eV<br>erg<br>ft-lb force<br>cal<br>Btu | $1.60210 \times 10^{-19}$<br>$10^{-7}$<br>1.3558<br>4.1868<br>1055.04 | $6.24181 \times 10^{18}$<br>$10^7$<br>0.73755<br>0.23885<br>$9.478 \times 10^{-4}$ |
| $\mathscr{Q}$ | electrical charge | C | electrons<br>statcoulomb | $1.60210 \times 10^{-19}$<br>$(0.1/c) = 3.33564 \times 10^{-10}$ | $6.24181 \times 10^{18}$<br>$2.9979 \times 10^9$ |
| $\mathscr{Q}_s$ | surface charge concentration | C/meter$^2$ | electron/cm$^2$<br>electron/micron$^2$<br>statcoulombs/cm$^2$<br>coulombs/cm$^2$ | $1.60210 \times 10^{-15}$<br>$1.60210 \times 10^{-7}$<br>$3.3356 \times 10^{-6}$<br>$10^4$ | $6.24181 \times 10^{14}$<br>$6.24181 \times 10^6$<br>$2.9979 \times 10^5$<br>$10^{-4}$ |

*Continued*

CONVERSION FACTORS FOR COMMONLY USED UNITS (Continued)

| Quantity Symbol | Description | To convert to the MKSA units listed below | From the units listed below[a] | Multiply by | Or divide by |
|---|---|---|---|---|---|
| $R_c$ | charge-to-mass ratio | C/kg | statcoulomb/gm | $3.3356 \times 10^{-7}$ | $2.9979 \times 10^6$ |
| | | | electrons/$\mu\mu$g | $1.60210 \times 10^{-4}$ | 6,241.81 |
| | | | $\mu$C/gm | $10^{-3}$ | $10^3$ |
| $u$ | velocity | meter/sec | cm/sec | $10^{-2}$ | $10^2$ |
| | | | ft/sec | 0.30480 | 3.2808 |
| | | | mile/hr | 0.4470 | 2.237 |
| $V$ | volume | meter$^3$ | cm$^3$ | $10^{-6}$ | $10^6$ |
| | | | U.S. gal | 0.003785 | 264.2 |
| | | | cu ft | 0.028316 | 35.316 |
| $\mu$ | viscosity | (kg)/(meter)(sec) | centipoise | $10^{-3}$ | $10^3$ |
| | | | poise or (gm)/(cm)(sec) | $10^{-1}$ | 10 |
| | | | (lb)/(ft)(sec) | 1.4882 | 0.6720 |
| $\rho$ | density | kg/meter$^3$ | lb/ft$^3$ | 16.019 | 0.06243 |
| | | | gm/cm$^3$ | $10^3$ | $10^{-3}$ |
| $\sigma$ | surface tension | N/meter or J/meter$^2$ | dyn/cm or erg/cm$^2$ | $10^{-3}$ | $10^3$ |

[a] Calories are gram calories. All units in each group are arranged in the order of increasing magnitude of the unit.

## Universal Constants and Defined Values

| Constant | Value |
|---|---|
| c | speed of light = $2.997925 \times 10^8$ meter/sec. <br> [$1/c = 3.3356405 \times 10^{-9}$ sec/meter; $c^2 = 8.98755 \times 10^{16}$ (meter/sec)$^2$; <br> $1/c^2 = 1.112650 \times 10^{-17}$ (sec/meter)$^2$] |
| g | standard gravitational acceleration = 9.8067 meter/sec$^2$ |
| h | Planck's constant = $6.6256 \times 10^{-34}$ (J)(sec) |
| k | $R/N$ = Boltzmann constant = $1.38054 \times 10^{-23}$ J/°K |
| $m_e$ | mass of electron = $9.1091 \times 10^{-31}$ kg |
| $m_p$ | mass of proton = $1.67252 \times 10^{-27}$ kg |
| N | Avogadro's number = $6.023 \times 10^{26}$ molecules/kg mole |
| Q | charge on electron = $1.60210 \times 10^{-19}$ C |
| R | gas constant = 8314 J/(°K)(kg mole) |
| $R_{ce}$ | charge-to-mass ratio for electron = $1.75880 \times 10^{11}$ C/kg |
| $R_{cp}$ | change-to-mass ratio for proton = $9.5790 \times 10^{7}$ C/kg |
| ε | permittivity of free space = $10^7/4\pi c^2$ <br> = $8.854 \times 10^{-12}$ (C)$^2$/(meter)$^2$ (N) <br> [$4\pi\varepsilon = 1.11265 \times 10^{-10}$ (C)$^2$/(meter)$^2$(N)] |

## Properties of Standard Air

Conditions

Temperature, $T = 25°C = 298°K = 77°F = 539°R$

Pressure, $p = 1$ atm $= 1.013 \times 10^6$ dyn/cm$^2$ = 14.7 lbs/in.$^2$
$= 1.013 \times 10^5$ N/meter$^2$

Molecular weight, $M$ = 29.0 gm/gm mole = 29.0 kg/kg mole

Properties at Above Conditions

Density, $\rho = 1.185 \times 10^{-3}$ gm/cm$^3$ = 0.0740 lb/ft$^3$
= 1.185 kg/meter$^3$

Viscosity, $\mu = 1.840 \times 10^{-4}$ poise = $1.840 \times 10^{-5}$ (kg)/(meter)(sec)

Mean molecular velocity, $\bar{u}_m = (8RT/\pi M)^{1/2}$ = 46,600 cm/sec
= 1530 ft/sec = 466.0 meters/sec

Mean free path, $\lambda_m$ (simple kinetic theory) = $3\mu/\rho\bar{u}_m$ = 0.1000 microns
= $1.000 \times 10^{-7}$ meters

Smoluchowski flocculation function $K_S/k_C = 4kT/3\mu = 2.980 \times 10^{-10}$ cm$^3$/sec
= $1.052 \times 10^{-14}$ ft$^3$/sec
= $2.980 \times 10^{-16}$ m$^3$/sec
= 298.0 micron$^3$/sec

## Acknowledgments

The investigation upon which this chapter is based was sponsored by the U.S. Army, Edgewood Arsenal, under Contract DA-18-035-AMC-122(A). The following individuals contributed to specific portions of this work: D. E. Blake, who conducted the literature search on electrostatic properties, checked many of the calculations, and summarized the methods of charge analysis; K. G. Dedrick, who analyzed the problem of maximum stable charge of fine particles; A. P. Brady, G. L. Pressman, and R. L. Kiang, who reviewed and analyzed the literature related to charging by interface and phase change mechanisms.

## Nomenclature

All equations are given on a dimensionally consistent basis and can be used with any dimensionally consistent units. The illustrative units given in the following are based on the MKSA system, using the rational basis for electrical units. For the cgs–esu (irrational) system, the corresponding cgs–esu units would be used and the permittivity, $\varepsilon$, would have a value of $1/(4\pi)$. Universal constants and defined values are represented by a symbol in Gothic (sans serif) or in bold face type.

| | | | |
|---|---|---|---|
| $a$ | Acceleration, meters/sec$^2$ | $E_{ps}$ | Specific particle potential $= \mathscr{Q}_p/2\pi\varepsilon\delta\, D_p$, V |
| $A$ | Surface area of a bounded volume $V$, meters$^2$ | $\mathscr{E}$ | Potential gradient or field intensity, V/meter |
| $A_b$ | Ground area covered by cloud in time $t$, meters$^2$ | $\mathscr{E}_c$ | Field intensity at surface of wire or cylinder, V/meter |
| $\mathbf{c}$ | Speed of light, $2.9979 \times 10^8$ meters/sec | $\mathscr{E}_0$ | Uniform external field intensity or that at surface of an electrode, V/meter |
| $c$ | mass concentration, kg/meters$^3$ | $\mathscr{E}_{\max}$ | Maximum field intensity, V/meter |
| $c_p$ | Mass concentration of particles, kg/meters$^3$ | $\mathscr{E}_{ps}$ | Specific particle surface gradient for particle of size $D_p$, V/meter, $\mathscr{Q}_p/\pi\varepsilon\delta\, D_p^2$ |
| $c_{p0}$ | Initial mass concentration of particles, kg/meters$^3$ | | |
| $(ct)_c$ | Critical dosage (kg)(sec)/(meters$^3$) | $\mathscr{E}_{ps1}$ | Specific particle surface gradient for particle of size $D_{p1}$, V/meter |
| $\mathscr{C}$ | Capacitance, farads | $\mathscr{E}_{ps2}$ | Specific particle surface gradient for particle of size $D_{p2}$, V/meter |
| $C_D$ | Drag coefficient, dimensionless | | |
| $d$ | Differential of | $\mathscr{E}_{ps\max}$ | Maximum stable value of $\mathscr{E}_{ps}$, V/meter |
| $D$ | Diameter; diameter of spherical cloud, meters | | |
| $D_c$ | Diameter of wire or cylinder, meters | $\mathscr{E}_{ps\mathrm{opt}}$ | Optimum value of specific particle surface gradient, V/meter |
| $D_p$ | Particle diameter, meters | $f$ | Frequency, cps |
| $D_{pf}$ | Diameter of floc, meters | $f_p$ | Probability of an electron occupying a given quantum energy level, dimensionless |
| $D_{p1}$ | Diameter of particles of one size, meters | | |
| $D_{p2}$ | Diameter of particles of another size, meters | $F$ | Force, N |
| $D_{p0}$ | Initial diameter of drop, meter | $F_A$ | Force of attraction or adhesion, N |
| $D_{psfa}$ | Apparent Stokes settling diameter of floc, meters $\{18\mu\, u_{tg}/[k_C\, \mathbf{g}\,(\rho_p - \rho)]\}^{1/2}$ | $F_e$ | Force due to electric field, N |
| | | $F_r$ | Resisting force due to fluid drag, N |
| $\mathscr{D}$ | Diffusion coefficient, meters$^2$/sec | $F_R$ | Force of repulsion, N |
| $\mathscr{D}_p$ | Diffusion coefficient for particles, of diameter $D_p$, meters$^2$/sec | $\mathbf{g}$ | Standard gravitational acceleration, 9.8067 meters/sec$^2$ |
| $\mathscr{D}_{p1}$ | Diffusion coefficient for particles of diameter $D_{p1}$, meters$^2$/sec | $G$ | Mass velocity, (kg)/(meter$^2$)(sec) |
| $\mathscr{D}_{p2}$ | Diffusion coefficient for particles of diameter $D_{p2}$, meters$^2$/sec | $G_{de}$ | Mass deposition rate due to electrostatic forces, (kg)/(meter$^2$)(sec) |
| $e$ | Natural logarithmic base, $2.718\cdots$ | $G_{dg}$ | Mass deposition rate due to gravity, (kg)/(meters$^2$)(sec) |
| $E$ | Potential or potential difference, V | | |
| $E_1$ | Ionization potential, V | $\mathbf{h}$ | Planck's constant $= 6.6256 \times 10^{-34}$ (J)(sec) |

| | | | |
|---|---|---|---|
| $j$ | Current, amp | $m$ | Total mass of particles, kg |
| $j_L$ | Current per unit length (of wire or tube), amp/meter | $m_e$ | Mass of electron = $9.1091 \times 10^{-31}$ kg |
| $j_A$ | Current per unit area, amp/meter² | $m_i$ | Mass of ion, kg |
| $k$ | Boltzmann constant = $R/N$ = $1.38054 \times 10^{-23}$ J/°K | $m_p$ | Mass of proton = $1.67252 \times 10^{-27}$ kg |
| $k_C$ | Stokes–Cunningham correction factor, dimensionless | $m_p$ | Mass of particle of diameter $D_p$, kg |
| $k_{em}$ | Correction factor on flocculation rate to allow for both particle charge and particle mean free path, dimensionless | $m_{p1}$ | Mass of particle of diameter $D_{p1}$, kg |
| | | $m_{p2}$ | Mass of particles of diameter $D_{p2}$ kg |
| $k_d$ | Distribution factor, dimensionless | $m_{pf}$ | Mass of floc, kg |
| $k_{F\mathscr{E}}$ | Factor relating attractive force to maximum field intensity for proximate bodies of opposite charge, dimensionless | $M$ | Molecular weight of molecule, kg/kg mole |
| | | $M_i$ | Molecular weight of ion, kg/kg mole |
| $k_s$ | Sticking factor, dimensionless | $n$ | Order of surface zonal harmonic, dimensionless |
| $k_{sA}$ | Factor to allow for inductive shifting of charges on proximate conducting, oppositely charged particles, dimensionless | $n_e$ | Concentration of electrons in space, electrons/meters³ |
| | | $n_i$ | Ion concentration in space, ions/meters³ |
| $k_{s\mathscr{C}}$ | Factor to allow for proximity effects on capacitance between two charged bodies, dimensionless | $n_p$ | Number concentration of particles, particles/meters³ |
| | | $n_{p1}$ | Number of particles of size $D_{p1}$, dimensionless |
| $k_{s\mathscr{E}}$ | Factor to obtain the maximum field intensity from the specific particle surface gradient of proximate charged bodies, dimensionless | $N$ | Avogadro's number = $6.023 \times 10^{26}$ molecules/kg mole |
| | | $N_m$ | Parameter to allow for slip between particles during collision, dimensionless (see Eq. 20) |
| $k_{sE}$ | Factor to obtain the maximum field intensity from the nominal voltage gradient between proximate charged bodies, dimensionless | $N_{me}$ | Parameter that measures effect of particle slip as corrected for particle charge, dimensionless (see Eq. 19) |
| $k_{sR}$ | Factor to allow for inductive shifting of charges on proximate conducting particles of like charge, dimensionless | $N_q$ | Parameter that measures effect of charge on flocculation, dimensionless (see Eq. 18) |
| $K_S$ | Smolukowski flocculation coefficient, meters³/sec | $N_{Re_p}$ | Particle Reynolds number = $D_p u_r \rho/\mu$, dimensionless |
| $K_W$ | Proportionality factor in equation for force of adhesion by van der Waal's forces, N/meter | $p$ | Pressure, N/meter² |
| | | $p_e$ | Pressure due to electrostatic stress in a fluid, N/meter² |
| $K_\delta$ | Proportionality factor, C/meter² | $p_p$ | Equilibrium vapor pressure from particle of size $D_p$, N/meter² |
| $K_{\delta\mathscr{E}}$ | Proportionality factor, V/meter | | |
| ln | Natural logarithm of | $p_T$ | Total pressure, N/meter² |
| $L$ | Distance between centers of two particles or bodies, meters | $p_\sigma$ | Constraining pressure due to surface tension, N/meter² |
| $L_m$ | Migration distance, meters | | |
| $L_s$ | Clearance between proximate surfaces of two bodies, meters | $p_\infty$ | Vapor pressure from a flat surface, N/meter² |

| | | | |
|---|---|---|---|
| $Q$ | Charge on the electron $= 1.60210 \times 10^{-19}$ C | $u_R$ | Radial outward velocity of expanding cloud, meters/sec |
| $\mathcal{Q}$ | Charge or charge at time $t$, C | $V$ | Volume, meters$^3$ |
| $\mathcal{Q}_1$ | Magnitude of arbitrary point charge, C | $x$ | Distance, meters |
| | | $u_{tg}$ | Terminal gravitational settling velocity of particle, meters/sec |
| $\mathcal{Q}_2$ | Magnitude of another arbitrary point charge, C | $\alpha$ | First Townsend ionization coefficient, meters$^{-1}$ |
| $\mathcal{Q}_0$ | Initial charge, C | | |
| $\mathcal{Q}_p$ | Particle charge or charge on particle of diameter $D_p$, C | $\beta_i$ | Ion mobility, (meters/sec)/(V/meter) |
| $\mathcal{Q}_{p1}$ | Charge on particle of diameter $D_{p1}$, C | $\beta_p$ | Mechanical particle mobility $= u_m/F_r$, (meters/sec)/(N) |
| $\mathcal{Q}_{p2}$ | Charge on particle of diameter $D_{p2}$, C | $\beta_{pe}$ | Electrostatic particle mobility $= u_m/\mathscr{E}$, (meters/sec)/(V/meter) |
| $\mathcal{Q}_{pa}$ | Average charge on each of other particles, C | $\delta$ | Dielectric constant of medium surrounding particle, dimensionless |
| $\mathcal{Q}_{ps}$ | Surface concentration of particle charge, $= \mathcal{Q}_p/\pi D_p^2$, C/meter$^2$ | | |
| $\mathcal{Q}_s$ | Surface charge concentration, C/meter$^2$ | $\delta_p$ | Dielectric constant of particle, dimensionless |
| $R$ | Universal gas constant $= 8314$ (J)/($^\circ$K)(kg mole) | $\delta_v$ | Dielectric constant of a volume, dimensionless |
| $R$ | Radius, meters | $\delta_1$ | Dielectric constant of one material, dimensionless |
| $\mathscr{R}_0$ | Resistivity of bulk material, ($\Omega$)(meter) | $\delta_2$ | Dielectric constant of another material, dimensionless |
| $R_c$ | Charge-to-mass ratio $= \mathcal{Q}_p/m_p$, C/kg | $\epsilon$ | Permittivity of free space $=(10^7/4\pi c^2) = 8.854 \times 10^{-12}$ (C)$^2$/(meter)$^2$(N) |
| $t$ | time, sec | | |
| $t_a$ | Time for objects at edge of cloud to receive critical dosage of agent, sec | $\lambda_i$ | Mean free path of ions, meters |
| | | $\lambda_m$ | Mean free path of electrons, ions, or molecules, meters |
| $t_{ps}$ | Specific particle stopping time $= k_c \rho_p D_p^2/18\mu$, sec | $\sigma$ | Surface tension of liquid or surface energy of solid, N/meter or J/meter$^2$ |
| $t_R$ | Relaxation time $= \epsilon \delta_v \mathscr{R}_0$, sec | | |
| | | $\rho$ | Fluid density, kg/meter$^3$ |
| $t_{sd}$ | Specific dilution time $= \mu \rho_p / 2k_c \epsilon \delta_v c_p \mathscr{E}_{ps}^2$, sec | $\rho_p$ | Particle density, kg/meter$^3$ |
| | | $\rho_g$ | Density of pure gas phase, kg/meter$^3$ |
| $T$ | Absolute temperature, $^\circ$K | | |
| $u$ | Fluid velocity, meters/sec | $\mu$ | Absolute gas viscosity, (kg)/(meter)(sec) or dekapoise |
| $\bar{u}_i$ | rms velocity of ions $= (3kT/m_i)^{1/2} = (3RT/M_i)^{1/2}$, meters/sec | $\zeta$ | Probability factor for producing ionizing photon, dimensionless |
| $u_m$ | Particle migration velocity, meters/sec | $\xi$ | Probability factor for producing photoelectrons, dimensionless |
| $\bar{u}_m$ | Mean molecular velocity $= [8kT/(\pi M/N)]^{1/2} = [8RT/(\pi M)]^{1/2}$ meters/sec | $\tau_y$ | Tensile strength of liquid, N/meter$^2$ |
| | | $\varphi$ | Work function for surface, J |
| $u_r$ | Relative velocity between particle and fluid, meters/sec | $\varphi_F$ | Fermi energy level, J |
| | | $\varphi_T$ | Quantum energy level, J |

## References

A1. Arabadzhi, V. I., The electrification of particles in clouds, *Meteorol. i Gidrol.* No. 6, 37 (1955).
A2. Arabadzhi, V. I., On some electrical properties of water and ice, *Zh. Eksperim. Teor. Fiz.* **30**, 193 (1956).
A3. Arendt, P., and Kallmann, H., The mechanism of the electrification of small particles in clouds, *Z. Physik* **35**, 421 (1926).
B1. Ballou, J. W., Static electricity in textiles, *Textile Res. J.* **24**, 146 (1954).
B2. Beach R., Preventing static electricity fires. Part II, *Chem. Eng.* **72**, No. 1, 63 (1965).
B3. Berg, T. G. O., Final report: Dissemination and use of CW agents, Aerojet-General Corp., Contr. DA-18-108-405-CML-829 (1963).
B4. Berg, T. G. O., The mechanism of the tribo effect, Aerojet-General Corp. Rept. R-444, Contr. DA-18-108-405-CML-829 (1961).
B5. Berg, T. G. O., Fernish, G. C., and Flood, W. J., Investigation of the electrification of powders in flow through tubes and nozzles. Aerojet-General Corp. Rept. 0395-04(08)SP, Contr. DA-18-108-405-CML-829 (1963).
B6. Berg, T. G. O., Fernish, G. C., and Flood, W. J., Charge analyzer for aerosols and spray, *Rev. Sci. Instr.* **35**, 719–23 (1964).
B7. Berg, T. G. O., Fernish, G. C., and Gaukler T. A., The mechanism of the coalescence of liquid drops, *J. Atmospheric Sci.*, **20**, N0. 2 153 (1963).
B8. Berg, T. G. O., and Flood, W. J., Investigations of the electrification of powders in flow through tubes and nozzles. II. Charge analysis of deagglomerated powders, Aerojet-General Corp. Rept. 0395-04(14)SP, Contr. DA-18-108-405-CML-829 (1963).
B9. Berg, T. G. O., and Gaukler, T. A., The mechanism of the tribo effect. III. The effect of the humidity of the air, Aerojet-General Corp. Rept. 0395-04(05)SP, Contr. DA-18-108-405-CML-829(1963).
B10. Blake, D. E., and Lapple, C. E., Bibliography on electrostatic phenomena in aerosol dissemination, Stanford Res. Inst., Special Tech. Rept. No. 1, Contr DA-18-035-AMC-122(A) 1965) (available to the public as Doc. No. AD-468 320, Clearinghouse for Fed. Sci. and Tech. Inform, U.S. Dept. of Commerce, Springfield, Va 22151).
B11. Bradley, R. S., The cohesion between smoke particles, *Trans. Faraday Soc.* **32**, 1088 (1936).
B12. Brasefield, C. J., Electrification of carbon black by contact with a metal surface, *J. Franklin Inst.* **270**, No. 4, 283 (1950).
B13. Bredov, M. M., and Kshemyanskaya, I. Z., The electricity observed after contact between two bodies, *Zh. Tekh. Fiz.* **27**, 921 (1957).
C1. Chalmers, J. A., Atmospheric electricity, *Rept. Prog. Mod. Phys.* **17**, 101 (1954).
C2. Cho, A. Y. H., Contact charging of micron-sized particles in intense electric fields, *J. Appl. Phys.* **35**, 2561 (1964).
C3. Cochet, R., Lois de charge des fines particules (submicronique). Edudes theoretiques —Controles Recents Spectre de Particules, *Colloq. Intern. Phys. Forces Electrostatiques et Leurs Appl.* p. 331. Centre Nat. Rech. Sci., Paris, 1961.
C4. Coehn, A., and Raydt, V., The quantitative validity of the law concerning the charging of dielectrics, *Ann. Physik* **30**, 777 (1909).
C5. Cohen, E., Research on charged colloid generation, Wright–Patterson AFB, Aero Propulsion Lab. Final Rept. Contr. AF33(657)-10999. ASTIA Doc. AD 601, 390, NASA Rept. No. N64-27961 (1964).

C6. Cooper, W. F., The electrification of fluids in motion, *Brit. J. Appl. Phys.*, *Suppl.* 2, S11 (1953).
C7. Corn, M., The adhesion of solid particles to solid surfaces. I. A review, *J. Air Pollution Control Assoc.* 11, No. 11, 523 (1961).
D1. Daniel, J. H., and Brackett, F. S., An electrical method for investigating the nature and behavior of small, airborne charged particles, *J. Appl. Phys.* 22, 542 (1951).
D2. Davies, C. N., Recent advances in aerosol research, Macmillan (Pergamon), New York, 1964.
D3. Dawkins, G. S., Electrostatic effects in the deposition of aerosols on cylindrical shapes, Univ. of Illinois Eng. Exptl. Sta. Tech. Rept. 15, AEC Rept. COO-1017, Contr. AT(11-1)-276 (1958).
D4. Deutsch, W., Bewegung and Ladung der Elektrizitätsträger im zylinder Kondensator, *Ann. Physik* 68, (4) 335 (1922).
D5. Dodd, E. E., The statistics of liquid spray and dust electrification by the Hopper and Laby method, *J. Appl. Phys.* 24, 73 (1953).
D6. Doyle, A., Moffett, D. R., and Vonnegut, B., Behavior of evaporating electrically charged droplets, *J. Colloid Sci.* 19, 136 (1964).
D7. Drozin, V. G., The electrical dispersion of liquids as aerosols, *J. Colloid Sci.* 10, 158 (1955).
D8. Dubois, J., Measurement of the electric charge on water and salt-solution droplets during evaporation and condensation, *J. Phys. (France)* 24, 661 (1963).
D9. Dunskii, V. F., and Kitaev, A. V., Electrostatic spraying, *Zashchita Rast. ot Vreditelei i Boleznei* 3, No. 4, 17 (1958).
E1. Earhart, R. F., The sparking distances between plates for small distances, *Phil. Mag.* 1, 147 (1901).
E2. Eigen, M., and DeMaeyer, L., Self-dissociation and protonic charge transport in water and ice, *Proc. Roy. Soc. (London)*, Ser. A, 247, 505 (1958).
E3. Einbinder, H., Generalized equations for the ionization of solid particles, *J. Chem. Phys.* 26, 948 (1957).
E4. Evans, D. G., and Hutchinson, W. C. A., The electrification of freezing water droplets and colliding ice particles, *Quart. J. Roy. Meteorol. Soc.* 89, 370 (1963).
F1. Foster, W. W., Deposition of unipolar charged aerosol particles by mutual repulsion, *Brit. J. Appl. Phys.* 10, No. 5, 206 (1959).
F2. Frickel, R., Containment of charged particles, paper presented at Fifth Coordination Seminar of U.S. Army (CRDL), at Illinois Institute of Technology Research Institute, Chicago, Ill., 1964.
F3. Fuchs, N. A., Uber die Stabilitat und Aufladung der Aerosole, *Z. Physik* 89, 736 (1934).
F4. Fuchs, N. A., "The Mechanics of Aerosols," Macmillan (Pergamon), New York, 1964.
F5. Fukada, E., and Fowler, J. F., Triboelectricity and electron traps in insulating materials: Some correlations, *Nature* 181, 693 (1958).
G.1 Gallily, I., and Ailam, G., On the vapor pressure of electrically charged drops, *J. Chem. Phys.* 36,, 1781 (1962).
G2. Geist, J. M., York, J. L., and Brown, G. G., Electronic spray analyzer for electrically conducting particles, *Ind. Eng. Chem.* 43, 1371 (1951).
G3. Germer, L. H., Electrical breakdown between close electrodes in air, *J. Appl. Phys.* 30, 46 (1959).
G4. Gillespie, T., and Langstroth, G. O., An instrument for determining the electric-charge distribution in aerosols, *Can. J. Chem.* 30, 1056 (1952).

G5. Göhlich, H., Investigations to improve precipitation of plant protectives by means of electrical charges, *Forsch. Gebiete Ingenieurw., VDI-Forschungsh.* 467 (1958).
G6. Gordieyeff, V. A., Some properties of unipolarly charged aerosols, *Arch. Ind. Health* **14**, 471 (1956).
G7. Goyer, G. G., Gruen, R., and LaMer, V. K., Filtration of monodisperse electrically charged aerosols, *J. Phys. Chem.* **58**, 137 (1954).
G8. Graf, P. E., Breakup of small liquid volumes by electrical charging, presented at *API Res. Conf. Distillate Fuel Combustion, Chicago, 1962* (paper CP-62-4).
G9. Guest, P. G., Static electricity in nature and industry, U.S. Bur. of Mines, Bull. 368 (1933).
G10. Guyton, A. C., Electronic counting and size determination of particles in aerosols, *J. Ind. Hyg. Toxicol.* **28**, 133 (1946).
H1. Harper, W. R., Liquids giving no electrification by bubbling, *Brit. J. Appl. Phys., Suppl. No.* 2, S19 (1953).
H2. Harper, W. R., Adhesion and charging of quartz surfaces, *Proc. Roy. Soc. (London)*, Series A3, **231**, 388 (1955).
H3. Harper, W. R., The generation of static charge, *Advan. Phys.*, **6**, 365 (1957).
H4. Harper, W. R., Electrification following the contact of solids, *Contemporary Phys.* **2**, No. 5, 345 (1961).
H5. Hendricks, C. D., Jr., Charged droplet experiments, paper presented at *Symp. Advan. Propulsion Concepts, 2nd*, ARDC and AVCO-Everett Res. Labs., Boston, Mass. (1959).
H6. Hendricks, C. D., Jr., Charged droplet experiments, *J. Colloid Sci.* **17**, 249 (1962).
H7. Henry, P. S. H., The role of asymmetric rubbing in the generation of static electricity, *Brit. J. Appl. Phys. Suppl.* 2, S31 (1953).
H8. Hersh, S. P., and Montgomery, D. J., Static electrification of filaments, *Textile Res. J.* **25**, No. 4, 279 (1955).
H9. Hertz, H., Uber die Berührung fester elastischer Körper, *J. Reine Angew. Math.* **92**, 156 (1882).
H10. Hewitt, G. W., "The charging of small particles for electrostatic precipitation," *Commun. Electron.*, No. 31, 300 (1957).
H11. Hinkle, B. L., Orr, C., and DallaValle, J. M., A new method for the measurement of aerosol electrification, *J. Colloid Sci.* **9**, 70 (1954).
H12. Hogan, J. J., Parameters influencing the charge-to-mass ratio of electrically sprayed liquid particles, Univ. of Illinois, Charged Particle Res. Lab., Rept. CPRL-2-63, NASA Rept. N64-20822 (1963).
H13. Hogan, J. J., and Hendricks, C. D., Investigation of the charge-to-mass ratio of electrically sprayed liquid particles, *AIAA J.* **3**, No. 2, 296 (1965).
H14. Hopper, V. D., and Laby, T. H., The electronic charge, *Proc. Roy. Soc. (London) Ser. A*, **178**, 243 (1941).
H15. Hurd, R. M., Schmid, G. M., and Snavely, E. S., Electrostatic fields: their effect on the surface tension of aqueous salt solutions, *Science* **135**, 791 (1962).
J1. Jordan, D. W., The adhesion of dust particles, *Brit. J. Appl. Phys., Suppl. 3*, S194 (1954).
K1. Kachurin, L. G., and Bekryaev, V. I., An investigation of the process of electrification of crystallizing water, Dokl. Akad. Nauk. SSSR, **130**, 57 (1960).
K2. Keily, D. P., Measurement of drop size distribution and liquid water content in natural clouds, Dept. of Meteorol., Mass. Inst. of Tech., Contr. No. AF19(628)-259, NASA Rept. No. N64-30005 (1964).

K3. Kisliuk, P., Electron emission at high fields due to positive ions, *J. Appl. Phys.* **30**, 51 (1959).
K4. Kitaev, A. V., Unipolar electrification of aerosols in the field of the corona discharge, *Vestn. Sel'skokhoz. Nauki.* **2**, No. 9, 127 (1957).
K5. Knoblauch, O., Experiments on contact electricity, *Z. Phys. Chem.* **29**, 225 (1902).
K6. Kraemer, H. F., Properties of electrically charged aerosols, Univ. of Illinois, Eng. Exptl. Sta. Tech. Rept. 12, AEC Rept. COO-1013, Contr. AT(11-1)-276 (1954).
K7. Kraemer, H. F., and Johnstone, H. F., Collection of aerosol particles in presence of electrostatic fields, *Ind. Eng. Chem.* **47**, 2426 (1955).
K8. Kraemer, H. F., and Ranz, W. R., Homopolar electrification of aerosols, Univ. of Illinois, Eng. Exptl. Sta. Tech. Rept. 7, AEC Rept. SO-1008, Contr. AT (30-3)-28 (1952).
K9. Kunkel, W. B., Growth of charged particles in clouds, *J. Appl. Phys.* **19**, 1053 (1948).
K10. Kunkel, W. B., Magnitude and character of errors produced by shape factors in Stokes' law estimates of particle radius, *J. Appl. Phys.* **19**, 1056 (1948).
K11. Kunkel, W. B., The static electrification of dust particles on dispersion into a cloud, *J. Appl. Phys.* **21**, 820 (1950).
K12. Kunkel, W. B., and Hansen, J. W., A dust electricity analyser, *Rev. Sci. Instr.* **21**, 308 (1950).
L1. Ladenburg, R., and Sachsse, H., Physical phenomena of electrical gas purification, *Ann. Physik* (5) **4**, 863 (1930).
L2. LaMer, V. K., Goyer, G., Gruen, R., and Kruger, J., Filtration of monodisperse electrically charged aerosols, AEC Rept. NYO-514, Contr. AT(30-1)-651 (1952).
L3. Langer, G., and Radnik, J. L., Development and preliminary testing of a device for the electrostatic classification of submicron airborne particles, *J. Appl. Phys.* **32**, 955 (1961).
L4. Latham, J., Electrification produced by the asymmetric rubbing of ice on ice, *Brit. J. Appl. Phys.* **14**, 488 (1963).
L5. Latham, J., and Mason, B. J., " Electric charge transfer associated with temperature gradients in ice," *Proc. Roy. Soc. (London), Ser. A* **260**, 523 (1961).
L6. Latham, J., and Mason, B. J., Generation of electric charge associated with the formation of soft hail in thunderstorms, *Proc. Roy. Soc. (London), Ser. A* **260**, 537 (1961).
L7. Lindblad, N. R., Effects of relative humidity and electric charge on the coalescence of curved water surfaces, *J. Colloid Sci.* **19**, 729 (1964).
L8. Loeb, L. B., The Threshold for the positive pre-onset burst pulse corona and the production of ionizing photons in air at atmospheric pressure, *Phys. Rev.* **73**, 798 (1948).
L9. Loeb, L. B., " Static Electrification," Springer, Berlin, 1958.
L10. Loeb, L. B., " Basic Processes of Gaseous Electronics " 2nd ed. Univ. of Calif. Press Berkeley, California, 1961.
L11. Loeb, L. B., Static electrification. I. in " Progress in Dielectrics " (J. B. Birks and J. Hart, eds.), Vol. IV, pp. 249–309. Heywood, London, 1962.
L12. Loeb, L. B., Static electrification. II. in " Progress in Dielectrics " (J. B. Birks and J. Hart, eds.), Vol. V, pp. 233–289. Heywood, London, 1963.
L13. Loeb, L. B., Parker, J. H., Dodd, E. E., and English, W. N., The choice of suitable gap formations for the study of corona breakdown and the field along the axis of hemispherically capped cylindrical point-to-plane gap, *Rev. Sci. Instr.* **21**, 42 (1950).
M1. Mason, B. J., and Maybank, J., The fragmentation and electrification of freezing water drops, *Quart. J. Roy. Meteorol. Soc.* **86**, 176 (1960).

M3. Masters, J. I., An aerosol analyser, *Rev. Sci. Instr.* **24**, 586 (1953).
M3. Matthews, J. B., and Mason, B. J., Electrification produced by the rupture of large water drops in an electric field, *Quart. J. Roy. Meteorol. Soc.* **90**, 275 (1964).
M4. Medley, J. A., "The electrostatic charging of some polymers by mercury," *Brit. J. Appl. Phys., Suppl.* **2**, S28 (1953).
M5. Mierdel, G., Migration of dust particles in electric filters, *Z. Tech. Physik* **13**, 564 (1932).
M6. Mitchell, R. I., Retention of aerosol particles in the respiratory tract, *Am. Rev. Respiratory Diseases* **82**, 627 (1960).
M7. Montgomery, D. J., Static electrification of solids, *Solid State Phys.* **9**, 139 (1959).
N1. Natanson, G. L., The electrification of drops during atomization of liquids as a result of fluctuations in the ion distribution, *Zh. Fiz. Khim.* **23**, No. 3, 304 (1949).
P1. Pauthenier, M., and Moreau-Hanot, M., Spherical particles in an ionised field, *J. Phys. Radium* **3**, 590 (1932).
P2. Penney, G. W., and Lynch, R. D., Measurements of charge imparted to fine particles by a corona discharge, *Commun. Electron.*, No. 31, 294 (1957).
P3. Peskin, R. L., and Raco, R. J., Some results from the study of ultrasonic and electrostatic atomization, presented at API Res. Conf. Distillate Fuel Combustion, paper CP 63-3, 1963.
P4. Peterson, J. W., Contact charging between a borosilicate glass and nickel, *J. Appl. Phys.* **25**, 501 (1954).
P5. Peterson, J. W., Contact charging between nonconductors and metal, *J. Appl. Phys.* **25**, 907 (1954).
R1. Randall, J. M., Marshall, W. R., and Tschernitz, J. L., The atomization of liquids by high voltage electrical energy, presented at *A.I.Ch.E. Meeting, Las Vegas, 1964*.
R2. Rayleigh, Lord, The influence of electricity on colliding water drops, *Proc. Roy. Soc. (London), Ser. A* **28**, 406 (1879).
R3. Rayleigh, Lord, Further observations upon liquid jets, *Proc. Roy. Soc. (London)* **34**, 130 (1882).
R4. Richards, H. F., The contact electricity of solid dielectrics, *Phys. Rev.* **22**, 122 (1923).
R5. Richardson, E. G. (ed.), "Aerodynamic Capture of Particles," Macmillan (Pergamon), New York, 1960.
R6. Robertson, A. J. B., Viney, B. W., and Warrington, M., The production of positive ions by the field ionization at the surface of a thin wire, *Brit. J. Appl. Phys.* **14**, 278 (1963).
R7. Rohmann, H., Messung der grösse von Schwebeteilchen, *Z. Physik* **17**, 253 (1923).
R8. Rose, G. S., and Ward, S. G., Contact electrification across metal-dielectric and dielectric-dielectric interfaces, *Brit. J. Appl. Phys.* **8**, 121 (1957).
R9. Russell, A., "A Treatise on the Theory of Alternating Currents," Vol. I, 2nd ed. Cambridge Univ. Press, London, 1914.
R10. Ryce, S. A., An equilibrium value for the charge-to-mass ratio of droplets produced by electrostatic dispersion, *J. Colloid Sci.* **19**, 490 (1964).
S1. Schonland, B. F. J., "Atmospheric Electricity," 2nd ed. Wiley, New York, 1953.
S2. Schultz, R. D., and Branson, L., The colloid rocket: Progress towards a charged-liquid-colloid propulsion system, presented at *Symp. Advan. Propulsion Concepts, 2nd, Boston, 1959*.
S3. Schultz, R. D., and Wiech, R. E., Electrical propulsion with colloidal materials, AGARD Combustion and Propulsion Panel, Tech. Meeting, *Advanced Propulsion Concepts, Pasadena, 1960*.
S4. Sergiyeva, A. P., The electrical charges of cloud particles, *Akad. Nauk. SSSR, Izv., Ser. Geofiz.*, No. 3, 247 (1958).

S5. Shaffer, R. E., Dissemination of aerosols by electrostatics, U.S. Army, Edgewood Arsenal, CRDL Rept. 3141, AD 294 639 (1962).
S6. Shapiro, A. R., and Watson, W. K. R., An electrodynamic foucault pendulum, *J. Appl. Phys.* **34**, 1553 (1963).
S7. Shashoua, V. E., Static electricity in polymers I. Theory and Measurement, *J. Polymer Sci.* **33**, 65 (1958).
S8. Shaw, P. E., and Jex, C. S., Tribo-electricity and friction, Parts II and III, *Proc. Roy. Soc.* **118**, 97 (1928).
S9. Silsbee, F. B., Static electricity, Nat. Bur. Stand. Circ. No. C-438 (1942).
S10. Solov'yev, V. A., A method for the measurement of charges and sizes of fog droplets, *Mezhduved. Konf. po Voprasam Issledovan. Oblakov, Osadkov i Grozovogo Elektr.* (*no.* 5), Leningrad, 170, 1957.
S11. Soo, S. L., Trezek, G. J., Dimick, R.C., and Hohnstreiter, G.F., Concentration and mass flow distributions in a gas-solid suspension, *Ind. Eng. Chem. Fundamentals* **3**, No. 2, 98 (1964).
S12. Straubel, H., The electrostatic atomization of liquids, *Z. Angew. Physik* **6**, No. 6, 264 (1954).
S13. Suzuki, S., and Tomura, M., Studies on the measurement of charges on fine particles. I. Charge measurement of tobacco and mosquito incense aerosols by a charge spectrometer, *Denshi Shashin* (*Japan*) **4**, No. 2, 20 (1962).
T1. Taneya, S., Electrification of powder, *Japan J. Appl. Phys.* **2**, 798 (1963).
T2. Thomas, D. G. A., The measurement of oxidation of coal by static electrification, *Brit. J. Appl. Phys., Suppl. 2*, S55 (1953).
V1. Van Ostenburg, D. O., and Montgomery, D. J., Charge transfer upon contact between metals and insulators, *Textile Res. J.* **28**, No. 1, 22 (1958).
V2. Vick, F. A., Theory of contact electrification, *Brit. J. Appl. Phys., Suppl. 2*, S1 (1953).
V3. Vonnegut, B., Moffett, D. R., Sliney, P. M., and Doyle, A. W., Electrical phenomena associated with aerosols, Arthur D. Little, Inc., Final Rept., Contr. DA-18-108-405-CML-852, AD 299 716 (1962).
V4. Vonnegut, B., and Neubauer, R. L., Production of monodisperse liquid particles by electrical atomization, *J. Colloid Sci.* **7**, 616 (1952).
W1. Wagner, P. E., Electrostatic charge separation at metal–insulator contacts, *J. Appl. Phys.* **27**, 1300 (1956).
W2. Wells, P. U., and Gerke, R. H., An oscillation method for measuring the size of ultramicroscopic particles, *J. Am. Chem. Soc.* **41**, 312 (1919).
W3. Whitby, K. T., Generator for producing high concentrations of small ions, *Rev. Sci. Instr.* **32**, 1351 (1961).
W4. White, H. J., "Industrial Electrostatic Precipitation." Addison–Wesley, Reading, Massachusetts, 1963.
W5. Whitman, V. E., Electrification of dust clouds, *Phys. Rev.* **28**, 1287 (1926).
W6. Wright, T. E., Stasny, R. J., and Lapple, C. E., High velocity air filters, Wright Air Developt. Center WADC Tech. Rept. 55-457, ASTIA No. AD-142 075 (1957).
Z1. Zebel, G., On the theory of coagulation of electrically uncharged aerosols, *Kolloid Z.* **156**, No. 2, 102 (1958).
Z2. Zebel, G., On the theory of the behavior of electrically charged aerosols, *Kolloid Z.* **157**, No. 1, 37 (1958).

# MATHEMATICAL MODELING OF CHEMICAL REACTIONS

## J. R. Kittrell

Chevron Research Company
Richmond, California

|  |  |
|---|---|
| I. Introduction | 98 |
|    A. Mechanisms and Models | 98 |
|    B. Reaction-Rate Models | 99 |
|    C. Integral and Differential Models | 101 |
| II. Linearly Reducible Models | 102 |
|    A. Power-Function Models | 102 |
|    B. Hyperbolic Models | 105 |
| III. Parameter Estimation | 110 |
|    A. Least Squares | 111 |
|    B. Reparameterization | 121 |
|    C. Confidence Intervals and Regions | 124 |
|    D. Multiple-Response Parameter Estimation | 129 |
| IV. Tests of Model Adequacy | 131 |
|    A. Analysis of Variance | 131 |
|    B. Residual Analysis | 137 |
| V. Use of Diagnostic Parameters | 142 |
|    A. Model Discrimination with Diagnostic Parameters | 142 |
|    B. Adaptive Model-Building with Diagnostic Parameters | 147 |
| VI. Empirical Modeling Techniques | 154 |
|    A. Response-Surface Methodology | 155 |
|    B. Transformations of Variables | 159 |
|    C. Empirical Model Tuning | 164 |
| VII. Experimental Designs for Modeling | 168 |
|    A. Model-Discrimination Designs | 171 |
|    B. Parameter-Estimation Designs | 173 |
|    Nomenclature | 178 |
|    References | 181 |

## I. Introduction

### A. Mechanisms and Models

For any physical system there exists a precise mathematical and physical representation of all of the phenomena that make up the system. A chemical reaction is no exception, since it does possess some *true* mechanism that is descriptive of every microscopic detail of the reaction. In practice, of course, a complete description of this mechanism cannot be obtained, and approximations must be made. For example, for studies of a fundamental nature on heterogeneously catalyzed reaction systems, this may entail reference to some average property, such as an average pore size, an average size of a particle of metal dispersed on a support, an average adsorbed state, or an average mode of reaction. For a broader problem such as the modeling of the operation of an entire plant, we must make more extensive approximations. For industrial reactions generally we are fortunate if we can characterize the reactor feed in terms of a limited number of major components, if we can precisely define the reaction kinetics of these major components in the industrial scale, and if we can evaluate the residence time distribution of the reactor. Consequently, we will often be concerned with models of the system that are at best quasimechanistic, simply because our reacting system is chosen on the basis of economics rather than simplicity or convenience. This review presents statistical methods for developing mathematical descriptions (either theoretical or empirical) of reacting systems.

In contrast to the above allusion to a mechanism, the term model has been used to describe a wide range of chemical engineering subjects. Chemical engineering modeling activities in general can be thought of as belonging to one of three members of a nested set. The outermost member is plant modeling. This entails the description of, for example, an entire refinery complex, taking into full account topics ranging from crude resources to geographical product distributions. Economic models of such a complex obviously require a knowledge of the behavior of the several processes making up the plant. Hence, modeling activity within the plant-wide modeling is required, that is, process modeling. Process models are capable of predicting the steady-state level of operation at the settings of the process variables, and perhaps the dynamic response of the process to any disturbance. Several process models must be included in the plant model. The development of the process model, in turn, cannot be carried out without recourse to phenomenological or mechanistic modeling. The mechanistic model must be capable of describing adequately the basic physical and chemical steps that take place within the processes, under the conditions imposed by the settings of the process variables.

Considerable unification of the concepts of plant modeling and process modeling has taken place in recent years, for example, as described in References (F4) and (R3) for process and plant concepts, respectively. Attention has been devoted only intermittently, however, to the development of methodology for mechanistic modeling. For reaction-rate modeling, in particular, there was considerable activity in the 1930's and 1940's and again in the late 1950's and 1960's. The latter activity apparently resulted in large part from the introduction of computer techniques into chemical engineering technology. This review will be concerned specifically with methodology useful in the elucidation of such reaction rate models, treated artificially as two distinct stages. In the specification stage, the problem is to determine the appropriate functional form of a rate model that adequately describes the reaction. Although parameter estimates are often necessary to test the adequacy of any particular functional form, these estimates often need not be very precise. After a model has been specified, it is often advantageous to give particular attention to the estimation of the parameters within the adequate functional form. Once the functionality of the rate equation is known, the estimation of the parameter values is relatively straightforward.

Although examples of the methodology will utilize entirely reaction rates or reactant concentrations, the procedures are equally valid for other model responses. They have been used, for example, with responses associated with catalyst deactivation and diffusional limitations as well as with copolymer reactivity ratios and average polymer molecular weights.

## B. Reaction-Rate Models

In the model-specification stage, we concentrate upon two types of reaction-rate models, the power-function model

$$r = k C_A{}^a C_B{}^b \tag{1}$$

and the hyperbolic model exemplified by

$$r = \frac{k C_A C_B}{1 + K_A C_A + K_B C_B} \tag{2}$$

Here, $r$ is a reaction rate, $C_A$ and $C_B$ are measures of concentration, and the remaining terms represent parameters to be estimated.

Equation (1), of course, is simply a representation of the law of mass action, stating that the reaction rate is proportional to a concentration driving force. Such an equation has been applied to many reacting systems, particularly those of considerable complexity.

Equation (2) is generally derived based upon some form of steady state hypothesis. Such models have been applied to gaseous reactions (Lindemann theory), to enzymatic reactions (Michaelis–Menten mechanism), and to gaseous reactions on solid surfaces (Langmuir–Hinshelwood mechanism). Considerable discussion of the derivation of these equations has been presented elsewhere (L2). In general, we refer to this model type as a hyperbolic model or, in the case of solid catalyzed gaseous reactions, as a Hougen–Watson model (to remove restrictions implicit in a Langmuir–Hinshelwood mechanism).

The use of any model more complex than Eq. (1) has frequently been criticized as an attempt to read too much into a set of kinetic data. The implicit results of such a misapplication would be a proliferation of meaningless constants; this would thus lead to unnecessary effort in data fitting and interpretation, or to misleading interpolation or extrapolation.

The latter danger is, of course, potentially present any time any data interpretation is attempted, particularly if nature is assumed always to follow Eq. (1). The only course of action is to attempt to include as much theory in the model as possible, and to confirm any substantial extrapolation by experiment. It is erroneous, however, to presume that kinetic data will always be so imprecise as to be misleading. The use of computers and statistical analyses for any linear or nonlinear reaction rate model allows rather definite statements about the amount of information obtained from a set of data. Hence, although imprecision in analyses may exist, it need not go unrecognized and perhaps become misleading.

However, the amount of error in the data is not generally the limiting factor in data interpretation. Rather, the locations at which the data are taken most severely hinder progress toward a mechanistic model. Reference to Fig. 1 indicates that the decision between the dual- and single-site models would be quite difficult, even with very little error of measurement, if data are taken only in the 2- to 10-atm range. However, quite substantial error can be tolerated if the data lie above 15 atm total pressure (assuming data can be taken here). Techniques are presented that will seek out such critical experiments to be run (Section VII).

The concept of the reaction-rate model should be considered to be more flexible than any mechanistically oriented view will allow. In particular, for any reacting system an entire spectrum of models is possible, each of which fits certain overlapping ranges of the experimental variables. This spectrum includes the purely empirical models, models accurately describing every detail of the reaction mechanism, and many models between these extremes. In most applications, we should proceed as far toward the theoretical extreme as is permitted by optimum use of our resources of time and money. For certain industrial applications, for example, the closer the model approaches

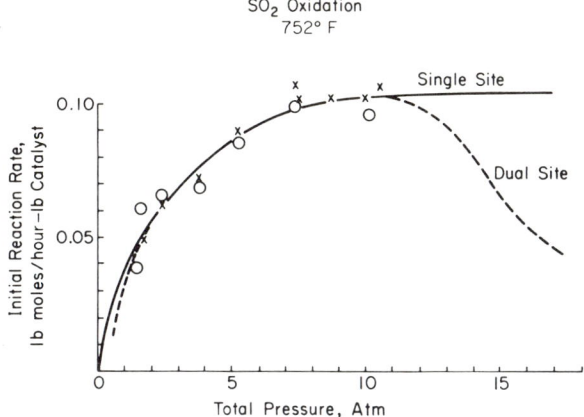

FIG. 1. Initial rate dependence on total pressure for $SO_2$ oxidation (M3).

the true mechanism, the better the final plant design and operation may become. For example, for a high-throughput process of which many exist and many more are expected to be built, a precise mechanistic model can often be justified. In this area, in fact, recent developments in computing, process dynamics, and large-system design are increasingly demanding more sophisticated models of the mechanistic and transport phenomena of the plant in general and reaction kinetics in particular. On the other hand, the construction of a precise mathematical model can require considerable experimentation and analysis. It would often be more profitable to get a new process on stream with a grossly nonoptimum design than to delay construction for a precise definition of a kinetic model. For most applications, an economic balance must be achieved to allow flexibility for modeling as well as rapid progress toward the process design.

C. INTEGRAL AND DIFFERENTIAL MODELS

Equations (1) and (2) represent reaction rates and, as such, can represent directly only data from a differential reactor. In many cases, however, data are obtained from an integral reactor. Are the data to be differentiated and compared directly to Eqs. (1) or (2), or are the equations to be integrated with the conservation equations and compared to the integral data?

If the modeling process has entered the estimation stage, then Eqs. (1) or (2) should clearly be integrated and fitted to the integral data. This may be accomplished quite easily, through analytical or numerical integration of the conservation equations combined with one of the estimation procedures of Section III. General programs providing such combinations have been described (H6).

In the specification stage, however, gross elimination of rival models is difficult to accomplish using such generalized techniques with a set of existing data (cf. the example of Section III,A,3). Instead, careful comparisons of the features of the data and the theoretical model surface are required; these are most easily compared in the simplest possible model form, i.e., the differential form (see, for example, Section V). Furthermore, the experimental design procedures of Section VII can consume substantial computer time, a problem made more severe by iterative numerical integration. With the exception of purely empirical modeling (as in Section VI, which is always in the estimation stage), then, it is preferred to carry out the initial model specification with the simplest form of the model with which the data can be made consistent. For example, factors such as severe variation in catalyst-bed temperature can require the integrated equations even in the specification stage.

## II. Linearly Reducible Models

Most of the modeling procedures commonly used require that the model first be reduced to a form which is linear in the unknown parameters. This procedure represents very good tactics; the technique will be exploited frequently in this review, particularly in Section V. If the scope of the models to be used or the range of experimental variables to be explored is not limited when applying this philosophy, the procedure also represents good modeling strategy.

For these reasons, a proper balance must be achieved between the linearization tactics and the overall modeling strategy. Both linear and nonlinear methods will be illustrated in the review, along with the problems encountered when relying too heavily on either single approach. First, however, some of the more common linear procedures will be discussed.

### A. Power-Function Models

For a single reacting component, Eq. (1) reduces to a form that can easily be compared to rate data to determine (a) if the model is adequate and, if so, (b) the best estimates of the rate constant and the reaction order. For this case, Eq. (1) becomes

$$r = -dC_A/dt = kC_A^a \qquad (3)$$

Here, $t$ is a measure of reaction time. To analyze data from certain reactor types using this model, one can use one of the following methods.

## 1. The Method of Integration

Here, the rate equation, such as Eq. (3), is integrated to yield

$$C_A^{1-a} - C_{A0}^{1-a} = (a-1)kt \qquad a \neq 1$$
$$\ln(C_A/C_{A0}) = -kt \qquad a = 1 \qquad (4)$$

Now, appropriate plots of the data are made, which, if linear, would indicate that the assumed model of Eq. (3) is adequate. For example, if $\ln(C_A/C_{A0})$ were linear with $t$, a first-order model would be adequate. Alternatively, one could assume a model (including the value of the parameter $a$), calculate the rate constant $k$ at each data point, and tabulate the constants. If these "constants" remain constant, or if there is a reasonable trend of the constants with any independent variable, then the data do not reject the assumed model. For example, the value of $\ln k$ would be expected to be independent of the value of the reaction time and to change linearly with the reciprocal of the absolute temperature.

## 2. The Method of Differentiation

In this method, the reaction rate is read directly, say, as the slope of the concentration–time data. Then, the logarithm of the rate is plotted versus the logarithm of the concentration; if the data lie along a straight line, the slope is equal to the reaction order.

## 3. Fractional-Life Methods

Models may also be tested by utilizing the time required for a given fraction of a reactant to disappear, since this varies with the initial concentration in a fashion characteristic of the reaction order. For example, if the half-life of a reaction is defined as the time required for one-half of the initial amount of reactant to be consumed, then Eq. (4) may be written

$$\ln t_{1/2} = \ln\left(\frac{2^{a-1}-1}{k(a-1)}\right) - (a-1)\ln C_{A0} \qquad a \neq 1$$
$$t_{1/2} = 0.693/k \qquad a = 1 \qquad (5)$$

Here $t_{1/2}$ is the half-life and $C_{A0}$ the initial concentration of reactant. Now, half-life data (versus $C_{A0}$) may be analyzed by the same techniques used in the method of integration.

Several other methods are available for analyzing reaction-rate data, such as utilizing linear least squares in the above methods, or such as the method of dimensionless curves. The procedures and their advantages and disadvantages have been adequately discussed elsewhere (L2); numerous examples of the techniques have also been set forth (B3, F5, H7, L2, W2) and are not discussed further here.

### 4. Multicomponent Cases

For the more usual multicomponent case, as described by Eq. (1), variations of the above methods are used to allow a data analysis. One method is to take raction-rate data at such large concentrations of one reactant that this concentration is effectively a constant during the entire course of the reaction:

$$r = kC_A^a C_{B0}^b = k'C_A^a \qquad (6)$$

The order of the other reactant can be determined by any of the previously discussed methods. This technique, called the isolation method, will allow a determination of the component reaction orders, but it should be kept in mind that a very limited region of the experimental space has been covered in determining these orders. Thus, because the model has not been tested for conditions in which both concentrations are varying, the model should be used with caution here.

A second method of analyzing data to be described by Eq. (1) is to use initial reaction rates. Here, one can vary the initial concentrations of the individual reactants (holding all other reactant concentrations constant), measure the rate at zero time for each case, and analyze the data by the previous methods. An advantage of this technique over the isolation method is that the concentrations of the reactants can be nearly equal instead of some concentrations being in large excess. The method allows reaction rates to be obtained over the entire range of composition with respect to known major reaction participants. However, these rates may not be equal to those obtained from experiments in which extensive conversions are allowed, due to the presence of trace byproducts generated during the reaction that affect reaction rate.

For more complex reacting systems, the simple modeling procedures described above are not sufficient. In these more complex systems, the kinetic model describing the system will more generally consist of a group of coupled, nonlinear differential equations. For specific systems, standard analysis techniques are available (W1, B5); more general procedures have also been made available through recent developments of computing hardware and software, and are discussed in the Sections III–V. One very real danger in the treatment of any such system is the inclusion of extraneous and unnecessary parameters, simply because the descriptive equations become so complex that the contributions of the parameters are hidden. Also, with such multiparameter systems, convergence of nonlinear estimation programs can become difficult. Techniques reducing such problems are described.

## B. Hyperbolic Models

The hyperbolic model types have very commonly been used in the analysis of kinetic data, as discussed in Section I. Such applications are sometimes justified on the theoretical bases already alluded to, or simply because models of the form of Eq. (2) empirically describe the existing reaction-rate data. Considerably more complex models are quite possible under the Hougen–Watson formalism, however. For example, Rogers, Lih, and Hougen (R1) have proposed the competitive–noncompetitive model

$$r = \frac{\gamma_1 K_1 K_2 p_1 p_2}{(1 + K_1 p_1 + K_2 p_2)^2} + \frac{\gamma_2 K_1 K_2 p_1 p_2}{(1 + K_1 p_1)(1 + K_1 p_1 + K_2 p_2)} \quad (7)$$

for cases in which large and small molecules react over a solid catalyst.

Shabaker (S1), for the hydrogenation of propylene over a platinum alumina catalyst, selected a Hougen–Watson model of the form

$$r = k_1 K_1 K_2 L^2 (C_L/L)^3 p_1 p_2 + k_2 K_1 L (C_L/L)^2 p_1 p_2 \quad (8)$$

where

$$\frac{C_L}{L} = -\frac{(1 + K_2 p_2)}{4 K_1 p_1} + \left[ +\frac{(1 + K_2 p_2)^2}{16 K_1^2 p_1^2} + \frac{1}{2 K_1 p_1} \right]^{1/2}$$

The adsorption and rate constants in each case exhibit an exponential temperature dependence. The study of these models, in addition to the linearizable models of the form of Eq. (2), have become possible through the use of nonlinear least squares (see Section III).

The usual data analysis procedures for the linearizable models typified by Eq. (2) consist of (1) isolating a class of plausible rival models by means of plots of initial reaction-rate data as a function of total pressure, feed composition, conversion, or temperature; (2) fitting the models passing the screening requirements of the initial rates by linear least squares, and further rejecting models based upon physical grounds.

### 1. Dependence of Reaction Rate on Experimental Variables

The examination of the initial rate dependence upon total pressure is by far the most common means of examining the various classes of rival models in terms of their ability to fit the observed data. Yang and Hougen (Y1) have presented a classification of a multitude of possible models in terms of the dependence of the predicted rate upon total pressure. Generally speaking, the analysis may be performed by writing each of the candidate models in terms of its dependence upon total pressure. For an alcohol dehydrogenation, for example, two models that might be considered are the dual site

$$r = \frac{k\bar{K}_E(p_E - p_A p_H/K)}{(1 + \bar{K}_E p_E + \bar{K}_A p_A + \bar{K}_H p_H)^2} \tag{9}$$

and the single site

$$r = \frac{k\bar{K}_E(p_E - p_A p_H/K)}{(1 + \bar{K}_E p_E + \bar{K}_A p_A)} \tag{10}$$

For initial reaction rates $r_0$ with a pure alcohol feed, all product partial pressures are zero, and $p_{AO} = y_{AO}\pi$, so that for the dual site

$$r_0 = \frac{kK_E \pi}{(1 + K_E \pi)^2} \tag{11}$$

and for the single site

$$r_0 = \frac{kK_E \pi}{(1 + K_E \pi)} \tag{12}$$

At low $\pi$, the denominator simplifies to unity in each case and both models are linear in $\pi$. For sufficiently high $\pi$, the parenthesis in the denominator approaches $K_E \pi$; the initial rate for the dual-site model then approaches zero, and that of the single-site model approaches a constant value. Thus the plot of the experimental data will indicate that the dual-site model is preferable if a maximum exists in the data, or that the single-site model is preferable if a horizontal high-pressure asymptote exists. Hence, for the data of Franckaerts and Froment (F1) shown in Fig. 2, the dual-site model is preferred over the single-site model.

In addition to the maximum point, inflection points of the rate data can be used for testing model adequacy (M6).

The more commonly applied procedure, however, is to linearize the models of Eqs. (11) and (12) to yield

$$\left(\frac{\pi}{r_0}\right)^{1/2} = \frac{1}{(kK_E)^{1/2}} + \frac{K_E}{(kK_E)_{1/2}}\pi \tag{13}$$

$$\left(\frac{\pi}{r_0}\right) = \frac{1}{kK_E} + \frac{1}{k}\pi \tag{14}$$

The data should lie along a straight line when plotted as $(\pi/r_0)^{1/2}$ vs $\pi$, if the dual-site model is adequate. If, in addition, the dual-site model is to be preferred over the single-site model, the plot of $\pi/r_0$ vs $\pi$ should be curved.

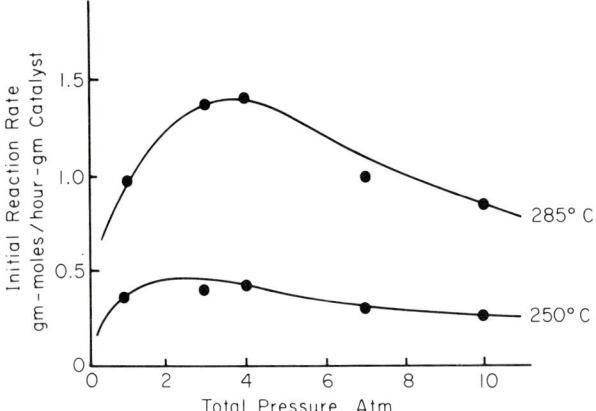

FIG. 2. Initial reaction rate versus total pressure for alcohol dehydrogenation, 250 and 285°C.

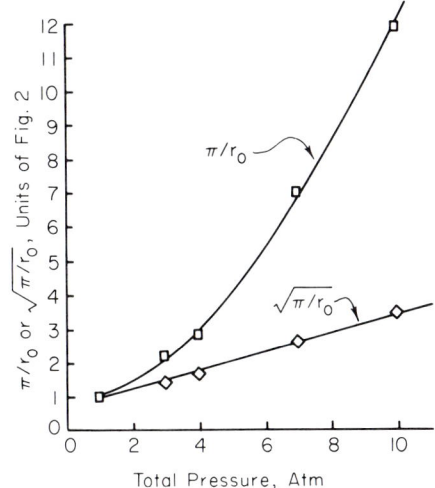

FIG. 3. Linearized initial rate plots for alcohol dehydrogenation, 285°C.

(cf., for example, Section V). The data of Fig. 3, replotted from Fig. 2 according to Eqs. (13) and (14), suggest that the dual-site model is adequate and is preferred over the single-site model.

Often it is advisable to utilize the slopes and intercepts of such plots as well as the above described curvature of the data. Kabel and Johanson (K1), for example, could eliminate an adsorption-controlling model with only 70%

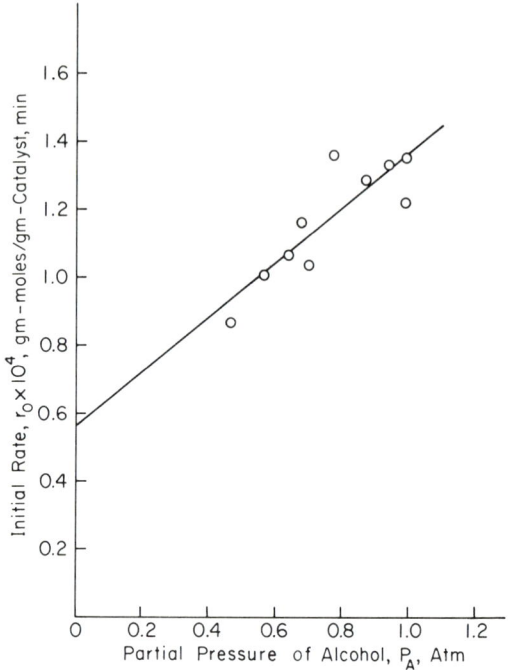

FIG. 4. Linearized initial rate plot for alcohol dehydration, Eq. (16).

certainty using the line curvature for the model representing dehydration of ethanol to ether over an ion-exchange-resin catalyst:

$$r = \frac{kL(p_A - p_E p_W/K)}{1 + [(K_A/K)^2 p_E p_W]^{1/2} + K_W p_W} \quad (15)$$

For ten data points reported at varying alcohol partial pressures ($p_A$) and zero water partial pressure ($p_W$), this model becomes

$$r_0 = kLp_A \quad (16)$$

This suggests that a plot of $r_0$ vs $p_A$ (Fig. 4) or $r_0/p_A$ vs $p_A$ should be linear. It is very difficult to reject this model on the basis of data curvature, even though it is evident that some curvature could exist in Fig. 4. However, Eq. (16) demands that Fig. 4 also exhibit a zero intercept. In fact, the 99.99% confidence interval on the intercept of a least-square line through the data does not contain zero. Hence the model could be rejected with 99.99% certainty.

The dependence of the reaction rate upon conversion, temperature, and composition has not been used as widely as the pressure dependence in model screening. The conversion behavior is discussed in detail in Section V.

The utility of the overall dependence of the reaction rate upon temperature appears to be slight, perhaps in some cases allowing a discrimination between adsorption and surface reaction classes of models. The study of the residuals of various estimated parameters as a function of temperature can clearly indicate model inadequacies (Section IV) and, in some cases, can lead to model modifications correcting these model inadequacies (Section V).

## 2. Magnitude of Estimated Parameters

The second stage of this analysis involves the fitting of the model to the data and examination of the parameter values thus obtained to ensure their physical reasonableness. In particular, (a) the estimated adsorption and rate constants should be positive; (b) a plot of the logarithm of the rate constant with reciprocal absolute temperature should be linear with a negative slope; (c) a plot of the logarithm of the adsorption constant with reciprocal absolute temperature should be linear with a positive slope (exothermic adsorption) and generally a negative intercept (decrease in entropy with adsorption); (d) the model should adequately fit the data (Section IV) with the best estimates of the parameter values. These criteria should, of course, be used with caution. For example, it is possible that adsorption could be endothermic if some of the adsorbed molecules dissociate. The use of these criteria has been discussed at length elsewhere (K7). Much of the present discussion is necessarily deferred until Section III. However, general cautionary measures must include an examination of the confidence interval for any parameter estimate before one rejects a given model. For example, it is illustrated in Section III that although an estimated adsorption constant is negative, its confidence region can easily contain positive values. Also, constants estimated to be negative by unweighted linear least squares can be estimated to be positive by nonlinear least squares or weighted linear least squares (Section III). The general utility of examining the magnitude of the estimated parameters can be illustrated by the data of Kabel and Johanson (K1), which, at zero water partial pressure, can be shown to reject the single-site model

$$r = \frac{kK_A L(p_A^2 - p_E p_W/K)}{(1 + K_A p_A + K_W p_W)} \qquad (17)$$

The appropriate linear plot of $p_A^2/r_0$ vs $p_A$ is shown in Fig. 5. With the amount of scatter in the data for this plot, there is no justification for rejecting the model because of curvature of correlation in the locus of data points. However, the intercept of this graph suggests that the adsorption constant is significantly negative. If a model that possesses theoretical insight into the reacting system is desired, then this model must be rejected.

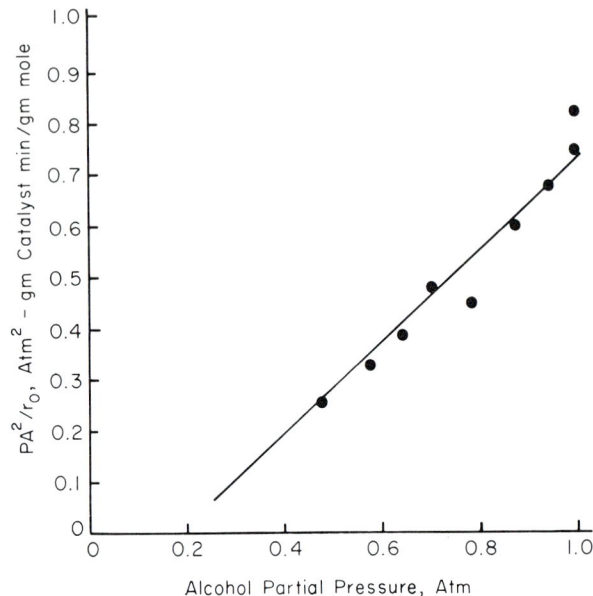

FIG. 5. Linearized initial rate plot for alcohol dehydration, Eq. (17).

An example of a model rejection based upon temperature coefficients can be obtained from the single-site model considered by Ayen and Peters (A3) for the reduction of nitric oxide over a commercial catalyst:

$$r = \frac{kK_{NO} p_{NO} p_{H_2}}{1 + K_{NO} p_{NO} + K_{H_2} p_{H_2}} \qquad (18)$$

This model was fitted to the data of all three temperature levels, 375, 400, and 425°C, simultaneously using nonlinear least squares. The parameters were required to be exponentially dependent upon temperature. Part of the results of this analysis (K6) are reported in Fig. 6. Note the positive temperature coefficient of this nitric oxide adsorption constant, indicating an endothermic adsorption. Such behavior appears physically unrealistic if NO is not dissociated and if the confidence interval on this slope is relatively small. Ayen and Peters rejected this model also.

## III. Parameter Estimation

In any modeling procedure, values of the predicted rate must approximate the values of the observed rate before an adequate model has been established. A simple indication of the comparative shapes of the predicted and

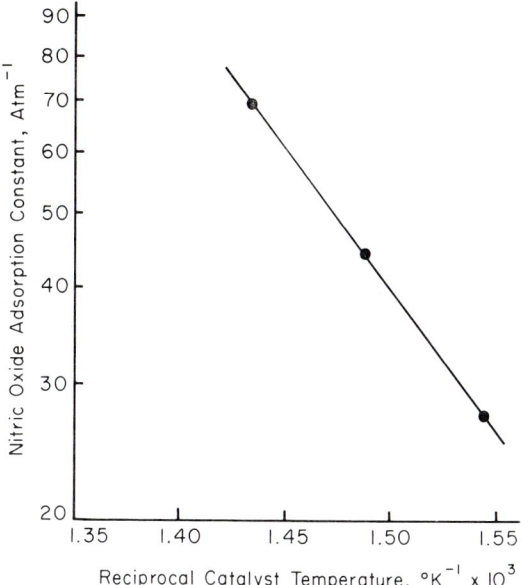

Fig. 6. Temperature dependence of adsorption constant for nitric oxide reduction, Eq. (18).

observed rate surfaces can be obtained by procedures described in Section II. For the final check of the agreement of the model and the data as well as for further use of the model, estimates of the parameters within the model must be obtained. Linear and nonlinear least squares have proven to be of significant value for this purpose. For effective utilization of the parameter estimates thus obtained, the precision of the estimates should also be measured.

A. Least Squares

1. *Linear Least Squares*

Parameter values producing predicted-model results near the observed results have frequently been selected by minimizing $S$:

$$S = \sum_{u=1}^{N} (y_u - \hat{y}_u)^2 \tag{19}$$

This method of *least squares* is not only intuitively desirable, but also provides estimates having desirable properties, if certain assumptions are met (D4).

For models that are linear in the parameters, that is, models of the form

$$\hat{y} = \sum_{i=1}^{p} b_i x_i \tag{20}$$

the parameters may be estimated analytically by linear least squares. In using unweighted linear least squares, however, it is desirable that the $x$'s be free of error, that the response $y$ have constant error variance, and that this error be independent and of mean zero.

For a simple linear model

$$\hat{y}_u = b_0 + b_1 x_u \tag{21}$$

the least-squares estimates $b_0$ and $b_1$ may be obtained by substituting Eq. (21) into Eq. (19), differentiating, and setting the first derivatives to zero:

$$\frac{\partial S}{\partial b_0} = -2 \sum_{u=1}^{N} (y_u - b_0 - b_1 x_u) = 0 \tag{22}$$

$$\frac{\partial S}{\partial b_1} = -2 \sum_{u=1}^{N} x_u (y_u - b_0 - b_1 x_u) = 0 \tag{23}$$

Solving these two equations in two unknowns,

$$b_1 = \frac{N \sum x_u y_u - \sum x_u \sum y_u}{N \sum x_u^2 - (\sum x_u)^2} \tag{24}$$

$$b_0 = \frac{\sum x_u^2 \sum y_u - \sum y_u \sum x_u y_u}{N \sum x_u^2 - (\sum x_u)^2} \tag{25}$$

where the summations range from $u = 1$ to $u = N$. In the more general matrix notation, the least-squares estimates for Eq. (20) become

$$\mathbf{b} = (\mathbf{X}'\mathbf{X})^{-1}\mathbf{X}'\mathbf{Y} \tag{26}$$

where $\mathbf{b}$ is a $p \times 1$ column vector, $\mathbf{Y}$ is an $N \times 1$ column vector, and $\mathbf{X}$ is an $N \times p$ matrix of the levels of $x$. For Eq. (21),

$$\mathbf{b} = \begin{bmatrix} b_0 \\ b_1 \end{bmatrix} \tag{27}$$

$$\mathbf{Y} = \begin{bmatrix} y_1 \\ y_2 \\ \vdots \\ y_N \end{bmatrix} \tag{28}$$

$$\mathbf{X} = \begin{bmatrix} 1 & x_1 \\ 1 & x_2 \\ 1 & x_3 \\ \vdots & \vdots \\ 1 & x_N \end{bmatrix} \tag{29}$$

The equations represented by Eq. (26) are termed the normal equations; they provide parameter estimates possessing the minimum distance from the data $Y$ to the model surface, thus representing a line that is perpendicular (or normal) from the point $Y$ to the surface.

Many kinetic equations can be suitably linearized to the form of Eq. (20). For example, Eq. (1) can be transformed logarithmically, or Eq. (2) can be transformed reciprocally. Two equations proposed for describing pentane-isomerization data (C1, J1) are the single site

$$r_1 = \frac{kK_2(p_2 - p_3/K)}{(1 + K_1p_1 + K_2p_2 + K_3p_3)} \tag{30}$$

and the dual site

$$r_2 = \frac{kK_2(p_2 - p_3/K)}{(1 + K_1p_1 + K_2p_2 + K_3p_3)^2} \tag{31}$$

where $p_1$, $p_2$, and $p_3$ are the partial pressures of hydrogen, normal pentane, and isopentane. Rearranging and redefining parameters,

$$\left(\frac{p_2 - p_3/K}{r_1}\right) = b_0 + b_1p_1 + b_2p_2 + b_3p_3 \tag{32}$$

$$\left(\frac{p_2 - p_3/K}{r_2}\right)^{1/2} = b_0 + b_1p_1 + b_2p_2 + b_3p_3 \tag{33}$$

Unweighted linear least-squares estimates of these parameters are shown in Table I. The negative parameter estimates of the single-site model are of no

TABLE I

PARAMETER ESTIMATES FOR PENTANE ISOMERIZATION

| Parameters | Least-square technique | | |
|---|---|---|---|
| | Unweighted linear | Weighted linear | Unweighted nonlinear |
| Single Site | | | |
| $k$ | 189 | 30.5 | 36.2 |
| $K_1$ | −0.4 | 2.0 | 0.9 |
| $K_2$ | −0.03 | 1.3 | 0.5 |
| $K_3$ | −1.08 | 4.3 | 2.1 |
| Dual Site | | | |
| $k$ | 289 | 73.9 | 133 |
| $K_1$ | 0.03 | 0.03 | 0.03 |
| $K_2$ | 0.005 | 0.02 | 0.02 |
| $K_3$ | 0.09 | 0.07 | 0.07 |

great concern, as they arise from a negative estimate of $b_0$ in Eq. (32) having a confidence interval (Section III,C) so large that positive values of $b_0$ may be chosen without significantly harming the fit of the data.

Several variations of these concepts (e.g., stepwise regression) have also been proposed (D4).

### 2. Weighted Linear Squares

Thus far, we have assumed that all of the observations are independent and have the same inherent uncertainty. Deviations from such assumptions have been discussed in a generalized fashion elsewhere (H8).

Taking into account deviations from constant variance is a particularly straightforward matter. If, for example, the variable $y$ of Eq. (19) possesses a variance $\sigma_i^2$ which varies from data point to data point, then Eq. (19) divided by $\sigma_i^2$ is the appropriate sum of squares to minimize. If, in general, we define

$$\mathbf{C} = \begin{bmatrix} \sigma_i^2 & 0 & \cdots & 0 \\ 0 & \sigma_2^2 & \cdots & 0 \\ \vdots & \vdots & & \vdots \\ 0 & 0 & \cdots & \sigma_N^2 \end{bmatrix} \tag{34}$$

then Eq. (26) becomes

$$\mathbf{b} = (\mathbf{X}^1 \mathbf{C}^{-1} \mathbf{X})^{-1} \mathbf{X}^1 \mathbf{C}^{-1} \mathbf{Y} \tag{35}$$

or if $\hat{y} = \hat{b}x$, this is simply

$$\hat{b} = \frac{\sum (x_i y_i / \sigma_i^2)}{\sum (x_i^2 / \sigma_i^2)} \tag{36}$$

where the sum is taken over all observations. Now, if replication (Section IV) has been carried out at each data point, then the estimate of error variance at each data point, $\sigma_i^2$, can be inserted into Eq. (34) and the least-square estimate calculated from Eq. (35).

More typically, we have an indication that a transformed variable $f(y)$ has constant error variance and will wish to use this information to weigh $y$ appropriately. For example, we may suspect log $y$ has constant error variance and wish to fit $y$. More typically, we might feel that $y$ has constant error-variance and wish to fit $1/y$.

This problem was solved approximately in 1947 (B2, K9, J1), wherein was suggested that the transformed function of $y$ be expanded in a linear Taylor series to provide

$$\text{variance } [f(y)] = (\partial f(y)/\partial y)^2 \cdot \text{variance } (y) \tag{37}$$

For example, if the rate $r$ is thought to have constant error variance and we wished to fit $f(y) = (a/r)^n$ as in Eq. (33), then the quantity

$$\sigma_i^2 = [-nf(y_i)/r_i]^2 \sigma_r^2 \tag{38}$$

should be used in the weighting matrix **C** of Eq. (34) (the constants $n$ and $\sigma_r^2$ are unnecessary). For comparison to the results of the unweighted least-squares analysis of the pentane isomerizations data just discussed, the second column of Table I should be examined. Here, the weighting function of Eq. (38) was used, assuming the rate $r$ had constant error variance [in contrast to the function of $r$ defined in Eqs. (32) and (33) in the unweighted case]. Note the significant shift in the weighted estimates from the unweighted estimates, including the positive single-site parameter estimates. Since the nonlinear least-squares results also assume $r$ has constant error variance, there is a close correspondence in the weighted linear and the nonlinear results.

Equations (34) and (35) can also be generalized to include correlation between observations. In such a case

$$\mathbf{C} = \begin{bmatrix} \sigma_{11}^2 & \sigma_{12}^2 & \cdots & \sigma_{1N}^2 \\ \sigma_{21}^2 & \sigma_{22}^2 & \cdots & \sigma_{2N}^2 \\ \vdots & \vdots & & \vdots \\ \sigma_{N1}^2 & \sigma_{N2}^2 & \cdots & \sigma_{NN}^2 \end{bmatrix} \tag{39}$$

where the $\sigma_{ij}^2 (i \neq j)$ are the covariances between the observation errors. This variance–covariance matrix is again applied through Eq. (35).

## 3. Nonlinear Least Squares

Many of the models encountered in reaction modeling are not linear in the parameters, as was assumed previously through Eq. (20). Although the principles involved are very similar to those of the previous subsections, the parameter-estimation procedure must now be iteratively applied to a nonlinear surface. This brings up numerous complications, such as initial estimates of parameters, efficiency and effectiveness of convergence algorithms, multiple minima in the least-squares surface, and poor surface conditioning.

*Iterative Techniques.* In estimating parameters in a model that is nonlinear in the parameters

$$r = f(\mathbf{x}; \mathbf{K}) \tag{40}$$

we still minimize $S$ of Eq. (19). Since the partial derivatives similar to Eqs. (22) and (23) do not generally lead to equations that are easily solved, we turn to one of three basic types of methods (K8), even though a wide variety of specific minimization techniques are available (S2).

In the *linearization method*, the nonlinear model of Eq. (40) is linearized by a truncated Taylor expansion:

$$y_u = r_u - f(\mathbf{x}_u ; \mathbf{K}) = \sum_{i=1}^{p} f'_{iu} b_i \qquad (41)$$

where

$$f'_{iu} = \frac{\partial f(\mathbf{x}_u ; \mathbf{K})}{\partial K_i} \bigg|_{\mathbf{K} = \mathbf{K}^\circ} \qquad (42)$$

Here $\mathbf{K}^\circ$ is the set of initial parameter estimates. Since Eq. (41) is linear, estimates of the correction vector **b** may be obtained through Eq. (26), where $y_u$ is now given by Eq. (41) and **X** is given by

$$\mathbf{X} = \{f'_{iu}\} \qquad (43)$$

Using estimates thus obtained, improved estimates of $K_i$ are given by

$$K_i^{(1)} = b_i + K_i^\circ \qquad (44)$$

The procedure is repeated until the procedure converges, that is until the correction vector **b** approaches zero. Methods of modifying the size of the correction vector have been developed to improve convergence (B5) and this method, in theory, should always converge (H1). In practice, nonlinearities in the model and poor parameter estimates can prevent convergence. A modification of this method linearizes the normal equations; higher derivatives have also been used.

In the *steepest-descent method* (B10, B18, K8, S2, R2), the sums-of-squares surface is visualized to be a response surface in which the parameters are variables. Steepest-descent procedures are then applied to determine the minimum of the surface. The procedure is, then, to (a) select initial parameter estimates; (b) set up a first-order design (B18) in the parameter space about these latter estimates, calculating the sum of squares at each point (c) determine the direction of steepest descent; (d) proceed in this direction in the parameter space until the sums of squares begin increasing; (e) if the minimum is not near, go back to step (b); otherwise (f) set up a second-order design for precise location of the minimum, and continue toward the minimum, reparameterizing (Section III,B) as necessary to improve surface conditioning. Although this method converges for nearly any set of initial estimates, its convergence can be agonizingly slow for multiparameter problems (K8).

Experience with fitting many models indicates that the steepest descent is very stable for the initial iterations, while the linearization method more efficient for the final iterations. A *compromise method* has been suggested (L5, M1, M2) which tends to emphasize steepest descent at the outset ar linearization in the final stages. In this method, the correction vector used in Eq. (44) is calculated from

$$\mathbf{b} = (\mathbf{X}^1 \mathbf{X} + \psi \mathbf{I})^{-1} \mathbf{X}^1 \mathbf{Y} \tag{45}$$

where $y_u$ is given by Eq. (41) and $\mathbf{X}^1 \mathbf{Y}$ is the vector of steepest descent. Thus, a large value of $\psi$ makes the correction vector near the vector of steepest descent while a small $\psi$ makes the correction vector near that obtained by the linearization method. The compromise method, then, incorporates a continually decreasing value of $\psi$ as the iterations proceed.

Numerous examples of applications of nonlinear least squares to kinetic-data analysis have been presented (K7, K8, L3, L4, M7, P2); an exhaustive tabulation of references would, at this point, approach 100 entries. Typical results of a nonlinear estimation and comparison to linear estimates are shown in Table I and discussed in Section III,A,2. Many estimation problems exist, however, as typified in part by Fig. 7. This is the sum-of-squares surface obtained at fixed values of $K_s$ and $K_u$ in the rate equation used for the catalytic hydrogenation of mixed isooctenes (M7)

$$r = \frac{k K_H K_U p_H p_U}{(1 + K_H p_H + K_U p_U + K_S p_S)^2} \tag{46}$$

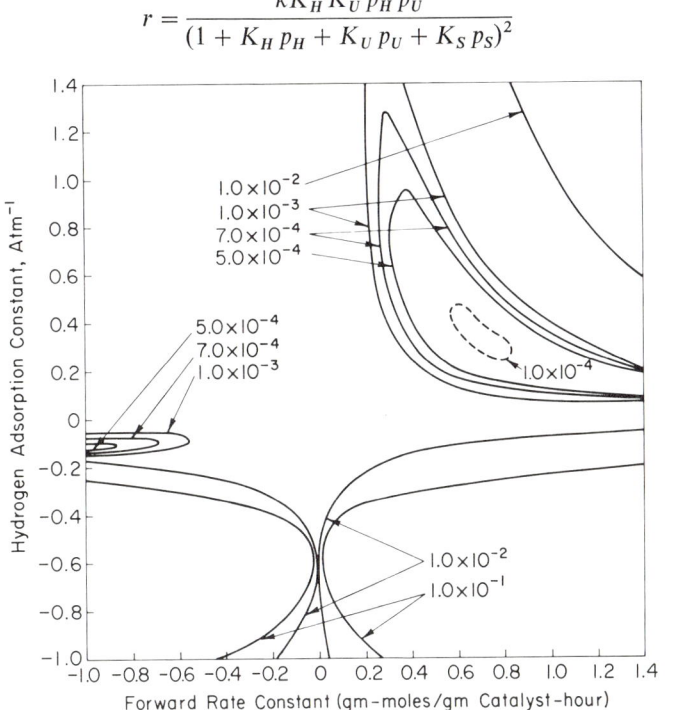

FIG. 7. Contours of sums of squares of residual rates for isooctene hydrogenation, Eq. (46).

First, this figure shows long skewed contours that frequently occur with such models. Also, it should be noted that two minima exist in this plot, in the first and third quadrants. Accordingly, in the estimation of the rate parameters for this model with the standard nonlinear estimation techniques, it would be possible to obtain certain negative parameter estimates. In fact, it can be seen from Eq. (46) that any two of the parameters $k$, $K_H$, or $K_U$ can simultaneously be negative without greatly hindering the fit of the equation to the data owing to the fact that they are multiplied together in the numerator. These considerations are of importance in the estimation of the parameter values, since negative estimates of any adsorption constants are generally used as a justification for rejecting the model under consideration. Therefore, one should attempt to examine the sums-of-squares surface in enough detail to ensure that the model should truly be rejected and not just refitted. Measures of nonlinearity of the sum-of-squares surface are available (B3, B8, M2, M7).

Part of the problems encountered with sums-of-squares surfaces such as shown in Fig. 7 can be relieved by obtaining precise initial-parameter estimates for the nonlinear estimation. Employing a pilot or first-stage estimation procedure to obtain values $\mathbf{K}°$ to use in an iterative nonlinear estimation program is often more efficient than simply making rough guesses, even though at first sight it may appear to be more time-consuming. Numerous examples of this type of pilot estimation have been described (K8). Figure 8, for example, presents the sums-of-squares surface for alcohol dehydration data and the model of Eq. (47). As indicated in the figure, any trial value of the two parameters in the first quadrant will, if restricted to the first quadrant, cause the nonlinear estimation procedure to diverge (M7):

$$r = \frac{kK_A(p_A^2 - p_E p_W/K)}{(1 + K_A p_A + K_E p_E + K_W p_W)} \qquad (47)$$

A brief survey of the sums-of-squares surface, such as shown in this figure can eliminate many computer hours of fruitless iteration with repeated initial parameter estimates. Other problems related to estimation from Figs. 7 and 8 can be reduced by reparameterization, as discussed in Section III,B.

If the nonlinear estimation procedure is carefully applied, a minimum in the sums-of-squares surface can usually be achieved. However, because of the fitting flexibility generally obtainable with these nonlinear models, it is seldom advantageous to fit a large number of models to a set of data and to try to eliminate inadequate models on the basis of lack of fit (see Section IV). For example, thirty models were fitted to the alcohol dehydration data just discussed (K2). As is evident from the residual mean squares of Table II, approximately two-thirds of the models exhibit an acceptable fit of the data

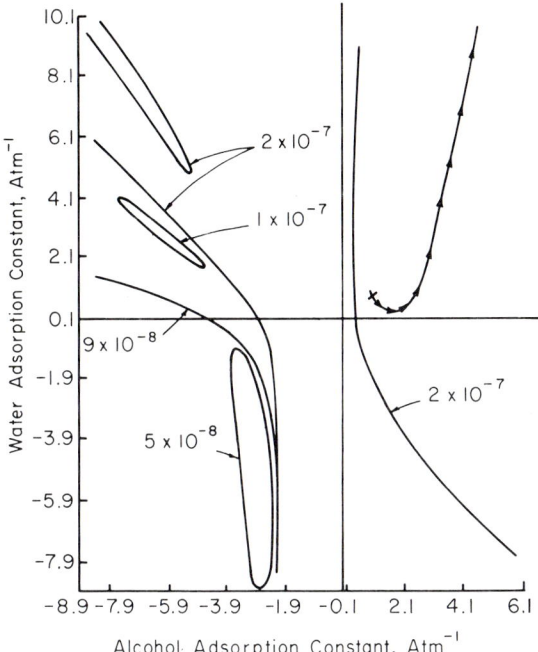

FIG. 8. Contours of sums of squares of residual rates for alcohol dehydration, Eq. (47).

points used (rms $\leq 1.2 \times 10^{-6}$). The experimental design, as is shown later, plays a major role in determining the success of such a project (Section VII).

For fitting such a set of *existing* data, a much more reasonable approach has been used (P2). For the naphthalene oxidation system, major reactants and products are symbolized in Table III. In this table, letters in bold type represent species for which data were used in estimating the frequency factors and activation energies contained in the body of the table. Note that the rate equations have been reparameterized (Section III,B) to allow a better estimation of the two parameters. For the first entry of the table, then, a model involving only the first-order decomposition of naphthalene to phthalic anhydride and naphthoquinone was assumed. The parameter estimates obtained by a nonlinear-least-squares fit of these data, $\hat{\theta}$, are seen to be relatively precise when compared to the standard errors of these estimates, $s_{\hat{\theta}}$. The residual mean square, using these best parameter estimates, is contained in the last column of the table. This quantity should estimate the variance of the experimental error if the model adequately fits the data (Section IV). The remainder of Table III, then, presents similar results for increasingly complex models, each of which entails several first-order decompositions.

TABLE II: SUMMARY OF PARAMETER ESTIMATES FOR ALCOHOL DEHYDRATION

| Model No. | $k \times 10^4$ | $K_A$ | $K_E$ | $K_W$ | $K_{E1}$- | $K_{W1}$- | $K_{A1}$- | $L$ | Residual mean square $\times 10^{10}$ |
|---|---|---|---|---|---|---|---|---|---|
| 1 | 1.49 | 0 | 0 | 4.09 | — | — | — | — | 1.09 |
| 2 | 2.26 | 3.39 | 0.13 | 7.31 | — | — | — | — | 0.70 |
| 3 | 8.63 | $10^{-8}$ | — | 3.68 | $10^{-8}$ | — | — | 0.17 | 1.13 |
| 4 | 5.28 | 46.5 | — | 91.4 | 666 | — | — | 0.51 | 0.75 |
| 5 | 8.98 | 0 | 0 | — | — | 40.3 | — | 0.17 | 1.36 |
| 6 | 9.72 | 0 | 0.19 | 2.71 | $10^{-7}$ | — | — | 0.15 | 1.67 |
| 7 | 16.2 | 4.26 | 0.04 | 1.34 | $10^{-6}$ | — | — | 0.08 | 6.71 |
| 8 | 15.8 | 0 | 0 | 8.40 | — | 8.40 | — | 0.10 | 1.45 |
| 9 | 5.0 | 54.3 | 0 | 4.0 | — | $10^3$ | — | 0.50 | 1.45 |
| 10 | 1.44 | 0 | 0 | 4.51 | — | — | — | — | 1.05 |
| 11 | 7.04 | 2.89 | 0 | 6.30 | — | — | — | — | 0.70 |
| 12 | 1.48 | $10^4$ | 0 | $10^4$ | — | — | — | — | 1.12 |
| 13 | 7.6 | 0 | — | 4.7 | $10^{-7}$ | — | — | 0.19 | 1.08 |
| 14 | 56.0 | 15.4 | — | 30.6 | 52.9 | 45.7 | — | 0.64 | 0.77 |
| 15 | 8.9 | 0 | 0 | — | — | 253 | — | 0.17 | 1.29 |
| 16 | 197 | 4.4 | 0.01 | 4.03 | $10^{-5}$ | — | — | 0.21 | 0.61 |
| 17 | 5.8 | 0 | 2.98 | 4.71 | $10^{-5}$ | — | — | 0.25 | 1.16 |
| 18 | 2.77 | 2.74 | 0 | 0.44 | — | $10^{-8}$ | — | 0.06 | 78.0 |
| 19 | 8.16 | 0 | 0 | 2.01 | — | 16.6 | — | 0.14 | 5.75 |
| 20 | 9.95 | 3.69 | 0 | 1.86 | — | — | — | 0.90 | 0.63 |
| 21 | 16.6 | — | 0 | $10^4$ | 0 | — | $10^{-3}$ | 0.30 | 1.06 |
| 22 | 9.94 | — | 0 | 1.74 | 0 | — | $10^9$ | 0.15 | 1.15 |
| 23 | 34.4 | — | — | $10^4$ | — | — | 0 | 0.21 | 35.3 |
| 24 | 9.94 | — | — | — | — | 5.4 | 109 | 0.15 | 1.15 |
| 25 | 6.5 | — | 0 | — | — | 0 | 0 | 0.48 | 1.12 |
| 26 | 17.4 | — | 0 | 1.03 | $10^{-6}$ | — | 8.51 | 0.17 | 5.7 |
| 27 | 38.1 | — | 0.05 | 0 | $10^{-9}$ | — | 0 | 0.19 | 1.50 |
| 28 | 17.4 | — | 0 | | | | 8.49 | 0.17 | 5.9 |

By slowly increasing the complexity of the models in this fashion, it was hoped that a model could be obtained that was just sufficiently complex to allow an adequate fit of the data. This conscious attempt to select a model that satisfies the criteria of adequate data representation and of minimum number of parameters has been called the principle of parsimonious parameterization. It can be seen from the table that the residual mean squares progressively decrease until entry 4. Then, in spite of the increased model complexity and increased number of parameters, a better fit of the data is not obtained. If the reaction order for the naphthalene decomposition is estimated, as in entry 5, the estimate is not incompatible with the unity order of entry 4. If an additional step is added as in entry 6, no improvement of fit is obtained. Furthermore, the estimated parameter for that step is negative and poorly defined. Entry 7 shows yet another model that is compatible with the data. If further discrimination between these two remaining rival models is desired, additional experiments must be conducted, for example, by using the model discrimination designs discussed later. The critical experiments necessary for this discrimination are by no means obvious (see Section VII).

## B. Reparameterization

The problems described in the last subsection can frequently be reduced by improving the surface conditioning through reparameterization. For example, if the two parameters in the first-order model

$$C_B = \exp\{-k_0 \exp(-E/RT)\} \tag{48}$$

are estimated from a set of data (H9), then one contour of the sums-of-squares surface can be represented as shown in Fig. 9. With such "natural" parameterization, an iterative nonlinear least-squares routine will generally converge to the minimum sum of squares quite slowly or not at all. However, writing the model in an equivalent mathematical form, but with different parameters, can lead to better sums-of-squares surface conditioning.

Considerable discussion of reparameterization and examples of its usefulness have been published (B3, B8, B12, G1, G2, M7). Although several specific techniques are useful, one reparametrization of kinetic models often necessary is a redefinition of the independent variables so that the center of the new coordinate system is near that of the experimental design. In particular, the exponential parameter

$$k = k_0 \exp(-E/RT) \tag{49}$$

TABLE III

SUMMARY OF NONLINEAR FITTING OF NAPHTHALENE OXIDATION DATA

| Entry | Mechanism[a] | | $k_0' = k_0 \exp(-E/RT)$ | | | | | | $E' = E/R$ | | | | | | $\alpha$ | $s^2 \times 10^3$ |
|---|---|---|---|---|---|---|---|---|---|---|---|---|---|---|---|---|
| | | | 1 | 2 | 3 | 4 | 5 | 6 | 1 | 2 | 3 | 4 | 5 | 6 | | |
| 1 | $N-1\to P$ | $\hat\theta$ | 8.79 | 0.842 | — | — | — | — | 14.5 | 12.4 | — | — | — | — | — | 1.06 |
| | $N-2\to Q$ | $s_\theta$ | 0.26 | 0.162 | — | — | — | — | 0.4 | 2.8 | — | — | — | — | — | |
| 2 | $N-1\to P$ | $\hat\theta$ | 9.23 | 0.873 | 0.833 | — | — | — | 14.6 | 12.4 | 9.68 | — | — | — | — | 0.772 |
| | $N-2\to Q$ | $s_\theta$ | 0.26 | 0.146 | 0.151 | — | — | — | 0.4 | 2.8 | 2.34 | — | — | — | | |
| | $N-3\to G$ | | | | | | | | | | | | | | | | |
| 3 | $N-1\to P$ | $\hat\theta$ | 8.10 | 2.16 | 0.882 | 18.8 | — | — | 15.0 | 11.9 | 9.80 | 7.41 | — | — | — | 0.671 |
| | ↑4 | | | | | | | | | | | | | | | | |
| | $N-2\to Q$ | $s_\theta$ | 0.61 | 0.59 | 0.142 | 7.8 | — | — | 1.0 | 3.2 | 2.09 | 5.60 | — | — | — | |
| | $N-3\to G$ | | | | | | | | | | | | | | | | |
| 4 | $N-1\to P-5\to M$ | $\hat\theta$ | 8.10 | 3.00 | 0.881 | 28.0 | 0.831 | — | 15.6 | 13.0 | 8.86 | 9.69 | 16.8 | — | — | 0.471 |
| | ↑4 | | | | | | | | | | | | | | | | |
| | $N-2\to Q$ | $s_\theta$ | 0.73 | 0.74 | 0.124 | 8.9 | 0.221 | — | 1.1 | 2.6 | 3.96 | .80 | 3.5 | — | — | |
| | $N-3\to G$ | | | | | | | | | | | | | | | | |

| Scheme | | | | | | | | | | | | | | | |
|---|---|---|---|---|---|---|---|---|---|---|---|---|---|---|---|
| 5  N—1→P—5→M<br>    ↑4<br>N—2→Q<br>N—3→G | $b$<br>$s_\theta$ | 8.26<br>0.88 | 3.55<br>1.11 | 0.926<br>0.142 | 31.3<br>10.4 | 0.750<br>0.236 | —<br>— | 15.9<br>1.3 | 13.5<br>2.8 | 9.93<br>1.85 | 17.6<br>3.9 | 9.05<br>3.8 | —<br>— | 1.13<br>0.15 | 0.475<br>— |
| 6  N—1→P—5→M<br>    ↑4<br>N—2→Q<br>    ↓6<br>N—3→G | $\theta$<br>$s_\theta$ | 8.09<br>0.72 | 2.76<br>0.74 | 1.20<br>0.48 | 29.2<br>9.0 | 0.861<br>0.227 | −4.15<br>6.10 | 15.6<br>1.1 | 12.8<br>2.7 | 11.1<br>3.2 | 16.6<br>3.6 | 8.84<br>3.90 | 10.9<br>12.1 | —<br>— | 0.480<br>— |
| 7  N—1→P—5→M<br>    ↑4<br>N—2—Q—6→G | $\theta$<br>$s_\theta$ | 7.62<br>1.05 | 4.16<br>1.10 | —<br>— | 32.0<br>13.1 | 0.787<br>0.234 | 10.7<br>1.9 | 15.4<br>1.5 | 12.6<br>2.5 | —<br>— | 9.60<br>4.64 | 17.8<br>3.9 | 3.58<br>2.33 | —<br>— | 0.497<br>— |

[a] N, naphthalene; P, phthalic anhydride; M, maleic anhydride; Q, naphthoquinone; G, off gases.
[b] $\propto$ denotes reaction order for naphthalene decomposition steps.

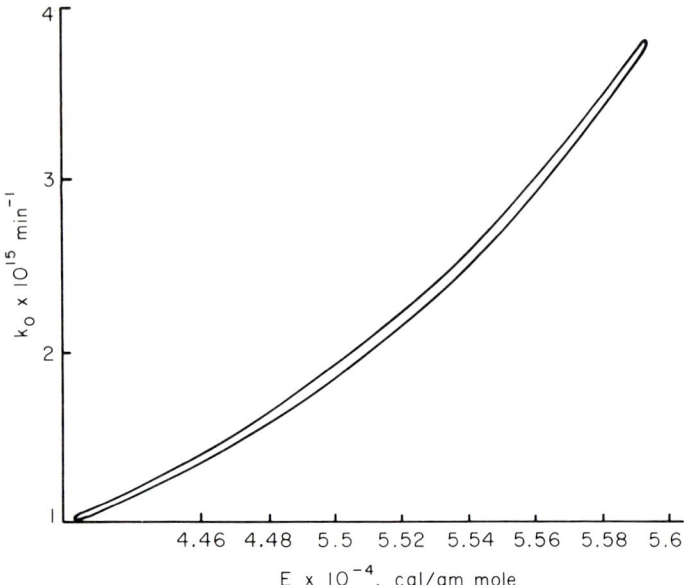

FIG. 9. Approximate 95% confidence region for first-order decomposition model.

should almost invariably be redefined as

$$k = k_0' \exp\left\{-\frac{E}{R}\left(\frac{1}{T}-\frac{1}{\overline{T}}\right)\right\} \quad (50)$$

where

$$k_0' = k_0 \exp(-E/R\overline{T}) \quad (51)$$

In the example above, the sums-of-squares surface is transformed to that shown by Fig. 10. The best point estimates of the parameters of $k_0'$ and $E$ are thus more readily obtained than $k_0$ and $E$; the initial-parameter estimates are less critical, and the estimation routine converges more rapidly to the minimum. Although the correlation between the parameter estimates has been reduced by this reparameterization, the size of the confidence region of the original parameters $k_0$ and $E$ will not change.

## C. Confidence Intervals and Regions

In reaction-rate modeling, precise parameter estimates are nearly as essential as the determination of the adequate functional form of the model. For example, in spite of imprecisely determined parameters, an adequate model will still predict the data well over the range that the data are taken.

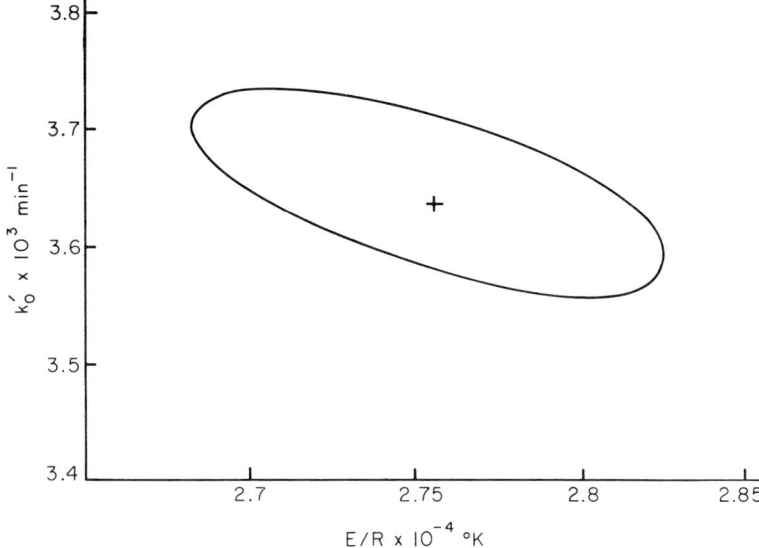

Fig. 10. Approximate 95% confidence region for reparameterized first-order decomposition model.

as shown in Fig. 11 (K11). However, if these parameter estimates are to be subjected to mechanistic interpretation or are to be used in reactor designs (with slight extrapolation), then clearly additional information about the precision of the estimates is required. An example of the first situation is provided by the negative parameter estimates of Table I. As is shown clearly by the confidence region for this entry, positive parameter estimates are quite compatible with the data (K3).

The amount of uncertainty in parameter estimates obtained for the hyperbolic models is particularly large. It has been pointed out, for example, that parameter estimates obtained for hyperbolic models are usually highly correlated and of low precision (B16). Also, the number of parameters contained in such models can be too great for the range of the experimental data (W3). Quantitative measures of the precision of parameter estimates are thus particularly important for the hyperbolic models. (C2).

. *Linear Models*

For models of the form of Eqs. (32) and (33) [or, more generally, Eq. (20)], the $(1 - \alpha)\,100\%$ confidence interval for the parameters $b_i$ (not $K_i$) is given by

$$b_i \pm t_{v,\,\alpha/2}[v(b_i)]^{1/2} \qquad (52)$$

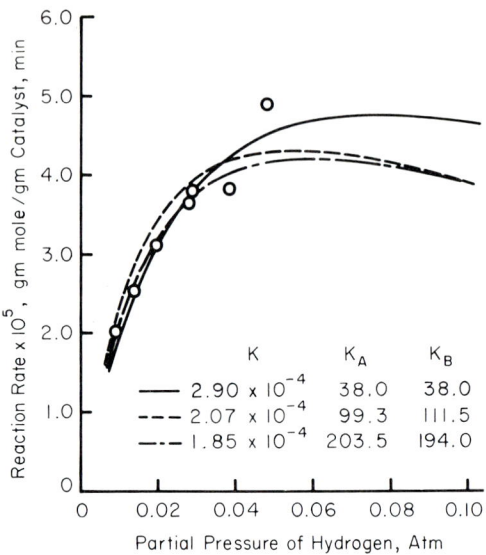

FIG. 11. Dependence of predicted reaction rate upon assumed parameter estimates Eq. (149).

where $t_v, \alpha/2$ is the $(1 - \alpha/2)$ 100% point of the $t$ distribution (D1) with degrees of freedom. If the errors in the dependent variable [e.g., left-hand side of Eq. (32)] are normally and independently distributed with constant variance $\sigma^2$, then $v(b_i)$ is the $i$th diagonal element of

$$v(\mathbf{b}) = (\mathbf{X}^1\mathbf{X})^{-1}\sigma^2 \qquad (53)$$

If, on the other hand, weighted linear least squares is required, Eq. (53) becomes

$$v(\mathbf{b}) = (\mathbf{X}^1\mathbf{C}^{-1}\mathbf{X})^{-1}\mathbf{X}^1\mathbf{C}^{-1}\sigma^2 \qquad (54)$$

The use of these equations allow the calculation of confidence intervals to assess, for example, the significance of negative parameter estimates or the additional experimentation required to estimate the parameters with particular precision.

As has been described, the correlation of the parameter estimates is also of great interest. This correlation is measured by the off-diagonal terms $v(\mathbf{b})$, completely ignored by the above equations. If an analysis of variance conducted, one finds that a joint $(1 - \alpha)$ 100% confidence region is defined

$$(\boldsymbol{\beta} - \mathbf{b})^1\mathbf{X}^1\mathbf{X}(\boldsymbol{\beta} - \mathbf{b}) = s^2 p F_\alpha(p, v) \qquad (5$$

Where $\boldsymbol{\beta}$ is a vector of parameter values being estimated by $\mathbf{b}$, $s^2$ is an independent estimate of the experimental error variance $\sigma^2$, $p$ is the number parameters, and $F_\alpha(p, v)$ is the $(1 - \alpha) \, 100\%$ point of the $F$ distribution. Alternatively,

$$S_c = S_{\min} + s^2 p F_\alpha(p, v) \tag{56}$$

where $S_c$ is the value of the sum of squares on the $100(1 - \alpha)\%$ confidence contour, and $S_{\min}$ is the minimum value of Eq. (19). As is described in Section IV, $s^2$ can be estimated from replicated data or, if the fitted model is adequate, from

$$s^2 = S_{\min}/(N - p) \tag{57}$$

For two parameters in the model, Eq. (55) simplifies to

$$(\beta_1 - b_1)^2 \sum x_1^2 + 2(\beta_1 - b_1)(\beta_2 - b_2) \sum x_1 x_2 + (\beta_2 - b_2)^2 \sum x_2^2 = s^2 p F_\alpha(p, v) \tag{58}$$

where the summations range over $N$ observations. For fixed-data $x_i$, this equation describes an ellipse in the parameter space. Thus, confidence regions are elliptical (or more generally, ellipsoidal) for linear models.

The covariance (off-diagonal) terms of Eq. (53) are also needed to calculate approximate intervals for parameters in the hyperbolic models. For a model such as

$$\frac{1}{r} = \frac{1}{k} + \frac{K_c}{k} p_c + \frac{K_D}{k} p_D \tag{59}$$

Eq. (53) provides the variance–covariance matrix for the parameter ratios. Letting $b_1$, $b_2$, and $b_3$ denote the respective ratios of Eq. (59), then Eq. (37) may be applied to find

$$v(k) = k^4 v(b_1)$$

where $v(b_1)$ is available from the usual linear least squares Eq. (53). Using a generalized form of Eq. (37), we further obtain

$$v(K_c) = \left(\frac{K_c}{b_2}\right)^2 v(b_2) + \left(\frac{K_c}{b_1}\right)^2 v(b_1) + \left(\frac{K_c}{b_1}\right)\left(\frac{K_c}{b_2}\right) \text{cov}(b_1, b_2) \tag{60}$$

$$v(K_D) = \left(\frac{K_D}{b_3}\right)^2 v(b_3) + \left(\frac{K_D}{b_1}\right)^2 v(b_1) + \left(\frac{K_D}{b_3}\right)\left(\frac{K_D}{b_1}\right) \text{cov}(b_1, b_3) \tag{61}$$

Confidence intervals thus calculated for the rate and adsorption constants have been reported (K7).

## 2. Nonlinear Models

The general approach used with nonlinear models, such as Eq. (40) is to linearize by a Taylor expansion [Eq. (41)] and apply the linear theory of Section III,C,1.

The more approximate method of measuring the precision of parameter estimates is to use confidence intervals, as defined by Eqs. (52) and (53). Here X is again given by Eq. (43). This estimate will be erroneous because the Taylor expansion does not perfectly describe the nonlinear model throughout the confidence intervals. Hence, although the intervals are calculated assuming the contours of the sums-of-squares surface are ellipsoidal, they deviate from this shape. An example comparing such contours is shown in Fig. 12, for estimates of the two rate constants in the sequence $A \to B \to C$ (B14). In fact, the deviation of the actual sums of squares from an ellipsoidal shape can be used as a measure of the nonlinearity of the model (B3, B8), using latent roots of the $(X^1 X)$ matrix or other indicators (G1, G2, M7).

A better measure of the precision of the parameter estimates for nonlinear models is provided by Eq. (56), which can take into account deviations of the sums-of-squares contours from the ellipsoidal shape. In fact, this equation does provide a rigorous confidence contour for nonlinear models; it is only the exact confidence level, $\alpha$, associated with this contour that is approximated.

Numerous applications of this theory have been made in calculating confidence intervals for parameter estimates in nonlinear kinetic models, such as typified in Table III (P2). The use of confidence regions is typified in Fig. 13 (M7) for the alcohol dehydration model

$$r = \frac{kK_A(p_A^2 - p_E p_W / K)}{(1 + K_A p_A + K_W p_W)^2} \tag{62}$$

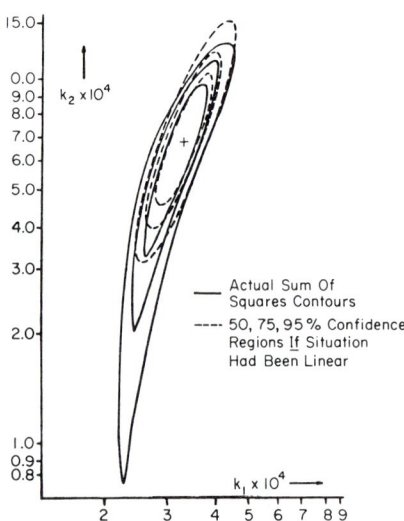

FIG. 12. Contours of actual sum of squares surface and of corresponding linear theo surface, $A \xrightarrow{k_1} B \xrightarrow{k_2} C$.

and Fig. 32 for the nitric oxide reduction model of Section VII. Note the large volume of these regions and the long skewed relationship to the axes (high correlation of estimates).

D. MULTIPLE-RESPONSE PARAMETER ESTIMATION

Parameter estimation through Eq. (19) has been widely used for experimental situations in which only a single response is being measured. Frequently, however, measurements can be made on two or more responses

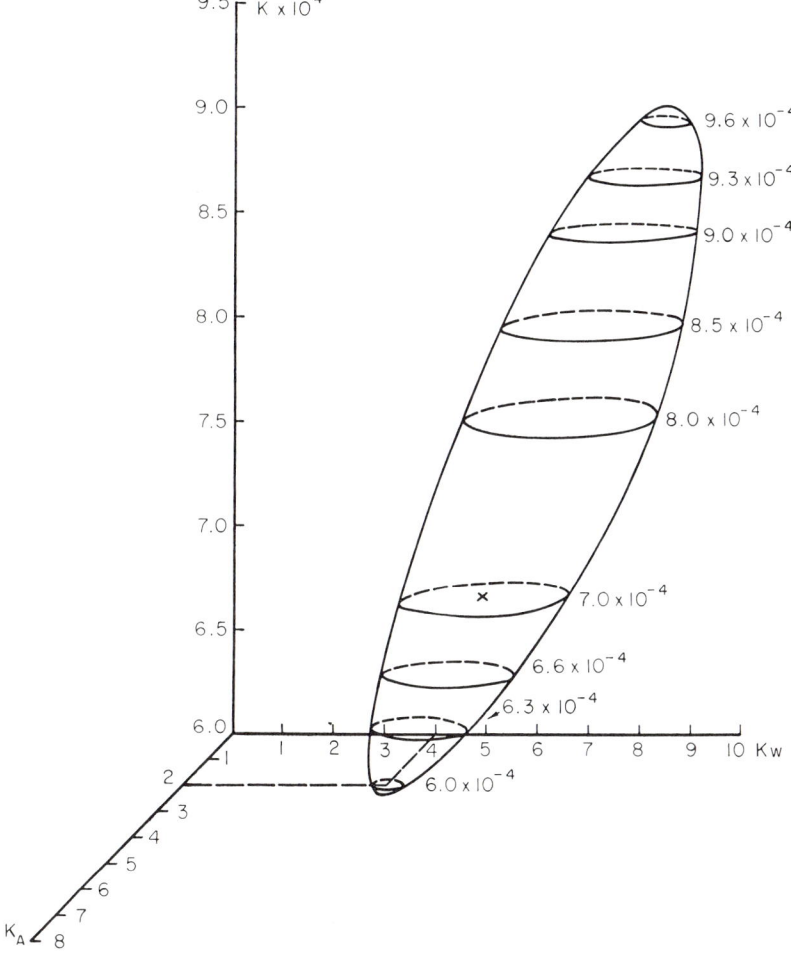

FIG. 13. Approximate 95% confidence region for alcohol dehydration, Eq. (62).

simultaneously. For example, Eq. (19) could be used if only the concentration of $B$ could be measured in the reaction sequence

$$A \rightarrow B \rightarrow C \tag{63}$$

If independent data were available for all three responses, however, all of these data should be used in estimating the two rate constants. One obvious extension of Eq. (19) to this case is to choose parameter estimates minimizing

$$S_r = \sum_{i=1}^{N}(C_{Ai} - \hat{C}_{Ai})^2 + \sum_{i=1}^{N}(C_{Bi} - \hat{C}_{Bi})^2 + \sum_{i=1}^{N}(C_{Ci} - \hat{C}_{Ci})^2 \tag{64}$$

This has been applied previously (B1).

This criterion, and others, can be derived using maximum likelihood arguments (H8). It has been shown that Eq. (64) is applicable (a) when each of the responses has normally distributed error; (b) when the data on each response are equally precise, and (c) when there is no correlation between the measurements of the three responses. These assumptions are rather restrictive.

A much more general criterion has been developed (B12, H8) for which these assumptions have been relaxed. The criterion developed when fitting a model, $r$, to observations, $y$, and for $v$ responses is to select parameter values minimizing the determinant

$$S_\Delta = \begin{vmatrix} \sum (y_u^1 - r_u^1)^2 & \sum (y_u^2 - r_u^2)(y_u^1 - r_u^1) & \cdots & \sum (y_u^v - r_u^v)(y_u^1 - r_u^1) \\ \sum (y_u^1 - r_u^1)(y_u^2 - r_u^2) & \sum (y_u^2 - r_u^2)^2 & \cdots & \sum (y_u^v - r_u^v)(y_u^2 - r_u^2) \\ \vdots & \vdots & & \vdots \\ \sum (y_u^1 - r_u^1)(y_u^v - r_u^v) & \sum (y_u^2 - r_u^2)(y_u^v - r_u^v) & \cdots & \sum (y_u^v - r_u^v)^2 \end{vmatrix} \tag{65}$$

where all sums range from $u = 1$ to $u = N$ observations. For the three response example above, $v = 3$, and the parameters would best be estimated by minimizing the 3 × 3 determinant of Eq. (65).

The advantages to be gained by following such a procedure are quite striking. The great improvement in the size of the confidence region for the two rate constants (reparameterized logarithmically) of Eq. (63) is shown in Fig. 14 (B12), with the measured responses reproduced thereon. Such precision of parameter estimates not only greatly improves the efficiency of the estimation procedure, but it can also assist in providing hard-to-obtain agreement between thermodynamic and kinetic estimations of kinetic parameters (M4). An application of the procedure is discussed in detail elsewhere (M4).

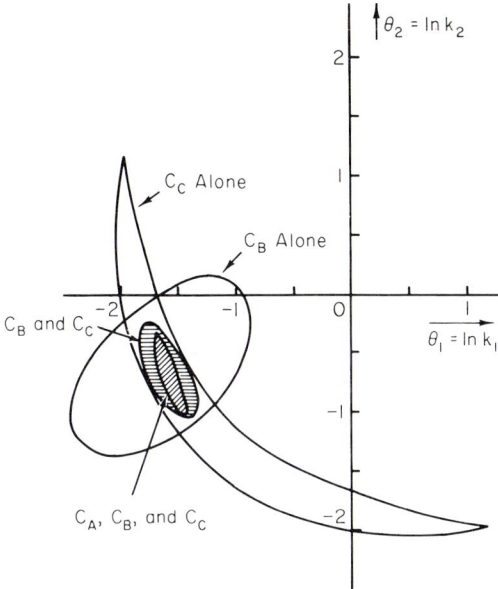

FIG. 14. Approximate 99.75% confidence regions for $A \xrightarrow{k_1} B \xrightarrow{k_2} C$.

## IV. Tests of Model Adequacy

The several modeling methods discussed in the accompanying sections are quite useful in testing the ability of a model to fit a particular set of data. These methods do not, however, supplant the more conventional tests of model adequacy of classical statistical theory, i.e., the analysis of variance and tests of residuals.

### A. ANALYSIS OF VARIANCE

The analysis of variance is used to compare the amount of variability of the differences of predicted and experimental rates with the amount of variability in the data itself. By such comparisons, the experimenter is able to determine (a) whether the overall model is adequate and (b) whether every portion of the model under consideration is necessary.

#### *Significance of the Overall Regression*

For every set of reaction-rate data, a total amount of variability in the data may be expressed as

$$\sum_{i=1}^{N} y_i^2$$

Also, by a least-squares analysis or some other suitable means, estimates of the parameters within a proposed model may be obtained. This allows the calculation of predicted reaction rates at each experimental point and thus an assessment of the total amount of variability which can be accounted for by the proposed model

$$\sum_{i=1}^{N} r_i^2$$

The difference in the predicted and observed rates $(y_i - r_i)$, is termed a residual and is a measure of the inability of the model to describe exactly the experimental data. If the model is entirely correct, in fact, the residual will be a measure of experimental error. A measure of the total amount of variation unaccounted for by the proposed model, then, is

$$\sum_{i=1}^{N} (y_i - r_i)^2$$

In statistical terminology, these quantities are called the crude sum of squares,

$$\sum_{i=1}^{N} y_i^2$$

the sum of squares due to regression,

$$\sum_{i=1}^{N} r_i^2$$

and the residual sum of squares,

$$\sum_{i=1}^{N} (y_i - r_i)^2$$

It is a direct result of the orthogonality property of linear least squares (Section III,A,1) that the crude sum of squares may be broken into two parts as follows:

$$\sum_{i=1}^{N} y_i^2 = \sum_{i=1}^{N} r_i^2 + \sum_{i=1}^{N} (y_i - r_i)^2 \qquad (66)$$

Associated with each data point is a certain degree of freedom, which will be used to attribute more information to, say, 500 data points than to 5 data points. In particular, if $N$ data points are used, the total sum of squares is said to possess $N$ degrees of freedom. The predicted rates estimated from a model containing $p$ parameters have $p$ degrees of freedom, and the remaining $N - p$ degrees of freedom are possessed by the residual sum of squares.

If several data points have been taken at the same settings of the independent variables, i.e., replicated data are available, then a knowledge of the inherent amount of error in the data is measured by the pure-error sum of squares,

$$\sum_{i=1}^{N}(y_i - \bar{y})^2.$$

Here, $\bar{y}$ is the average of all of the replicated data points. If the residual sum of squares is the amount of variation in the data as seen by the model, and the pure-error of squares is the true measure of error in the data, then the inability of the model to fit the data is given by the difference between these two quantities. That is, the lack-of-fit sum of squares is given by

$$\sum_{i=1}^{N}(\bar{y} - r_i)^2 = \sum_{i=1}^{N}(y_i - r_i)^2 - \sum_{i=1}^{N}(y_i - \bar{y})^2 \qquad (67)$$

If there are $\tilde{n}$ replications at $q$ different settings of the independent variables, then the pure-error sum of squares is said to possess $(\tilde{n} - 1)$ degrees of freedom (1 degree of freedom being used to estimate $\bar{y}$); while the lack-of-fit sum of squares is said to possess $N - p - q(\tilde{n} - 1)$ degrees of freedom, i.e., the difference between the degrees of freedom of the residual sum of squares and the pure-error sum of squares.

The sums of squares of the individual items discussed above divided by its degrees of freedom are termed mean squares. Regardless of the validity of the model, a pure-error mean square is a measure of the experimental error variance. A test of whether a model is grossly adequate, then, can be made by acertaining the ratio of the lack-of-fit mean square to the pure-error mean square; if this ratio is very large, it suggests that the model inadequately fits the data. Since an $F$ statistic is defined as the ratio of sum of squares of *independent normal* deviates, the test of inadequacy can frequently be stated

$$\frac{\text{lack-of-fit mean square}}{\text{pure-error mean square}} > F_\alpha[N - p - q(\tilde{n} - 1), q(\tilde{n} - 1)] \qquad (68)$$

The $F$ statistic is tabulated in many reference texts (D1). More rigorous discussions of the analysis of variance are, of course, available (D4).

One important application of analysis of variance is in the fitting of empirical models to reaction-rate data (cf. Section VI). For the model below, the analysis of variance for data on the vapor-phase isomerization of normal to isopentane over a supported metal catalyst (C1)

$$r = b_0 + b_1 x_1 + b_2 x_2 + b_3 x_3 + b_{11} x_1^2 + b_{22} x_2^2 + b_{33} x_3^2 \\ + b_{12} x_1 x_2 + b_{13} x_1 x_3 + b_{23} x_2 x_3 \qquad (69)$$

is reported in Table IV. Here, $x_1$ is the hydrogen partial pressure, $x_2$ the normal pentane partial pressure, and $x_3$ the isopentane partial pressure. It is evident from the $F$ ratios of Table IV that the model adequately fits these data.

TABLE IV

ANALYSIS OF VARIANCE FOR EQ. (69)

| Source | Sum of squares | Degrees of freedom | Mean squares | Ratios | F statistics 95% | 97.5% |
|---|---|---|---|---|---|---|
| Crude | 640.49 | 24 | — | — | — | — |
| Due to regression | 637.35 | 10 | 63.74 | — | — | — |
| Residual | 3.14 | 14 | 0.225 | — | — | — |
| Lack of fit | 2.85 | 11 | 0.260 | 2.73 | 8.8 | 14.3 |
| Pure error | 0.29 | 3 | 0.0954 | | | |

An analysis of variance can also be used to test the adequacy of more theoretical models. For example, two models considered in Section III for pentane isomerization are the single-site and dual-site models of Eqs. (30) and (31). These were linearized to provide Eqs. (32) and (33). The overall fit of these equations to the data may now be judged by an analysis of variance, reported in Tables V and VI (K3). It is seen that Eq. (33) fits the data quite

TABLE V

ANALYSIS OF VARIANCE FOR EQ. (32)

| Source | Sum of squares | Degrees of freedom | Mean squares | Ratios | F statistics 95% | 97.5% |
|---|---|---|---|---|---|---|
| Crude | 28894.02 | 24 | — | — | — | — |
| Due to regression | 26896.82 | 4 | 6,724.2 | — | — | — |
| Residual | 1997.20 | 20 | 99.86 | — | — | — |
| Lack of fit | 1964.56 | 17 | 115.56 | 10.62 | 8.7 | 14.3 |
| Pure error | 32.65 | 3 | 10.88 | | | |

TABLE VI

ANALYSIS OF VARIANCE FOR EQ. (33)

| Source | Sum of squares | Degrees of freedom | Mean squares | Ratios | F statistics 95% | 97.5% |
|---|---|---|---|---|---|---|
| Crude | 763.54 | 24 | — | — | — | — |
| Due to regression | 754.29 | 4 | 188.6 | — | — | — |
| Residual | 9.25 | 20 | 0.462 | — | — | — |
| Lack of fit | 8.98 | 17 | 0.528 | 5.9 | 8.7 | 14.3 |
| Pure error | 0.27 | 3 | 0.090 | | | |

well. However, there is some question as to whether Eq. (32) adequately represents the data, since the ratio of lack-of-fit to pure-error mean square is relatively large (although not so large as to allow a clear rejection of the model, particularly since the $y$ may not be independent normal deviates).

In some cases when estimates of the pure-error mean square are unavailable owing to lack of replicated data, more approximate methods of testing lack of fit may be used. Here, quadratic terms would be added to the models of Eqs. (32) and (33), the complete model would be fitted to the data, and a residual mean square calculated. Assuming this quadratic model will adequately fit the data (lack of fit unimportant), this quadratic residual mean square may be used in Eq. (68) in place of the pure-error mean square. The lack-of-fit mean square in this equation would be the difference between the linear residual mean square [i.e., using Eqs. (32) and (33)] and the quadratic residual mean square. A model should be rejected only if the ratio is very much greater than the $F$ statistic, however, since these two mean squares are no longer independent.

## 2. *Significance of Terms in the Regression*

Although the overall adequacy of a model to fit the data may be tested as described above, it is frequently desirable to test whether all of the terms in an adequate model need be included. This may be accomplished by a straightforward extension of Section IV,A,1, involving a decomposition of the regression sum of squares into the desired components. Suppose, for example, that we wished to test the necessity of including the quadratic and interaction terms $(b_{ij} x_i x_j)$ of Eq. (69). We do this by calculating the crude sum of squares and the regression sum of squares using the fitted linear model

$$r = b_0 + b_1 x_1 + b_2 x_2 + b_3 x_3 \tag{70}$$

This is, then, the regression sum of squares due to the first-order terms of Eq. (69). Then, we calculate the regression sum of squares using the complete second-order model of Eq. (69). The difference between these two sums of squares is the extra regression sum of squares due to the second-order terms. The residual sum of squares is calculated as before using the second-order model of Eq. (69); the lack-of-fit and pure-error sums of squares are thus the same as in Table IV. The ratio contained in Eq. (68) still tests the adequacy of Eq. (69). Since the ratio of lack-of-fit to pure-error mean squares in Table VII is smaller than the $F$ statistic, there is no evidence of lack of fit; hence, the residual mean square can be considered to be an estimate of the experimental error variance. The ratio

$$\frac{\text{extra mean square due to quadratic terms}}{\text{residual mean square}} > F_\alpha(p, N - p) \tag{71}$$

## TABLE VII

COMPLETE ANALYSIS OF VARIANCE FOR EQ. (69)

| Source | Sum of squares | Degrees of freedom | Mean squares | Ratios | F statistics 95% | 97.5% |
|---|---|---|---|---|---|---|
| Crude | 640.49 | 24 | — | — | — | — |
| Due to first order regression | 627.0 | 4 | 156.8 | — | — | — |
| Extra due to second order regression | 10.34 | 6 | 1.72 | — | — | — |
| Residual | 3.14 | 14 | 0.225 | 7.66 | 2.9 | 3.5 |
| Lack of fit | 2.85 | 11 | 0.260 | — | — | — |
|  |  |  |  | 2.73 | 8.8 | 14.3 |
| Pure error | 0.29 | 3 | 0.0954 | — | — | — |

then tests the necessity of the retaining the six quadratic terms in Eq. (69). Here, the ratio of quadratic to residual mean squares is so large that the quadratic terms must be used for an adequate fit of the rate data. This procedure may be carried out for any number of terms as long as we begin from the bottom of the table as was done here.

There are many cases of interest to kineticists in which the necessity for the inclusion of terms in nonlinear rate models must be tested. This may be done approximately by again calculating the regression sum of squares with and without the added term. Kabel and Johanson (K1), for example, considered the model

$$r = \frac{kK_A L(P_A^2 - P_E P_W/K)}{(1 + K_A P_A + K_E P_E + K_W P_W)^2} \quad (72)$$

in which they suspected $K_E$ was zero. This could be tested by comparing the residual mean squares to the pure-error mean square, with and without the inclusion of the $K_E$ term. Table VIII summarizes the information thus

## TABLE VIII

APPROXIMATE ANALYSIS OF VARIANCE FOR EQ. (72)

| Situation | $kL$ | $K_{A'}$ atm$^{-1}$ | $K_{W'}$ atm$^{-1}$ | $K_{E'}$ atm$^{-1}$ | Residual mean square |
|---|---|---|---|---|---|
| With $K_E$ | $7.02 \times 10^{-4}$ | 2.91 | 6.34 | $8.7 \times 10^{-3}$ | $7.0 \times 10^{-11}$ |
| Without $K_E$ | $7.04 \times 10^{-4}$ | 2.89 | 6.30 | — | $6.8 \times 10^{-11}$ |

obtained. From the closeness of the residual mean square, there is no reason to suspect that the $K_E$ term is needed. In this case, the parameter estimates foresee this situation, since the estimated value of $K_E$ is near zero *and* the parameter estimates obtained with the $K_E$ term excluded are near those with the $K_E$ term included. If similar results were reported with the $K_A$ term omitted, the residual mean square without $K_A$ would be greatly inflated over that including the $K_A$ term.

## B. Residual Analysis

The analysis of variance techniques of Section IV,A have been seen to provide information about the overall goodness of fit or about testing the importance of the contribution of certain terms in the model toward providing this overall fit of the data. Although these procedures are quite useful, more subtle model inadequacies can exist, even though the overall goodness of fit is quite acceptable. These inadequacies can often be detected through an analysis of the residuals of the model.

A residual is defined as the difference between the observed and predicted values of some response of interest, such as the reaction rate. For example, suppose that an experimentally observed reaction rate $y$ is linearly dependent on two partial pressures $x_1$ and $x_2$

$$y = \beta_0 + \beta_1 x_1 + \beta_2 x_2 + \varepsilon \qquad (73)$$

and that the predicted rate $r$ is obtained using the *correct* model,

$$r = b_0 + b_1 x_1 + b_2 x_2 \qquad (74)$$

If the residuals are estimated using some unbiased method such as linear least squares, the $b$'s would be expected to equal the $\beta$'s, so that the residual becomes

$$y - r = \varepsilon \qquad (75)$$

Since the difference between the predicted rate and the observed rate is attributable solely to experimental error, plots of this residual versus any independent variable should exhibit all of the characteristics of this error, such as being random with zero mean. If, on the other hand, the model of Eq. (74) is wrong in that the $b_2 x_2$ term is omitted, the residual should not be random when plotted versus either $r$ or $x_2$. Numerous suggestions have been made concerning the types of residual plots that are most revealing (A1, A2, B8, D4). These include plotting the residuals in overall plots, against predicted values $r$, against independent variables $x_i$, and against time.

### 1. *Overall Plots*

The overall plots consist of plotting the frequency of occurrence of the rounded values of the residual against the magnitude of the residual. These plots allow an approximate assessment of the normality of the error disbribution provided by $y - r$ if the model is correct. The plots are also a check of whether the mean of this error distribution is zero, as should be fufilled for linear and nonlinear least-squares applications. Hence, these plots allow an approximate check on some of the least-squares assumptions after the least-squares analysis has been completed. It should be noted that the mean of the residuals *must* be zero if the linear-least-squares calculations have been correctly carried out for linear models containing a constant [e.g., $b_0$ in Eq. (69)], so only a check on the algebra of linear least squares is obtained here.

## 2. *Predicted Value Plots*

A plot of the residual versus the predicted value, $r$, of a model can indicate whether the model truly represents the rate data. For example, residuals that are generally positive at low predicted rates and negative at high predicted rates can indicate a model inadequacy, even though the overall test of an analysis of variance indicates that the model is acceptable.

The analysis of variance for the model of Eq. (32), for example, for the data on the isomerization of normal pentane was shown in Table V; we concluded that the model was marginally acceptable. However, the plot of the residuals of Fig. 15 indicates that this overall fit is achieved by balancing predictions that are too low against predictions that are too high. Hence the

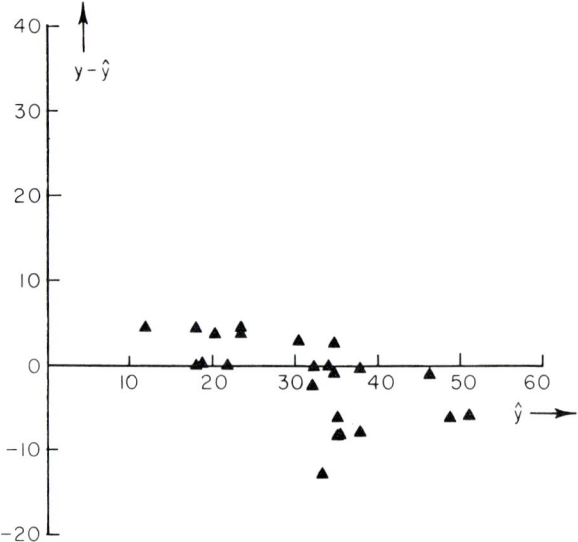

FIG. 15. Residual plot versus predicted response for Eq. (32).

model should be rejected.

These plots can also provide information about the assumption of constant error variance (Section III) made in the unweighted linear or nonlinear least-squares analyses. If the residuals continually increase or continually decrease in such plots, a nonconstant error variance would be evident. Here, either a weighted least-squares analysis should be conducted (Section III,A,2) or a transformation should be found to stabilize the error variance (Section VI).

Figure 16 presents residuals of 72 data points at three temperatures for the hydrogenation of propylene, reported by Shabaker (S1); the model was fitted by us to these data using unweighted nonlinear least squares with Eq. (8). The nonconstant error variance shown here indicates that a weighted nonlinear least-squares analysis would have been more appropriate. The residuals trends were largely removed by Shabaker by fitting the logarithm of the reaction rate rather than the rate itself (a transformation of coordinates such as discussed in Section VI).

### 3. *Independent Variable Plots*

The plot of the residuals versus the independent variables can yield information as to which of the variables in the equation is causing the residual

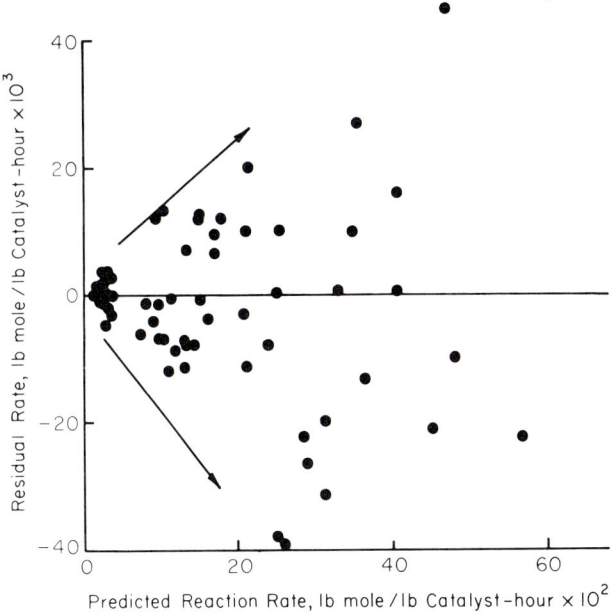

FIG. 16. Plot of residual rates versus predicted rate for Eq. (8).

trends that occur in the predicted value plots. The nonconstant error variance described in the above section also is exhibited in these plots, and can provide information useful in developing a weighting function.

Of particular value in kinetic studies are residual plots using the linearized form of the Hougen–Watson equation. For the model of Eq. (18), for example, we obtain

$$\frac{p_{NO}\,p_{H_2}}{r} = \frac{1}{kK_{NO}} + \frac{1}{k}\,p_{NO} + \frac{K_{H_2}}{kK_{NO}}\,p_{H_2} \qquad (76)$$

Reasoning analogous to that of Eqs. (73)–(75) indicates that if this model is wrong only in that adsorbed water should have been included, a plot of residuals versus water partial pressure should be linear. Or, if this model is erroneous only in that the hydrogen should have been considered to be dissociated, the residuals should be independent of the nitric oxide partial pressure *and* nonlinearly dependent upon the hydrogen partial pressure. It should be noted that these observations are strictly valid only if the data have been taken according to an orthogonal (e.g., factorial) design. This subject is treated more comprehensively in Section V; examples of these effects are included there.

### 4. Time-Residual Plot

The plot of residuals versus some measure of the time at which experiments were run can also be informative. If the number of hours on stream or the cumulative volume of feed passed through the reactor is used, nonrandom residuals could indicate improper treatment of catalyst-activity decay. In the same fashion that residuals can indicate variables not taken into account in predicting reaction rates, variables not taken into account as affecting activity decay can thus be ascertained.

### 5. Parametric Residuals

The residuals discussed thus far have been associated with some dependent variable, such as the reaction rate $r$. It is particularly advantageous in pinpointing the type of defect present in an inadequate model to expand this definition to include parametric residuals. The parametric residual, then, is simply the difference between a value of a given parameter estimated from the data and that predicted from a model. For example, the dots in Fig. 17 represent the logarithm of the alcohol adsorption constants measured in alcohol dehydrogenation experiments from isothermal data at each of several temperature levels (F1). The solid line represents the expectation that these points can be linearly dependent upon reciprocal temperature. If they do not, a defect exists in the model.

The plot of the residuals of this adsorption constant, using the data of Fig. 17, is shown in Fig. 18. A substantial trending effect is evident (it is also evident from Fig. 17). This is of interest in itself, but it is especially desirable to utilize the information of Fig. 18 to learn how the model must be modified to remove the observed defect. The other residual plots of the section can also assist this objective. This topic, called model-building, is treated in Section V.

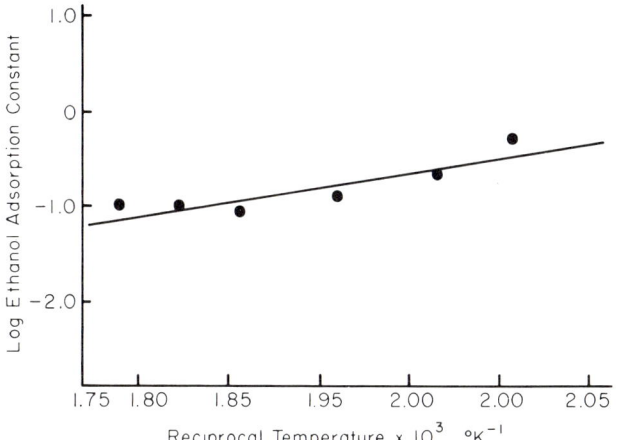

FIG. 17. Dependence of ethanol adsorption constant on temperature—alcohol dehydrogenation.

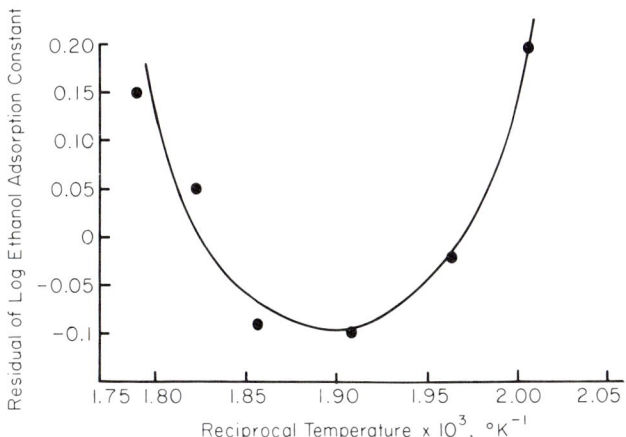

FIG. 18. Dependence of residuals of ethanol adsorption constant on temperature—alcohol dehydrogenation.

## V. Use of Diagnostic Parameters

Recently certain diagnostic parameters have been exploited to allow a discrimination among several rival models. These diagnostic parameters can be grouped into two broad classes—those that are inherently present in the model, and those that are introduced solely for the purpose of model discrimination.

There are two primary advantages to the use of diagnostic parameters in reaction-rate modeling. First, the use of these parameters allows an easy analysis of the adequacy of the model. This is accomplished by choosing the diagnostic parameters such that a linear test of a nonlinear model may be obtained. Second, in some cases, the diagnostic analysis will not only indicate a model inadequacy but also can suggest the precise nature of the inadequacy. Then, the model can be appropriately changed to remove this inadequacy.

### A. Model Discrimination with Diagnostic Parameters

#### 1. Nonintrinsic Parameters

A nonintrinsic parameter is one that is introduced into a model for the purpose of allowing the specification of a preferred model from a larger group of rival models. To be useful, such a parameter must simplify the analysis procedures, allow a broadening of the data base of the analysis, or both. One type of nonintrinsic parameter useful in this regard enters as a multiplier of each of a series of predicted responses (C3, C4). For two models, such a method reduces (W4, W5) to defining a dependent variable

$$z = y - 1/2(r_1 + r_2) \qquad (77)$$

This variable is then to be fitted using the equation

$$\hat{z} = \Lambda(r_2 - r_1) \qquad (78)$$

where $\Lambda$ is the nonintrinsic parameter used in specifying which of models $r_1$ and $r_2$ is to be preferred. It is to be noted that if

$$y = r_1 + \varepsilon \qquad (79)$$

then $\Lambda$ should be $-1/2$, while if $r_2$ were adequate $\Lambda$ should be $+1/2$.

The procedure to be followed, then, is to estimate the parameters **K** within each reaction-rate model by some appropriate technique (K8). The intrinsic parameter $\Lambda$ can then be estimated by linear least squares. Owing to experimental error in the data, this estimate of $\Lambda$ will typically be neither plus nor minus one-half. Hence the remaining portion of the analysis is to estimate the confidence interval for this discriminatory parameter estimate. If this confidence interval contains both possible values, this would suggest that no decision could be made about the most adequate model; if neither value is contained, both models should be viewed with suspicion.

In the above analysis, $y$ was considered to be a reaction rate. Clearly, any dependent variable can be used. Note, however, that if the dependent variable, $y$, is distributed with constant error variance, then the function $z$ will also have constant error variance and the unweighted linear least-squares analysis is rigorous. If, in addition, $y$ has error that is normal and independent, the least-squares analysis would provide a maximum likelihood estimate of $\Lambda$. On the other hand, if any transformation of the reaction rate is felt to fulfill more nearly these characteristics, the transformation may be made on $y$, $r_1$, $r_2$ and the same analysis may be applied. One common transformation will be logarithmic.

This method has been applied (M5) for modeling the vapor-phase rate of dehydration of secondary butyl alcohol to the olefin over a commercial silica-alumina cracking catalyst. Integral reactor data are available at 400, 450, and 500°F. Two models considered for describing this reaction are the single site

$$r_1 = \frac{kK_A p_A}{1 + K_A p_A + K_W p_W} \tag{80}$$

and the dual site

$$r_2 = \frac{kK_A p_A}{(1 + K_A p_A + K_{Wo} p_W)^2} \tag{81}$$

The usual plots of the initial reaction rate for 500°F are shown in Fig. 19. In

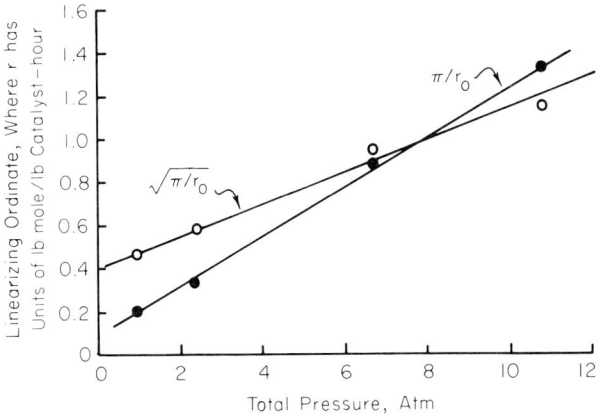

FIG. 19. Linearized initial rate plots for sec-butanol dehydration, Eqs. (80) and (81).

spite of this apparently good fit of the initial rate data by both models, the analysis of the intrinsic parameter $\Lambda$ will indicate that the single-site model cannot accurately represent the higher conversion data. In this case, the parameter $\Lambda$ is defined.

$$x - 1/2(\hat{x}_1 + \hat{x}_2) = \Lambda(\hat{x}_2 - \hat{x}_1) \tag{82}$$

where $x$ is the experimentally measured conversion, and $\hat{x}_1$ and $\hat{x}_2$ are the conversions predicted by the integrated forms of Eqs. (80) and (81). The parameters within these latter equations were estimated by nonlinear least squares (M5).

The values of $\Lambda$ thus obtained are reported in Table IX. It is clear that the

TABLE IX

ESTIMATES OF DISCRIMINATORY PARAMETER, $\Lambda$

| Temperature °F | Parameter estimate, $\Lambda^a$ |
|---|---|
| 400 | $0.36 \pm 0.28$ |
| 450 | $0.47 \pm 0.23$ |
| 500 | $0.46 \pm 0.14$ |

[a] The numbers following the $\pm$ signs define the 95% confidence interval on $\Lambda$, based upon residual mean squares.

preferred value of $\Lambda$ is plus one-half, and hence that the data do not reject the dual model $r_2$. However the probability that a $\Lambda$ of minus one-half could represent these data is less than 0.0001%. Thus, the single-site model should be rejected. The parameter estimates for the dual-site model are reported elsewhere (M5); they approximately exhibit an exponential temperature dependence, with slopes compatible with the usual predictions of kinetic theory.

## 2. *Intrinsic Parameters*

An intrinsic parameter is one that is inherently present in or arises naturally from a reaction-rate model. These parameters, which are of a simpler functional form than the entire rate model, facilitate the experimenter's ability to test the adequacy of a proposed model. Using these intrinsic parameters this section presents a method of preparing linear plots for high conversion data, which is entirely analogous to the method of the initial-rate plots discussed in Section II. Hence, these plots provide a visual indication of the ability of a model to fit the high conversion data and thus allow a more complete test of a model than does the initial-rate analysis alone. When used in conjunction with the linear initial-rate correlations, the high conversion plots provide estimates of all the parameters in models that exhibit only one adsorbed reactant and one adsorbed product.

When any hyperbolic model is written in terms of fractional conversions instead of partial pressures, two groupings of terms inherently arise within the denominator These two groupings will be called the intrinsic parameters $C_1$ and $C_2$. For example, when data are taken for the olefinic dehydration of a pure alcohol feed to a reactor, Eq. (80) becomes

$$r = \frac{kK_A[(1-x)/(1+x)]\pi}{1 + K_A[(1-x)/(1+x)]\pi + K_W[x/(1+x)]\pi} \tag{83}$$

In this formulation, the reaction is assumed to be essentially irreversible; the existence of product partial pressures in the numerator does not alter the method of analysis to be discussed here, however. Equation (83) may be written

$$r = \frac{(1-x)\pi}{(\hat{C}_1 + \hat{C}_2 x)} \tag{84}$$

$$\hat{C}_1 = \frac{1}{kK_A} + \frac{\pi}{k} \tag{85}$$

$$\hat{C}_2 = \frac{1}{kK_A} + \frac{(K_W - K_A)}{kK_A}\pi \tag{86}$$

The $\hat{C}_1$ term, then, is the collection of terms not multiplied by conversion while $\hat{C}_2$ is the collection multiplied by the conversion, $x$.

The use of the parameter $C_1$ in Eq. (84) was the subject of a portion of Section II, for Eq. (84) can be written at zero conversion in terms of the initial reaction rate:

$$C_1 = \frac{\pi}{r_0} \tag{87}$$

and Eq. (85) suggests that this variable should be linear with $\pi$. From Eq. (84), values of $C_2$ can be estimated by

$$C_2 = \frac{(1-x)\pi}{xr} - \frac{\pi}{xr_0} \tag{88}$$

where $C_1$ has been written as shown in Eq. (87). Here, the several values of $C_2$ (corresponding to several conversion points) at each pressure level can be averaged to give a single value of $C_2$ at each pressure. Equation (86) requires that these estimates be linear with pressure.

If this linear analysis is to be used, the experimental conversion–space-time data should first be taken at several pressure levels. Using the $C_1$ analysis alone, then, the plots of $C_1$ versus total pressure should be made for a preliminary indication of model adequacy. If several models are found to provide near-linear $C_1$ plots, the complete linear analysis using the $C_2$ plots should assist in the discrimination among the remaining rival models. If a model is adequate, both the $C_1$ and $C_2$ points should be correlatable by a straight line with a *common* intercept, as demanded by Eqs. (85) and (86). If only one model is found to be adequate following the initial $C_1$ analysis, the complete $C_1$ and $C_2$ analysis should still be carried out on this model to verify its ability to fit the high conversion data.

Consider again the 500°F data for the alcohol dehydration just discussed. Equations (84)–(88) arose from Eq. (80). A similar set of equations can be generated from Eq. (81) (K5). Let us examine the two models through an analysis of their intrinsic parameters.

The $C_1$ (initial reaction rate) plots were presented in Fig. 19. Note again that the data may be correlated well by straight lines for both models. The $C_2$ values are correlated by the solid lines of Fig 20. Note that the dual-site values can again be correlated by a straight line, but that the single-site values of $C_2$ show a definite curvature. Alternatively, the 0.975 atm value of the single-site $C_2$ could be rejected, and the three high-pressure points

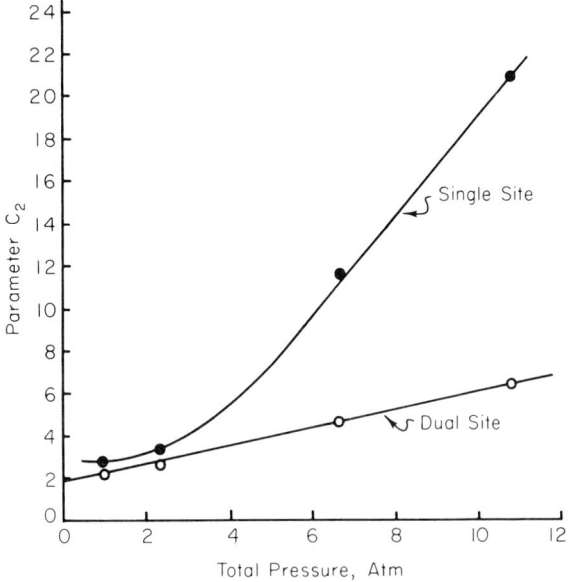

Fig. 20. Plot of parameter $C_2$ versus pressure for sec-butanol dehydration, Eqs. (80) and (81).

correlated by a straight line. Since the latter procedure results in a negative intercept for $C_2$ [cf. Eq. (85)], either approach eliminates the single-site model from further consideration. Although the intercept from the dual-site $C_2$ plot is somewhat higher than that of the dual-site $C_1$ plot, the 97.5% confidence intervals for these two intercepts overlap.

By using only simple hand calculations, the single-site model has been rejected *and* the dual-site model has been shown to represent adequately both the initial-rate and the high-conversion data. No replicate runs were available to allow a lack-of-fit test. In fact this entire analysis has been conducted using only 18 conversion–space-time points. Additional discussion of the method and parameter estimates for the proposed dual-site model are presented elsewhere (K5). Note that we have obtained the same result as available through the use of nonintrinsic parameters.

## B. Adaptive Model-Building with Diagnostic Parameters

The development of an adequate mathematical model representing a physical or chemical system is the object of a considerable effort in research and development activities. A technique has been formalized by Box and Hunter (B14) whereby the functional form of reaction-rate models may be exploited to lead the experimenter to an adequate representation of a given set of kinetic data. The procedure utilizes an analysis of the residuals of a diagnostic parameter to lead to an adequate model with a minimum number of parameters. The procedure is used in the *building* of a model representing the data rather than the postulation of a large number of possible models and the subsequent selection of one of these, as has been considered earlier. That is, the residual analysis of intrinsic parameters, such as $C_1$ and $C_2$, will not only indicate the inadequacy of a proposed model (if it exists) but also will indicate *how* the model might be modified to yield a more satisfactory theoretical model.

### 1. *Hyperbolic Models*

In the following discussion, we shall again separate the terms of a hyperbolic model and identify two parameters $C_1$ and $C_2$. As before, each of these two parameters will be a collection of terms, one of which is multiplied by conversion and one not multiplied by conversion. In previous formulations, however, we have oriented the discussion toward a familiar type of experimental design in kinetics: conversion versus space-time data at several pressure levels. Consequently, the parameters $C_1$ and $C_2$ were defined to exploit this data feature. Another type of design that is becoming more common is a factorial design in the feed-component partial pressures.

Consequently the definitions of the parameters $C_1$ and $C_2$ to be used here will reflect this type of design.

Let us first consider the type of analysis to be used. Suppose that the model, such as Eq. (83), were to predict that $\hat{C}_1$ should be related to the feed-component partial pressures as

$$\hat{C}_1 = b_0 + b_2 p_2 + b_4 p_4 \tag{89}$$

when in fact the true values of $C_1$ from the data were related as

$$C_1 = \beta_0 + \beta_2 p_2 + \beta_3 p_3 + \beta_4 p_4 + \varepsilon \tag{90}$$

Then, for an orthogonal design and zero expected value of error, the expected value of the residual of $C_1$ is (H5)

$$E(C_1 - \hat{C}_1) = \beta_3 p_3 \tag{91}$$

Hence, if a correlation of the residual with any independent variable can be detected (e.g., with $p_3$), then the model generating Eq. (89) is inadequate. Furthermore a modification of the manner in which $p_3$ enters the model is necessary.

The analysis could easily become more complicated. If, for example,

$$\hat{C}_1 = b_0 + b_1 p_1 + b_3 p_3 \tag{92}$$

then

$$E(C_1 - \hat{C}_1) = -\beta_1 p_1 + \beta_2 p_2 + \beta_4 p_4 \tag{93}$$

In this case, the residual will be correlated with the weighted sum of the three partial pressures. Although a trend may still be detectable, we will not generally be able to determine with which of the variables the residual is correlated and thus will not know how to correct the model; a careful initial model choice circumvents this problem.

*Application to Methane Oxidation.* This selection of an appropriate initial model can be accomplished as shown here for the complete oxidation of methane. A general representation of the surface reaction model is (K12)

$$r = \frac{k p_1 (1-x)(p_2 - 2 p_1 x)^2}{[1 + K_1 p_1 (1-x) + K_2(p_2 - 2 p_1 x) + K_3(p_3 + p_1 x) + K_4(p_4 + 2 p_1 x)]^n} \tag{94}$$

where $p_1$, $p_2$, $p_3$, and $p_4$ are the feed partial pressures of methane, oxygen, carbon dioxide, and water, respectively, and $x$ is conversion. Collecting terms and using the continuity equation for an integral reactor,

$$\frac{\partial x}{\partial W/F} = \frac{p_1(1-x)(p_2 - 2 p_1 x)^2}{(C_1 + C_2 x)^n} \tag{95}$$

For this generalized model, then,

$$C_1 = \left[p_1 p_2^2 \left(\frac{\partial W/F}{\partial x}\right)_{x=0}\right]^{1/n} \quad (96)$$

$$C_2 = \frac{p_1 p_2}{n C_1^{n-1}} \left[p_2 \left(\frac{\partial^2 W/F}{\partial x^2}\right)_{x=0} - \frac{C_1^n}{p_1 p_2^2}(p_2 + 4p_1)\right] \quad (97)$$

$$\hat{C}_1 = b_0 + b_1 p_1 + b_2 p_2 + b_3 p_3 + b_4 p_4 \quad (98)$$

$$\hat{C}_2 = \bar{b} p_1 \quad (99)$$

The problem of specifying an adequate model will now be to determine (1) the exponent $n$ by a $C_2$ analysis and (2) the denominator terms required by a $C_1$ analysis. Depending upon the particular surface-reaction model considered, the terms within Eq. (98) can change greatly. For any surface-reaction model, however, $\hat{C}_2$ remains the same. Equating Eqs. (97) and (99) provides an equation with two unknown parameters, $\bar{b}$ and $n$. Estimating $n$ in this way from the reaction data will specify the power of the denominator of Eq. (94). Generally, the selection of any of the models with the appropriate $n$ will eliminate the difficulties represented by Eq. (93) and hence allow an effective $C_1$ analysis.

For the complete vapor-phase oxidation of methane over a palladium alumina catalyst, conversion–space-time data were taken at 350°C and 1 atm total pressure; the fractional factorial design of Table X (H11) specified the settings of the feed partial pressures of the reacting species.

From these data, $n$ has previously been estimated to be 3 (K12). By starting with models of $n = 3$, the possibility of an ineffective $C_1$ analysis is diminished.

With $n = 3$, one model with a minimum number of parameters is model 1 of Table XI. By choosing a model with the smaller number of parameters, of course, we tend to prevent including parameters in the models unless their presence is absolutely necessary (principle of parsimonious parameterization). The residuals of the diagnostic parameter, $C_1$, for this model are shown in Table XII, allowing an analysis such as that suggested earlier. Note that the perfect correlation of the signs of residuals with the water partial pressure of Table X suggests that model 1 should be modified by changing the manner in which the effect of water is taken into account. To maintain $n = 3$, this requires the inclusion of an additional parameter, resulting in model 2 of Table XI. The residuals of model 2 of Table XII, being perfectly correlated with the carbon dioxide level of Table X, suggest that the effect of carbon dioxide is improperly described by model 2.

Keeping the number of parameters to a minimum, we are thus led to model 3 of Table XI. However, the residual trends of model 3 as shown in

## TABLE X

### EXPERIMENTAL DESIGN

| Run No. | $\bar{p}_1$ | $\bar{p}_2$ | $\bar{p}_3$ | $\bar{p}_4$ |
|---|---|---|---|---|
| 1 | −1 | −1 | −1 | +1 |
| 2 | +1 | −1 | −1 | −1 |
| 3 | −1 | +1 | −1 | −1 |
| 4 | +1 | +1 | −1 | +1 |
| 5 | −1 | −1 | +1 | −1 |
| 6 | +1 | −1 | +1 | +1 |
| 7 | −1 | +1 | +1 | +1 |
| 8 | +1 | +1 | +1 | −1 |

$$\bar{p}_1 = \frac{p_1 - 0.015}{0.005} \qquad \bar{p}_2 = \frac{p_2 - 0.120}{0.060}$$

$$\bar{p}_3 = \frac{p_3 - 0.065}{0.035} \qquad \bar{p}_4 = \frac{p_4 - 0.095}{0.055}$$

## TABLE XI

### MATRIX OF TERMS COMPRISING POSTULATED MODELS[a]

| Model No. | Methane $p_1$ | Oxygen $p_2$ | Carbon dioxide $p_3$ | Water, $p_4$ |
|---|---|---|---|---|
| 1 | 1 | 1 | 0 | 0 |
| 2 | 1 | 1 | 0 | 1 |
| 3 | 1 | 1 | 1 | 0 |
| 4 | 1 | 0 | 1 | 1 |
| 5 | 0 | 1 | 1 | 1 |

[a] 0 and 1. Gaseous and adsorbed state, respectively.

Table XII would indicate that the effect of water is not described adequately by the model. Utilizing the principle of parsimonious parameterization, one can consider both water and carbon dioxide to be adsorbed and oxygen to be nonadsorbed, resulting in the three-parameter model 4. The residuals in Table XII for model 4, however, are correlated with the oxygen level. Hence model 5 would perhaps be preferable, for it likewise contains only three parameters while allowing adsorbed oxygen. The random residuals of Table XII for model 5 indicate that this model cannot be rejected using the

MATHEMATICAL MODELING OF CHEMICAL REACTIONS 151

TABLE XII

Residuals of Parameter $C_1$ for the Postulated Models $\times\ 10^3$

| Run No. | Residuals for model | | | | |
|---|---|---|---|---|---|
| | 1 | 2 | 3 | 4 | 5 |
| 1 | +1.17 | −1.38 | +2.49 | −7.18 | −0.29 |
| 2 | −3.61 | −1.07 | −2.51 | −7.70 | +0.50 |
| 3 | −4.99 | −2.21 | −3.75 | +7.54 | −1.08 |
| 4 | +2.33 | −0.54 | +3.43 | +7.88 | +0.93 |
| 5 | −1.18 | +1.81 | −2.37 | −7.00 | +0.24 |
| 6 | +3.39 | +0.55 | +2.21 | −8.75 | −0.51 |
| 7 | +4.01 | +1.42 | +2.97 | +8.04 | −0.08 |
| 8 | −1.13 | +1.42 | −2.46 | +7.16 | +0.29 |

data presented. Thus, a reaction-rate model adequately describing these experimental data is model 5:

$$r = \frac{k_1 k_2 \, p_1 (1 - x)(p_2 - 2p_1 x)^2}{\{1 + K_2(p_2 - 2p_1 x) + K_3(p_3 + p_1 x) + K_4(p_4 + 2p_1 x)\}^3} \quad (100)$$

This model has previously been shown (H11, K12) to have a residual mean square comparing favorably with that expected from pure error, as discussed in Section IV. It is to be noted that we have been led logically from one model to another within the small class of models for which $n = 3$ by the above analysis. For these data, adsorbed methane is not required; however, for data with higher methane concentrations, the adsorbed-methane term may be needed.

The foregoing development represents a substantial deviation from, and in some cases a distinct improvement on, the more usual methods of obtaining a reaction model adequately describing a set of data. In particular, for the hyperbolic models, we have seen that it may be necessary to write a large number of possible models and to search extensively through these to find an adequate model. Common criteria for adequacy include the requirements that the rate data plot linearly with the proper choice of the ordinate, and that the parameter estimates be acceptable. For methane oxidation, more than eighty such models could be thus considered! By contrast, we started with the class of surface reaction controlling models, since experience has indicated that these models are generally found to represent reaction data. The analysis, then, provides first for an estimation of the number of sites participating in the reaction, $n$, *without* the necessity of the simultaneous estimation of the magnitudes of all of the individual adsorption constants in the model. Then, a model with the appropriate value of $n$ and with the minimum possible number of adsorption constants is considered, and the experimenter is led by the experimental data and the analysis to a model that, with a minimum number of parameter, adequately represents the data.

## 2. Power-Function Models

A similar procedure may also be used with the power-function models. For example, if we are considering the reaction (B14)

$$A + B \to F$$
$$A + F \to G$$

then an initial postulation of the form of the reaction rate model might be

$$-dC_A/dt = k_1 C_B + k_2 C_F \quad (101)$$

$$-dC_B/dt = k_1 C_B \quad (102)$$

$$dC_F/dt = k_1 C_B - k_2 C_F \quad (103)$$

$$dC_G/dt = k_2 C_F \quad (104)$$

These equations might be thought to apply for a high concentration of reactant $A$, at a given concentration of catalyst $C$, and at isothermal conditions. To describe measurements on the concentration of $F$ as a function of time at each of the initial conditions of the $2^4$ factorial design of Table XIII, we might tentatively entertain the integrated model:

$$C_F = \frac{C_{B0} k_1}{k_1 - k_2} (e^{-k_2 t} - e^{-k_1 t}) \quad (105)$$

This is simply the integrated form of the above equations.

Since several isothermal data points exist as a function of time for each of the 16 run conditions of Table XIII, we can estimate the values of the parameters $k_1$ and $k_2$ for each of these 16 conditions. This may be done using Eq. (105) and nonlinear least squares; the parameter estimates are shown in Table XIII (as taken from Reference B14).

Now, if the model of Eqs. (101)–(104) is oversimplified, the estimated values of the parameters $k_1$ and $k_2$ will likely be representable as

$$k = C_{AO}^m C_{CO}^n k_0 e^{-E/RT} \quad (106)$$

or

$$\ln k_1 = m_1 \ln C_{AO} + n_1 \ln C_{CO} + \ln k_{01} - E_1/RT \quad (107)$$

$$\ln k_2 = m_2 \ln C_{AO} + n_2 \ln C_{CO} + \ln k_{02} - E_2/RT \quad (108)$$

It should be noted at this point that we are confronted with a highly nonlinear problem involving extremely complex equations [if Eqs. (107) and (108) were

TABLE XIII

EXPERIMENTAL DESIGN AND ESTIMATED PARAMETER VALUES

| Run No. | $C_{AO}{}^a$ | $C_{BO}{}^a$ | $C_{CO}{}^a$ | $D^a$ | $-10 \ln k_1$ | $-10 \ln k_2$ |
|---|---|---|---|---|---|---|
| 1  | − | − | − | − | 79.74 | 72.68 |
| 2  | + | − | − | − | 61.99 | 62.09 |
| 3  | − | + | − | − | 77.49 | 68.03 |
| 4  | + | + | − | − | 64.56 | 62.04 |
| 5  | − | − | + | − | 74.80 | 67.25 |
| 6  | + | − | + | − | 59.91 | 62.19 |
| 7  | − | + | + | − | 75.99 | 69.35 |
| 8  | + | + | + | − | 61.01 | 62.40 |
| 9  | − | − | − | + | 69.48 | 65.09 |
| 10 | + | − | − | + | 55.70 | 58.05 |
| 11 | − | + | − | + | 68.78 | 64.31 |
| 12 | + | + | − | + | 54.82 | 58.19 |
| 13 | − | − | + | + | 65.71 | 63.42 |
| 14 | + | − | + | + | 52.04 | 57.86 |
| 15 | − | + | + | + | 65.22 | 64.44 |
| 16 | + | + | + | + | 51.76 | 57.01 |

[a] Values for pluses and minuses for $C_{AO}$, $C_{BO}$, $C_{CO}$, and $D$, respectively, are given below:

|   | $C_{AO}$ | $C_{BO}$ | $C_{CO}$ | $D$, temperature |
|---|---|---|---|---|
| + | 40 moles/liter | 2 moles/liter | 1 mM liter | 175°C |
| − | 20 moles/liter | 1 mole/liter | 0.5 mM liter | 165°C |

inserted into Eqs. (101)–(104)]. Rather, we have broken the problem into one easily solved equation [Eq. (105)] and two linear equations [Eqs. (107) and (108)]. In fact, the problem is entirely analogous to that of the previous subsection, except that the observed values of the parameters are contained in Table XIII and the predicted values are written as Eqs. (107) and (108). If we wished to adopt a solution similar to that used in model-building with hyperbolic models, we would assume the simplest form of Eqs. (107) and (108). Then we would predict parameter values $k_1$ and $k_2$ at each run, and compare the parameter residuals to the design of Table XIII. If the residuals were correlated with certain of the variables, we could introduce these variables, and proceed.

To illustrate a slightly different approach, however, let us simply fit Eqs. (107) and (108) to the observed values of the parameters $k_1$ and $k_2$ of Table XIII. Thus, we are able to estimate the magnitudes of the desired parameters of Eqs. (107) and (108). The results of such a linear least-squares fit are contained in Table XIV. This table suggests that Eq. (105) should be

TABLE XIV

PARAMETER ESTIMATE SUMMARY

| Parameter | $m^a$ | $n^a$ | $\ln k_0^a$ | $E$, kcal/gm-mole$^a$ |
|---|---|---|---|---|
| $k_1$ | $2.08 \pm 0.16$ | $0.47 \pm 0.16$ | $26.59 \pm 4.70$ | $35.2 \pm 4.2$ |
| $k_2$ | $0.99 \pm 0.20$ | $0.12 \pm 0.20$ | $11.26 \pm 0.65$ | $18.5 \pm 5.4$ |

$^a$ The numbers following $\pm$ signs represent 95% confidence intervals.

modified to read

$$C_F = \frac{C_{BO} C_{AO}^2 C_{CO}^{0.5} k_1'}{C_{AO}^2 C_{CO}^{0.5} k_1' - C_{AO} k_2'} (e^{-k_2' C_{AO} t} - e^{-k_1' {}_{AO}^2 C_{CO}^{0.5}}) \tag{109}$$

where

$$k_1' = 3.518 \times 10^{11} e^{-35200/RT}$$

$$k_2' = 7.795 \times 10^4 e^{-18500/RT}$$

That is, the data suggest that Eqs. (101)–(104) should be modified to allow formation of $F$ by a reaction second order in $A$, first order in $B$, and half-order in the catalyst concentration $C$. The disappearance of $F$ appears to be first order in $A$ and in $B$.

Before the analysis is complete, we must examine the adequacy of the model to fit the data. Since this procedure is described in Section IV as well as in Reference (B14), this aspect of the fitting will not be pursued here. We have simply demonstrated here how the experimenter can be *led* from an inadequate model to an adequate model by the data analysis.

## VI. Empirical Modeling Techniques

In every case in which a kinetic model is selected to represent adequately a reaction, the rate surface *predicted* by the model must be compared to the surface *observed* in the data. In the methods discussed in Section II, only one section through the entire rate surface was examined; for example, the dependence of initial rate on total pressure could be investigated when in fact the total rate surface constituted the dependence of rate on several component partial pressures and temperature. The misleading results obtainable by using sections of surfaces in this manner have been discussed (K3, K7).

In this section, methods are described for obtaining a quantitative mathematical representation of the entire reaction-rate surface. In many cases these models will be entirely empirical, bearing no direct relationship to the underlying physical phenomena generating the data. An excellent empirical representation of the data will be obtained, however, since the data are statistically sound. In other cases, these empirical models will describe the characteristic shape of the kinetic surface and thus will provide suggestions about the nature of the reaction mechanism. For example, the empirical model may require a given reaction order or a maximum in the rate surface, each of which can eliminate broad classes of reaction mechanisms.

## A. Response-Surface Methodology

The two primary features of response surface methodology are the experimental design and the method of data analysis (B7, D2, H3). In the design stages of the study, it is advantageous to use a minimum-variance design, such as the factorial or certain composite designs (H3). These designs not only cover a wide area of the kinetic surface, but they also can provide minimum-variance estimates of the parameters.

The importance of the parameter estimates becomes apparent from the data analysis. Suppose a nonlinear reaction-rate equation contains two independent variables and a set of unknown parameters:

$$r = f(x_1, x_2; \mathbf{K}) \tag{110}$$

The rate surface can be approximately represented, in the region where the data are taken, by a Taylor expansion

$$r = b_0 + b_1 x_1 + b_2 x_2 + b_{12} x_1 x_2 + b_{11} x_1^2 + b_{22} x_2^2 \tag{111}$$

Here, the coefficients **b** represent first and second partial derivatives of the rate expression $f(x_1, x_2; \mathbf{K})$ or functions thereof.

This is the general equation for an ellipse in two independent variables, as shown in Fig. 21. If a new coordinate system is defined that has its center at the point $S$ and axes directed along $X_1$ and $X_2$ of Fig. 21, then Eq. (111) reduces to

$$r - r_s = B_{11} X_1^2 + B_{22} X_2^2 \tag{112}$$

Examination of the coefficients $B$, then, characterizes the rate surface. If, for example, $B_{11}$ and $B_{22}$ were significantly positive, any movement along the axes $X_i$ would increase the rate $r$. Hence, the surface has a minimum at point $S$. Similarly, if the coefficients were negative, a maximum exists at point $S$.

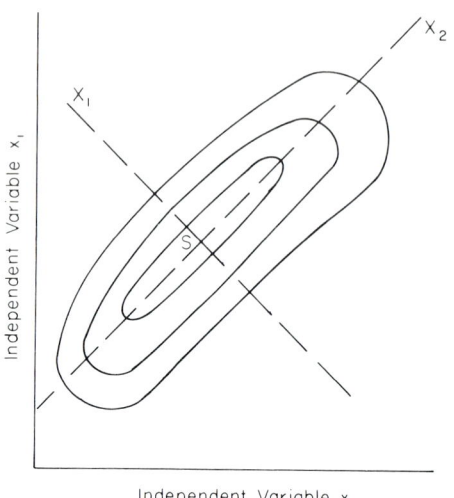

Fig. 21. Contours of constant reaction rate, Eqs. (111) and (112).

A saddle point exists if the coefficients are of unequal sign (B7, D2). Precise estimates of the parameters are of considerable importance.

The mathematical details of carrying out such a redefinition of coordinate system, termed a canonical transformation, have been presented elsewhere in detail (D2).

### 1. *Response-Surface Methods in Kinetic Modeling*

An augmented central composite design was used in obtaining reaction-rate data in a flow differential reactor; the reaction occurring was the isomerization of normal pentane to isopentane in the presence of hydrogen (C1). Using the subscripts 1, 2, and 3 for hydrogen, normal pentane, and isopentane respectively, an empirical rate equation can be written

$$r = b_0 + b_1 x_1 + b_2 x_2 + b_3 x_3 + b_{11} x_1^2 + b_{22} x_2^2 + b_{33} x_3^2 \\ + b_{12} x_1 x_2 + b_{13} x_1 x_3 + b_{23} x_2 x_3 \quad (113)$$

The parameter estimates obtained by a linear least-squares analysis of the original coded data (C1) are shown in Table XV. After a canonical analysis, this equation becomes

$$r - 2.11 = 0.155 X_1^2 - 0.292 X_2^2 + 0.419 X_3^2 \quad (114)$$

The standard errors of the coefficients are, respectively, 0.094, 0.108, and 0.095. The center of the coordinate system is at $x_1 = 1.32$, $x_2 = 1.43$, and $x_3 = 2.08$.

TABLE XV

Parameter Estimates for Eq. (113)

| Parameter | Estimate (with standard error) |
|---|---|
| $b_0$ | $3.06 \pm 0.20$ |
| $b_1$ | $-0.490 \pm 0.12$ |
| $b_2$ | $1.26 \pm 0.12$ |
| $b_3$ | $-1.50 \pm 0.12$ |
| $b_{11}$ | $-0.077 \pm 0.12$ |
| $b_{22}$ | $-0.208 \pm 0.07$ |
| $b_{33}$ | $0.257 \pm 0.12$ |
| $b_{12}$ | $-0.241 \pm 0.12$ |
| $b_{13}$ | $0.505 \pm 0.14$ |
| $b_{23}$ | $-0.158 \pm 0.14$ |

From a consideration of either Eqs. (113) or (114) (K3), it is evident that a saddle point is predicted from the fitted rate equation. This could eliminate from consideration any kinetic models not capable of exhibiting such a saddle point, such as the generalized power function model of Eq. (1) and the several Hougen-Watson models so denoted in Table XVI.

However, this particular experimental design only covered values of $x_3$ up to 1.68; consequently, the saddle point is only predicted by the model and not exhibited by the data. This is the reason the lack-of-fit tests of Section IV indicated neither model 3 nor model 4 of Table XVI could be rejected as inadequately representing the data. As is apparent, additional data must be taken in the vicinity of the stationary point to confirm this predicted nature of the surface and hence to allow rejection of certain models. This region of experimentation (or beyond) is also required by the parameter estimation and model discrimination designs of Section VII.

Response-surface methodology has also been used to gain theoretical insight into a reacting system (B19) and to determine the order of a reaction (P3).

### 2. *Response-Surface Methods in Process Exploitation*

Response-surface methodology has been used extensively for determining areas of process operation providing maximum profit. For example, the succinct representation of the rate surface of Eq. (114) indicates that increasing values of $X_3$ will increase the rate $r$. If some response other than reaction rate is considered to be more indicative of process performance (such as cost, yield, or selectivity), the canonical analysis would be performed on this response to indicate areas of improved process performance. This information

TABLE XVI

POSSIBLE RATE MODELS FOR PENTANE ISOMERIZATION

| Model | | Rate surface for model |
|---|---|---|
| 1. Adsorption controlled single site | $r = \dfrac{k(x_2 - x_3/K)}{[1 + K_1 x_1 + (K_2/K + K_3)x_3]}$ | No stationary point except at equilibrium; first derivative with $x_1$ zero only at equilibrium; first derivative with $x_2$ always positive and with $x_3$ always negative except at equilibrium. |
| 2. Adsorption controlled with dissociation | $r = \dfrac{k(x_2 - x_3/K)}{[1 + K_1 x_1 + (K_2/K x_3)^{\frac{1}{2}} + K_3 x_3]^2}$ | No stationary point except at equilibrium; first derivative with $x_1$ zero only at equilibrium; first derivative with $x_2$ always positive; first derivative with $x_3$ may be zero with second derivative positive. |
| 3. Single site, surface reaction controlled | $r = \dfrac{kK_2(x_2 - x_3/K)}{(1 + K_1 x_1 + K_2 x_2 + K_3 x_3)}$ | No stationary point except at equilibrium; first derivative with $x_1$ zero only at equilibrium; first derivative with $x_2$ always positive and with $x_3$ always negative except at equilibrium. |
| 4. Dual site, surface reaction controlled | $r = \dfrac{kK_2(x_2 - x_3/K)}{(1 + K_1 x_1 + K_2 x_2 + K_3 x_3)^2}$ | Can have saddle point; first derivative with $x_1$ zero at equilibrium; first derivative with $x_2$ may be zero with second derivative negative; first derivative with $x_3$ may be zero with second derivative positive. |
| 5. Dual site, surface reaction controlled, hydrogen dissociated | $r = \dfrac{kK_2(x_2 - x_3/K)}{[1 + (K_1 x_1)^{1/2} + K_2 x_2 + K_3 x_3]^2}$ | Same as model 4. |
| 6. Surface reaction controlled, dissociated $n$-pentane | $r = \dfrac{kK_2'(x_2 - x_3/K)}{[1 + K_1 x_1(K_2 x_2)^{1/2} + K_3 x_3]^2}$ | Same as model 4. |
| 7. Desorption controlled | $r = \dfrac{kK(x_2 - x_3/K)}{[1 + K_1 x_1 + (K_2 + KK_3)x_2]}$ | Same as model 1. |

is obtained in addition to the modeling information described above. Hence, the response surface approach can yield dividends in both areas. The use of response surface methods for general process improvement (B7, D2) and for investigating the influence of process variables and catalysts on reactions (F2, F3) has been discussed.

## B. Transformations of Variables

Particularly in empirical modeling, the transformation of a model can greatly assist in the modeling procedure. In some cases, a transformation can simplify the functional form of the model; in others, it can provide an improved fit of the data while requiring a previously specified function form (e.g., a power-function model). In still other cases, the transformations may ensure that certain assumptions are satisfied so that a simple and valid analysis may be performed. For example, it may be desirable to fit the logarithm of a dependent variable $y$ by unweighted least squares rather than fitting $y$ itself if the log function has constant variance. In this section, methods are discussed for determining transformations of the dependent variable, the independent variables, or both. Transformations of the parameters within a model were discussed in Section III.

### 1. Transformations of Dependent Variable

Theory for the transformation of the dependent variable has been presented (B11) and applied to reaction rate models (K4, K10, M8). In transforming the dependent variable of a model, we wish to obtain more perfectly (a) linearity of the model; (b) constancy of error variance, (c) normality of error distribution; and (d) independence of the observations to the extent that all are simultaneously possible. This transformation will also allow a simpler and more precise data analysis than would otherwise be possible.

Let us write an integrated $a$th-order rate equation as

$$C_{A0}^{1-a}[(1-x)^{1-a} - 1] = (a-1)kt \quad a \neq 1 \\ -\ln(1-x) = kt \quad a = 1 \tag{115}$$

If

$$\lambda = a - 1 \\ y = (1-x)^{-1} \tag{116}$$

then the rate equation may be written

$$y^{(\lambda)} = kt \tag{117}$$

where

$$v^{(\lambda)} = \begin{cases} (y^\lambda - 1)/\lambda C_{A0}^\lambda & \lambda \neq 0 \\ \ln y & \lambda = 0 \end{cases} \tag{118}$$

Now, if the transformed variable is normalized to $Z^{(\lambda)}$ by the Jacobian of the transformation, we obtain

$$Z^{(\lambda)} = \begin{cases} (y^\lambda - 1)/\lambda \dot{y}^{\lambda-1} & \lambda \neq 0 \\ \dot{y} \ln y & \lambda = 0 \end{cases} \quad (119)$$

where $\dot{y}$ is the geometric mean of the $y$'s. Then it has been shown (B11, K10) that the steps required for effective reaction-order estimation are:
1. Estimate, by unweighted linear least squares, the parameter $\tilde{b} = k/J^{1/N}$, which minimizes the sum of squares

$$S(\lambda) = \sum_{i=1}^{N} (Z_i^{(\lambda)} - \tilde{b} t_i)^2 \quad (120)$$

for a given $\lambda$, and calculate the sum of squares $S(\lambda)$.
2. Plot this minimum sum of squares for several $\lambda$.
3. Read off the minimum of this plot to obtain the best $\lambda$, $\hat{\lambda}$.
4. Calculate the 99% confidence interval for this $\hat{\lambda}$ by

$$\ln S(\lambda) - \ln S_{\min} < \chi_1^2(0.01)/N = 6.63/N \quad (121)$$

If this procedure is followed, then a reaction order will be obtained which is not masked by the effects of the error distribution of the dependent variables If the transformation achieves the four qualities (a–d) listed at the first of this section, an unweighted linear least-squares analysis may be used rigorously. The reaction order, $a = \lambda + 1$, and the transformed forward rate constant, $\tilde{b}$, possess all of the desirable properties of maximum likelihood estimates. Finally, the equivalent of the likelihood function can be represented b: the plot of the transformed sum of squares versus the reaction order. This provides not only a reliable confidence interval on the reaction order, but also the entire sum-of-squares curve as a function of the reaction order. Then, for example, one could readily determine whether any previously postulated reaction order can be reconciled with the available data.

This method has been discussed in detail (K10) and extended to cover other important cases elsewhere (K4, M8). It can logically be asked why such a precise reaction order is wanted. This and several other points in the analysis can best be shown by the following example.

Pannetier and Davignon (P1) studied the solid–solid reaction

$$NiS_2(s) \rightarrow NiS(s) + \tfrac{1}{2}S_2(g) \quad (122)$$

They presented seven determinations of the mass of unreacted solid at each of four temperatures. By the method of differentiation, Pannetier and Davignon (P1) found the reaction to be two-thirds order.

Typical sum-of-squares contours are shown in Fig. 22, where $Z^{(\lambda)}$ re presents the transformed dependent variable. The vertical broken line represent the 99% confidence interval. The results are shown below in complete form in Table XVII. Note that the two-thirds order is not compatible with the 415°C data. Although this order is adequate for the other data sets, a rather disturbing increase in the order with an increase of the temperature is present. In fact, the 395°C data alone would commonly be considered to be one-half order, with a relatively low probability of the two-thirds order being appropriate. This indication can be observed only for precise measures of the reaction order and its confidence interval. An analysis of variance for these data has been developed (K10).

TABLE XVII

ESTIMATES OF REACTION ORDER FOR EQ. (122)

| Temperature (°C) | Reaction order (with 99% confidence interval) |
|---|---|
| 395 | $0.53 \pm 0.25$ |
| 405 | $0.61 \pm 0.05$ |
| 415 | $0.91 \pm 0.13$ |
| 426 | $0.69 \pm 0.09$ |

If a single reaction order must be selected, an examination of the 95% confidence intervals (not shown) indicates that the two-thirds order is a reasonable choice. For this order, however, estimates of the forward rate constants deviate somewhat from an Arrhenius relationship. Finally, some trend of the residuals (Section IV) of the transformed dependent variable with time exists for this reaction order.

### 2. *Transformations of the Independent Variables*

In transforming the independent variables alone, it is assumed that the dependent variable already has all the properties desired of it. For example, if the $y$'s are normally and independently distributed with constant variance, at least approximately, then any transformations such as described in Section VI,B,1 would be unnecessary. Under such assumptions, Box and Tidwell (B17) have shown how to transform the *independent* variables to reduce a fitted linear function to its simplest form. For example, a function that has been empirically fitted by

$$y = b_0 + b_1 x + b_2 x^2 \tag{123}$$

might be fitted more simply by

$$y = b \ln x \tag{124}$$

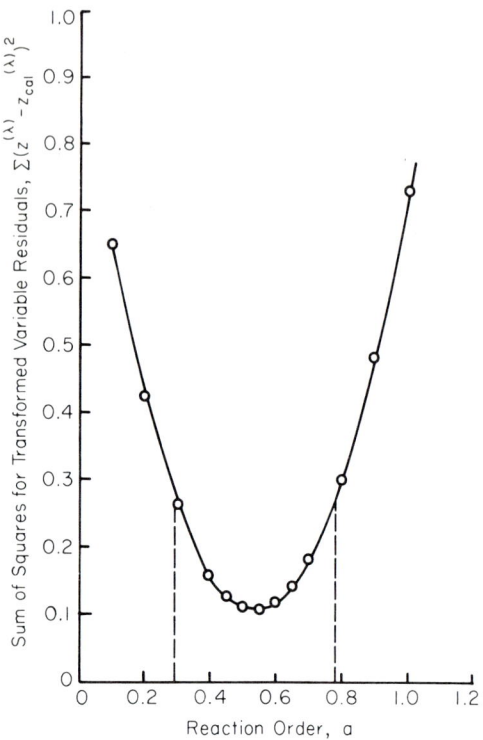

FIG. 22. Sum of squares curve for 395°C NiS$_2$ decomposition.

if the experimenter knew that it were preferable to transform $x$ to $\ln x$.

A type of transformation explored in detail is

$$U_i \begin{cases} = x_i^{a_i} & a_i \neq 0 \\ = \ln x_i & a_i = 0 \end{cases} \quad (125)$$

This transformation thus includes the familiar reciprocal, square-root, and logarithmic transformations. Box and Tidwell have shown how the functions

$$y = \sum b_i U_i \quad (126)$$

or

$$y = \sum \sum b_{ij} U_i U_j \quad (127)$$

can be fitted iteratively. We examine the usefulness of this approach more fully in Section VI,B,3, immediately following.

3. *Transformations of Independent and Dependent Variables*

Combinations of the above two methods lead to equations of the form

(H2, W6):

$$y^{(\lambda)} = b_0 + \sum b_i U_i \qquad (128)$$

where

$$y^{(\lambda)} \begin{cases} = (y^\lambda - 1)/\lambda & \lambda \neq 0 \\ = \ln y & \lambda = 0 \end{cases} \qquad (129)$$

and $U_i$ is given by Eq. (125). Here, if $y$ is taken to be the reaction rate, then any rate expression from

$$r = k \prod_{i=1}^{N} p_i^{a_i} \qquad (130)$$

to

$$r = \sum_i^N b_i p_i^{a_i} \qquad (131)$$

may be determined. If, on the other hand,

$$y = \frac{\text{reaction driving force}}{r} \qquad (132)$$

then, clearly, Eq. (128) can represent any of the hyperbolic models.

As an illustration, consider the hydrogenation of propylene over a platinum alumina catalyst discussed in Section II. These date were taken from 0 to 35°C, 1 to 4 atm total pressure, and 0 to 45% propylene. Equations (7) and (8) were obtained after considerable sifting and winnowing of rate equations. Both fit the observed data reasonably well.

An entirely empirical but very rapid technique is to fit

$$r^{(\lambda)} = b_0 + b_1 p_1^{a_1} + b_2 p_2^{a_2} + b_3 T^{a_3} \qquad (133)$$

where $p_1$ and $p_2$ are partial pressures of hydrogen and propylene, respectively. By first using nonlinear least squares to estimate the parameters on the right-hand side of Eq. (133) at each of several values of $\lambda$, it was found that $a_1$ and $a_2$ should be zero, but that the values $\lambda$ and $a_3$ were poorly defined. Consequently, the sums-of-squares contours similar to that of Fig. 22 were plotted as shown in Fig. 23 at the 95% confidence level. It is apparent that the effect of temperature on these data is not well defined, because a wide range of values of $a_3$ represents the data as well as the minimum sum-of-squares value. Choosing convenient values within the 95% confidence region, we obtain the rate equation

$$\ln r = 19.55 + 0.692 \ln p_1 - 0.226 \ln p_2 - 6.41 \times 10^3 T^{-1} \qquad (134)$$

or

$$r = a_0 p_1^{0.692} p_2^{-0.226} \exp(-6.41 \times 10^3 T^{-1}) \qquad (135)$$

A comparison of the predicted rates of Eqs. (7) and (135) is shown in Fig. 24. Since the bulk of the data were taken above 5% propylene, it is apparent that both models fit reasonably well. However, at lower concentrations of propylene, Eq. (135) will deviate widely from the data. Undoubtedly, deviations will also occur when extrapolating other directions from the data base as well. Checks on the adequacy of the transformation and calculations of the confidence regions for all parameters may also be carried out (B11, B17, H2).

The advantage of the approach is thus seen to be the speed and completeness with which a large number of empirical models can be tested and the best model of the group selected for closer examination. This empirical model may then be used as an end result, or it can serve as a starting point for more exhaustive analysis such as described in Sections III, V, and VII.

## C. Empirical Model Tuning

In the correlation of kinetic data, one may spend considerable time and effort obtaining a theoretical model using the techniques presented in the accompanying sections. Alternatively, one may simply fit an empirical function to the data, using the several techniques already discussed in this section. Many cases between these extremes are met in practice however. This subsection discusses procedures for empirically modifying an approximate mechanistic model such that (a) the function form of the mechanistic

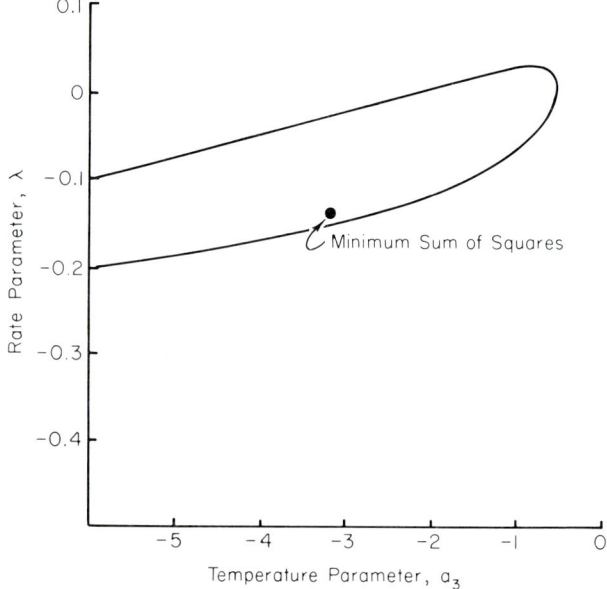

FIG. 23. Approximate 95% confidence region in $\lambda - a_3$ space for Eq. (133).

model can be retained; (b) the resulting model can describe the main features of the existing data; and (c) the model can possess a minimum number of parameters. This is done using the parametric residuals of Section IV,B,6, and already exploited in Section V.

Mickley, Sherwood, and Reed (M9) have discussed a useful method of empirically modifying a function to improve its fit of experimental data. The method would be applied when a particular rate equation

$$r_1 = f_1(\mathbf{x}; \mathbf{K}) \tag{136}$$

does not adequately describe the observed reaction rates, $y$. The procedure is to fit the residual of the dependent variable as an empirical function of the independent variables. If this empirical function were chosen to be linear, for example, we would estimate the empirical parameters $\mathbf{B}$ in

$$y - r_1 = \mathbf{x}\mathbf{B} \tag{137}$$

After estimating the parameters $\mathbf{B}$, the resulting equation

$$r_1 = f_1(\mathbf{x}; \mathbf{K}) + \mathbf{x}\mathbf{B} \tag{138}$$

may fit the data better than does Eq. (136). There are obvious disadvantages to this procedure for kinetic modeling. For example, we lose the advantage of the use of the functional form $f_1(\mathbf{x}; \mathbf{K})$, which would generally be selected

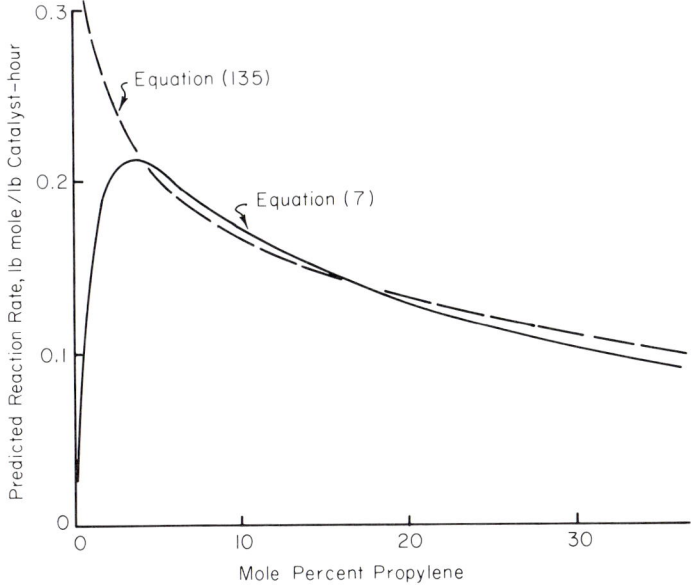

FIG. 24. Predicted reaction rates for Eqs. (7) and (135).

to provide reliable interpolation of kinetic data [e.g., Eqs. (1) or (2)]. Also, we add additional empirical constants **B** to Eq. (136), so that the residual mean square may actually be increased by following this procedure. Let us now describe how this general procedure may be utilized using the intrinsic parameters. In this way, the main behavior of, say, the Hougen–Watson model can be retained while making slight empirical modifications needed to fit the data. Also, the empirical modifications can be made without necessarily increasing the number of parameters in the model.

Let us consider the data taken by Laible (L1) on the dehydration of normal hexyl alcohol at 450°F over a silica alumina catalyst. The single- and dual-site surface reaction controlled models applying to alcohol dehydration were discussed in Section V,A,2. We now consider, however, the functional forms given, for example, by Eq. (84), as probably being capable of describing the data, but do not restrict the $C_1$ and $C_2$ plots to a linear pressure dependence as before. Rather, we obtain an empirical pressure dependence from the

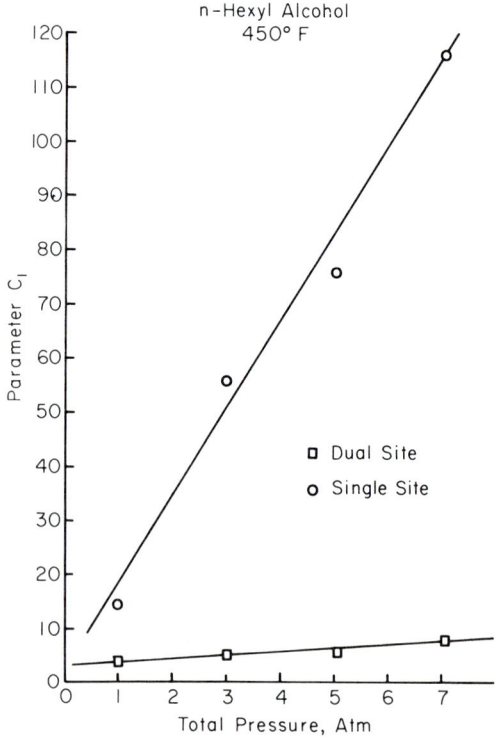

FIG. 25. Plot of parameter $C_1$ versus pressure for $n$-hexyl alcohol dehydration, Eqs (80) and (81).

data. Figure 25 presents the single- and dual-site $C_1$ plots; it can be seen that either model is well described by

$$\hat{C}_1 = \bar{a} + B_1\pi \tag{139}$$

The $C_2$ plots of Fig. 26, however, indicate that a linear $C_2$ dependence upon pressure will not be adequate. The use of the dual-site model will require three additional parameters for $C_2$, in order that the model approximately describe the data. The single-site model, however, can be used if we take

$$\hat{C}_2 = \bar{a} + B_2\pi^2 \tag{140}$$

where the parameter $\bar{a}$ is contained in both Eqs. (139) and (140). Thus, a model with only three parameters, which describes the main features of the data, is (K4)

$$r = \frac{(1-x)\pi}{(3+15\pi) + (3+15.5\pi^2)x} \tag{141}$$

Although we have intentionally chosen a simple example here to illustrate the method, we note that the same result could be obtained by a regression of any of the parameters of Sections IV,B or V on the independent variables. This procedure is then analogous to that described by Eqs. (136)–(138) except with the parameters as dependent variables.

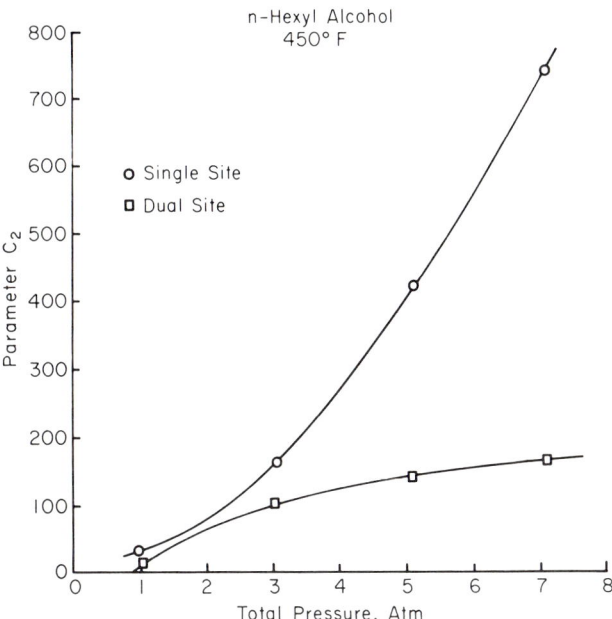

FIG. 26. Plot of parameter $C_2$ versus pressure for $n$-hexyl alcohol dehydration, Eqs. (80 and (81).

## VII. Experimental Designs for Modeling

The importance of a sound experimental design in modeling has already been emphasized. Normally, however, one cannot construct a detailed plan for the conduct of an entire future experimental program. At the outset of many kinetics experiments, we are not only unaware of the *ranges* over which certain variables should be studied but also of *which* variables should be studied. Often, it is only after observing unexpectedly wide data scatter or unacceptably rapid catalyst deactivation that we begin to appreciate the importance of certain variables. Consequently, we generally take a limited set of data and use the results to guide our next experiments.

Box recognized that not only do most kinetics experiments proceed in such a sequential fashion, but indeed most scientific experiments proceed iteratively. In his descriptions of the iterative nature of experimentation (B9), he has identified the stages of *conjecture, design, experiment,* and *analysis*. In such experimentation, one designs experiments to test a certain conjecture, conducts the experiment and analyzes the data. From this analysis, one gains information leading to new conjectures. The sequence is then repeated, making full use of all the information available at any given stage in planning experiments for the next stage. This strategy is particularly valuable in kinetic modeling studies.

It is partly the fault of statistics that experimenters have misconstrued the value of the *number* and *precision* of data points relative to the value of the location of the points. The importance of the location of the data in the model specification stage can be seen from Fig. 1, which represents literature data (M3) on sulfur dioxide oxidation. The dashed and solid lines represent the predicted rates of two rival models, and the points are the results of two series of experimental runs. It can be seen that neither a greater number of experimental points nor data of greater precision will be of major assistance in discriminating between the two rival models, if data are restricted to the total pressure range from 2 to 10 atm. These data simply do not place the models in jeopardy, as would data below 2 atm and greater than 10 atm total pressure. This is presumably the problem in the water–gas shift reaction, which is classical in terms of the number of models proposed, each of which adequately represent given sets of data.

The proper location of data is also important in parameter-estimation situations. For the nitric oxide reduction reaction (K11), for example, the relative sizes of the three-dimensional confidence regions calculated after each observation are shown in Fig. 27. The size of the confidence region after 12 points taken according to a one factor at a time variation of hydrogen and nitric oxide partial pressures is seen to be equivalent the size of the region after a $2^2$ factorial followed by one point taken by a minimum-volume design. The size of the region would have been substantially smaller had the first three points been taken by a Box–Lucas design (B16), followed by the minimum-volume design for the next two points. Note, from Fig. 27, that 18 times more information is obtained from 12 minimum-volume design points than from the 12 one-factor design points. This is a very significant increase in information; it has been shown, for example, that an increase in the precision of an estimated rate constant by only a factor of 4 can reduce the reactor overdesign required for such uncertainties by 15% of the reactor volume (K6).

Experimental design procedures exist for discriminating among $m$ rival models or estimating parameters in kinetic models. Figure 28 indicates three degrees of sophistication of such techniques that are available. At one extreme, the experimenter can simply use his intuition and experience, for example in suggesting that a tenfold variation from 1 to 10 atm total pressure should be adequate. Or, perhaps, high and low concentrations may be used. It is an unfortunate result of human nature that such experimental designs

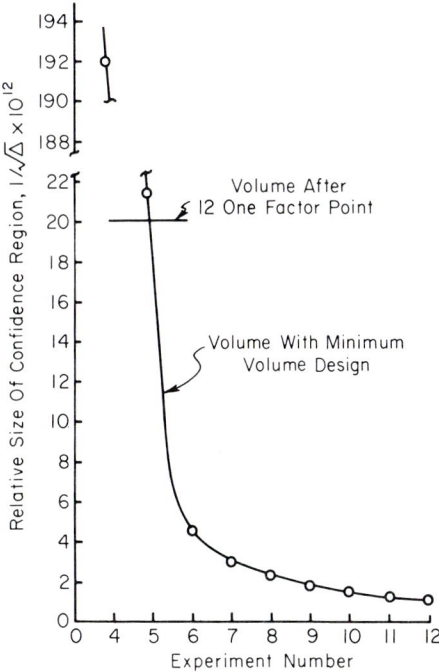

FIG. 27. Relative size of confidence region versus experiment number for nitric oxide reduction, Eq. (149).

may be influenced more by convenience than necessity; in any event, this method is best utilized by a gifted few. In very simple situations, such as the case in Fig. 1 with two models and one independent variable, a few hand calculations will suggest the proper regions of experimentation. For this example, given the experimental data of the figure and the scientific knowledge of two rival models, parameter estimates could be obtained and the predicted reaction rates sketched as shown in Fig. 1. These hand calculations, then, suggest that data at higher and lower pressure would be informative. The more usual situation, however, is where several complex models exist, each being influenced by several independent variables. In such cases, computer studies can best consider all of the available data and scientific knowledge, to evaluate the numerous possible alternatives.

The general scheme required here is shown in Fig. 29, a generalization of Fig. 28. With some ideas as to good areas of experimentation, the experimenter takes an initial set of data. These data are then analyzed to determine the best estimates of the parameters of the model or models under consideration. Since models that usually arise in these circumstances are nonlinear in the parameters, some version of nonlinear estimation will usually be employed in this analysis. Nonlinear estimation techniques, of course, almost always require the use of a computer.

Following this data analysis, the next step is the selection of a further set of runs to be performed; this also usually requires the use of a computer. For the two areas of modeling being discussed here, model discrimination and parameter estimation, this step involves maximization of the design criterion appropriate for each case. Settings of the independent variables, which extremize the criterion, are therefore determined, and these settings constitute the experimental design according to which the next set of runs

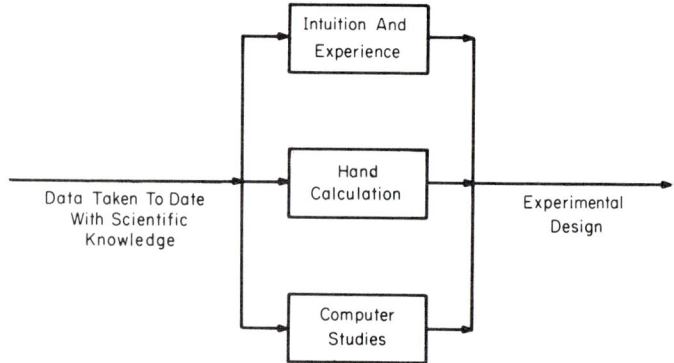

FIG. 28. General methods of selecting model discrimination or parameter estimation designs.

are performed. Once the data are collected, reestimation of the parameters follows. This cycle can be continued a sufficient number of times to accomplish the modeling objective (either model discrimination or parameter estimation). The strategy makes heavy use of the computer, there being a continual iteration between computer and laboratory.

## A. MODEL-DISCRIMINATION DESIGNS

Suppose that a given reacting system is about to be studied for which there exists a number of rival models; the object of the experimentation is here presumed to be the elimination of the inadequate models. Through this series of experiments, then, the experimenter is attempting to arrive at the best mathematical model for his system. The original paper (H13) in this area suggested sequential designs for discriminating between $m = 2$ models. Box and Hill (B13) later showed how to discriminate between an arbitrary number of rival models. In this later generalized approach, the

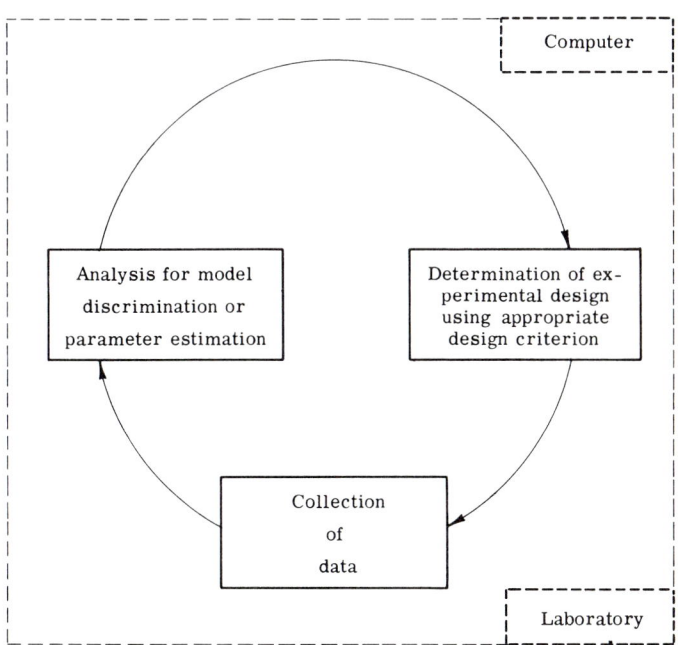

FIG. 29. Experimental strategy for mechanistic modeling.

experimenter iteratively selects those experimental conditions for the next stage of experimentation according to the following criterion:

$$D = \sum_{i=1}^{m} \sum_{j=1}^{m} \pi_i \pi_j D_{ij} \qquad (142)$$

where

$$D_{ij} = \frac{1}{2} \left[ \frac{(\hat{\sigma}_i^2 - \hat{\sigma}_j^2)^2}{(\sigma^2 + \hat{\sigma}_i^2)(\sigma^2 + \hat{\sigma}_j^2)} + (\hat{y}_i - \hat{y}_j)^2 \left\{ \frac{1}{\sigma^2 + \hat{\sigma}_i^2} + \frac{1}{\sigma^2 + \hat{\sigma}_j^2} \right\} \right]$$

The criterion $D$ is a measure of divergence among the models, obtained from information theory. The quantity $\pi_i$ is the prior probability associated with model $i$ after the $n$th observation is obtained; $\sigma^2$ is the common variance of the $n$ observations $y(1)$, $y(2)$, $\cdots$, $y(n-1)$, $y(n)$; $\hat{\sigma}_i^2$ is the variance for the predicted value of $y(n+1)$ by model $i$. When we have two models, $D$ simplifies to

$$D = \frac{(\hat{\sigma}_2^2 - \hat{\sigma}_1^2)}{(\sigma^2 + \hat{\sigma}_1^2)(\sigma^2 + \hat{\sigma}_2^2)} + (\hat{y}_1 - \hat{y}_2)^2 \left( \frac{1}{\sigma^2 + \hat{\sigma}_1^2} + \frac{1}{\sigma^2 + \hat{\sigma}_2^2} \right) \qquad (143)$$

Note from Eq. (142) that this discriminant is large when $\hat{y}_i$ is far from $\hat{y}_j$, $\hat{\sigma}_i^2$ and $\hat{\sigma}_j^2$ are small, and $\pi_i$ and $\pi_j$ are jointly large. This means that in order to decide which of several models is adequate, we should seek out settings of the independent variables for which the responses of the various models are expected to be quite different [$(\hat{y}_i - \hat{y}_j)$ large] and for which the responses can be predicted relatively precisely ($\hat{\sigma}_i^2$ small). Also, we should not give much weight to models that we suspect are relatively poor ($\pi_i \pi_j$ small), even though the two models predict responses far from one another. This discrimination procedure, then, utilizes all of these concepts to provide effective model discrimination designs. In particular, after $N$ observations become available, the quantity $D$ will be maximized with respect to the settings for the next experimental run, $N+1$, and these maximizing values will be used for the $(N+1)$th experiment. This procedure could be continued until one model was found to be clearly superior to the others. This iterative discrimination procedure can thus be represented by Fig. 29, where the appropriate design criterion is taken to be, at each stage, the maximization of $D$ as defined in Eq. (142).

As an illustration of the power of the method, two models known to fit the propylene hydrogenation data are shown in Eqs. (7) and (8). (See also Sections II,B and VI,B,3). Let us denote Eq. (7) by model 1 and Eq. (8) by model 2.

An initial nine data points were taken at 35°C in an adiabatic flow reactor The initial prior probabilities were taken to be equal (equal probability o adequacy) and $\sigma^2$ was estimated from experimental data (H12). Because of rapid catalyst deactivation and equipment limitations, total pressure was maintained at or below 3 atm.

Figure 30 portrays the grid of values of the independent variables over which values of $D$ were calculated to choose experimental points after the initial nine. The additional five points chosen are also shown in Fig. 30. Note that points at high hydrogen and low propylene partial pressures are required. Figure 31 shows the posterior probabilities associated with each model. The acceptability of model 2 declines rapidly as data are taken according to the model-discrimination design. If, in addition, model 2 cannot pass standard lack-of-fit tests, residual plots, and other tests of model adequacy, then it should be rejected. Similarly, model 1 should be shown to remain adequate after these tests. Many more data points than these 14 have shown less conclusive results, when this procedure is not used for this experimental system.

For the multiresponse situation, several measurable responses are implicit in each model under consideration. For example, for the reaction

$$A \begin{matrix} \to B \to D \\ \to C \end{matrix}$$

several models could be postulated, one of which could be that each component undergoes a first-order decomposition. For each such model, several responses could be measured as a function of time, e.g., the concentrations of components $A$, $B$, $C$, and $D$. Just as there is substantial advantage in making use of each of these responses in parameter estimation (Section II,D), considerable information about model discrimination will be lost unless all of the possible responses are utilized. The model-discrimination procedure for several responses has been developed (H10) and exemplified (H14).

## 3. Parameter-Estimation Designs

Once model discrimination has been accomplished (that is, from a large group of rival models one best model has been selected as being adequate), further experimentation can be conducted to improve parameter estimates. For such parameter-estimation experiments, one design criterion suggested is to choose experiments providing the most desirable posterior distribution of the parameters. Under certain assumptions (B15), the procedure reduces to the maximization of the determinant

$$\Delta = |\mathbf{X}'\mathbf{X}| \tag{144}$$

where the $(i, u)$th element of this $N \times p$ **X** matrix is

$$f'_{iu} = \frac{\partial f(\mathbf{x}_u ; \mathbf{K})}{\partial K_i}\bigg|_{\mathbf{K}=\mathbf{K}°} \tag{145}$$

It is to be noticed that for linear models, such as

$$f(\mathbf{x}_u ; \mathbf{K}) = K_1 x_{1u} + K_2 x_{2u}^2 \tag{146}$$

then Eq. (145) becomes

$$\begin{aligned} f'_{1u} &= x_{1u} \\ f'_{2u} &= x_{2u}^2 \end{aligned} \tag{147}$$

The determinant is then, for $N$ data points,

$$\Delta = \begin{vmatrix} \sum_{u=1}^{N} x_{1u}^2 & \sum_{u=1}^{N} x_{1u} x_{2u}^2 \\ \sum_{u=1}^{N} x_{1u} x_{2u}^2 & \sum_{u=1}^{N} x_{2u}^4 \end{vmatrix} \tag{148}$$

For this linear model, then, we simply find values of $x_1$ and $x_2$ that maximize $\Delta$. We do not need estimates of the parameters to define this minimum variance design. For nonlinear models, such as Eq. (40), this is not true. The

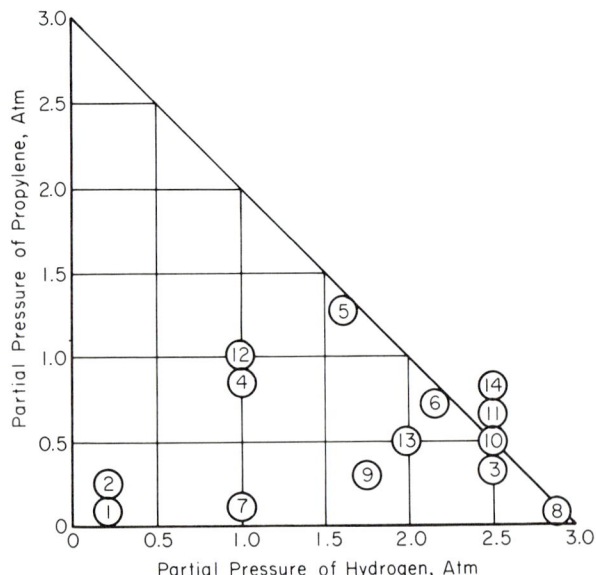

FIG. 30. Settings of independent variables for propylene hydrogenation for discriminating between Eqs. (7) and (8).

criterion $\Delta$ will be a function of the parameter values and, hence, the optimal design depends on these values; initial parameter estimates must be provided to use this criterion (B16).

By maximizing the determinant $\Delta$, we can expect to have a smaller confidence region of the parameter estimates than that obtained with any other experiment in the possible region of experimentation. The larger the confidence region, of course, the more uncertain is our knowledge concerning the estimated parameters; the smaller the confidence region, the more precise is our knowledge.

The iterative parameter-estimation procedure can also be represented as in Fig. 29. An experimenter performs an experiment, obtains the data and estimates the parameters, for example, by nonlinear least squares. Then he checks the confidence region of the parameter estimates, perhaps through $\Delta$ as shown in Fig. 27. If it is too large, the experimenter finds those experimental settings that maximize the parameter estimation criterion of Eq. (144), that is, experimental settings that will reduce the size of confidence region as

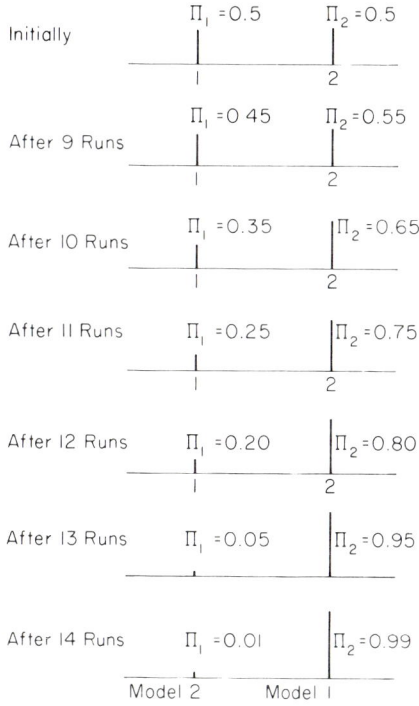

FIG. 31. Posterior probabilities in discrimination of propylene hydrogenation models Eqs. (7) and (8).

much as possible. He then conducts an experiment as indicated by this experimental design, analyzes the data, and checks whether the parameters have now been estimated with sufficient precision. If they have, he terminates the experimentation; if not, he repeats the cycle.

To illustrate this criterion, consider a simulation problem chosen to be similar to the catalytic reduction of nitric oxide (A3, K11):

$$NO + H_2 \rightleftharpoons H_2O + \tfrac{1}{2}N_2$$

In an experimental study (A3), the reaction model was found to be the surface reaction between an adsorbed nitric oxide molecule and one adjacentlyl adsorbed hydrogen molecule:

$$r = \frac{kK_{NO}K_{H_2}p_{NO}p_{H_2}}{(1 + K_{NO}p_{NO} + K_{H_2}p_{H_2})^2} \tag{149}$$

The confidence region obtained using a simulated one variable at a time design was first examined, since this was the design used by the original experimenters.

The approximate 95% confidence region for the non-linear least-squares estimates using these data is shown in Fig. 32. The surface represented in this figure is the contour of the sums of squares surface, which has the value $S_c$ given by Eq. (56).

In order to obtain more precise estimates of these parameters, the experimental design was also constructed using the above parameter estimation criterion. Using four preliminary observations taken according to a factorial design, the three parameters ($k$, $K_{NO}$, and $K_{H_2}$) were estimated. A fifth point was selected to minimize the joint confidence region of these parameters using the settings of the partial pressures maximizing $\Delta$. Following Fig. 29 the simulated reaction rate at this fifth point was used to reestimate the parameters by nonlinear least squares. Then the sixth point was chosen by again maximizing the determinant $\Delta$, and the entire procedure was repeated.

Since the square root of the determinant $\Delta$ is inversely related to the size of the confidence region, the rate of decrease in volume obtained by the design procedure may be represented by Fig. 27. The solid horizontal line is the size of the region obtained by the twelve one variable at a time experimental points. In contrast, the estimates from the minimum-volume design are equally precise after the fifth point (i.e., the first point chosen by the minimum volume design). A visual measure of the size of the confidence region after twelve well-designed points can be obtained from Fig. 3 which is one-eighteenth the size of the one variable at a time confidence region of Fig. 32.

This parameter-estimation technique has also been extended to the multiple-response case (D3). Just as was seen in the multiple-response

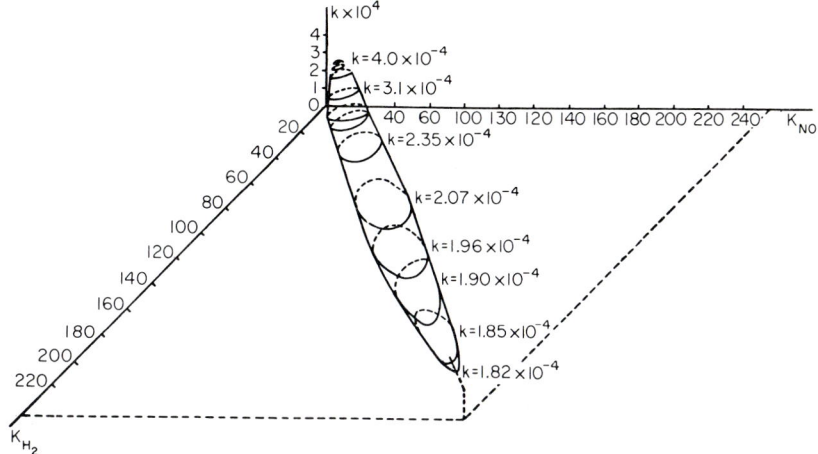

FIG. 32. Approximate 95% confidence region for Eq. (149) and one variable at a time design.

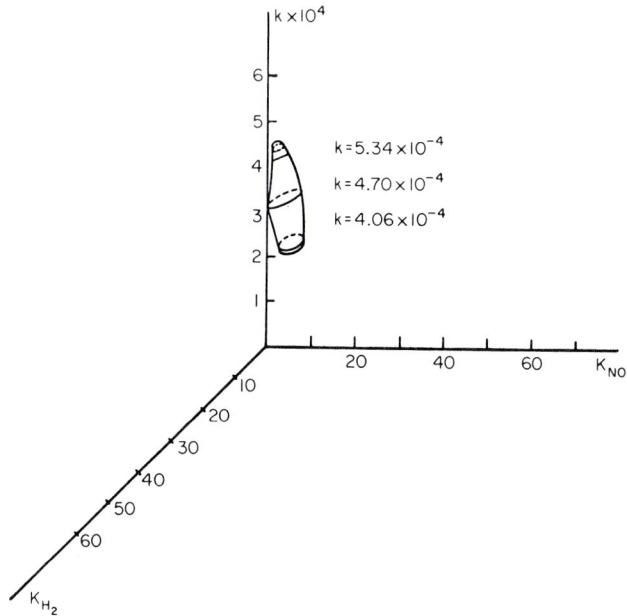

FIG. 33. Approximate 95% confidence region for Eq. (149) and sequential minimum volume design.

parameter-estimation example (B12), much greater power is available to the experimenter when all available responses are used. Also, a method has been suggested (H4) for combining the model-discrimination and parameter-estimation criteria, in order to increase the efficiency of the overall modeling program.

It will be noticed from Fig. 29 that there is a continual iteration between the laboratory and the computer. In the laboratory area, there could be a remote console linked directly to a time-shared digital computer, although immediate laboratory access is not essential. After obtaining his results, the experimenter would enter these into the digital computer, and the computer would perform routine analyses that have been programmed. For example, least-squares estimates would be calculated and the confidence region determined, for each model. Graphical output in the form of various plots of the residuals could also be obtained and examined. The computer could then be asked to determine those settings for a further run that would extremize the appropriate design criterion. The computer, having executed this instruction, would print out these optimal settings for consideration by the experimenter. The experimenter, in turn, could use these settings to proceed with further experimentation, or ask for other possible settings.

There is danger in describing this procedure in such an oversimplified way; it could be misconstrued that the investigator may only set variables and run experiments as the computer dictates. Experimentation in kinetics is far too complex for any such naive approach as this. In computer-assisted sequential experimentation, it is still necessary that the experimenter provide new ideas, intuition, and insights, which could possibly change the direction of the investigation in a fundamental manner. In short, the experimentation cannot be completely programmed ahead of time. Natural discoveries and surprises are the essence of research, and any experimental strategies for planning experiments must take them into account. However, the continual iteration between computer and laboratory, while not solving the problem completely, can contribute much in the way of solving modeling problems.

## Nomenclature

$A$. A component of a reacting mixture
$a$   A reaction order to be estimated for component $A$
$a_i$  An empirical exponent of Eq. (133)
$\bar{a}$  An empirical parameter in Eqs. (139) and (140)
$B$   A component of a reacting mixture

$B_i$  An empirical parameter Eqs. (139) and (140), $i = 1$,
$B_{ii}$  The canonical parameters defining the nature of the response surface in transformed coordinates, $i = 1, 2, 3, \cdots$
$\mathbf{B}$  A vector of empirical parameters of Eqs. (137) and (13?) obtained by linear regressi

of residuals on $x$
$b$ A reaction order to be estimated for component $B$
$b_i$ An estimated value of a parameter
$\bar{b}$ A combination of rate and absorption constants, defined by Eqs. (94), (95), and (99)
$\tilde{b}$ A transformed rate constant for Eq. (115), obtained by dividing the rate constant by $J^{1/N}$
$\hat{b}$ Least-squares estimate of a parameter
$\mathbf{b}$ A vector of parameters to be estimated
$C$ Designates a component of a reacting mixture
$C_1$ An observed value of an intrinsic parameter, which collects all denominator terms of hyperbolic models not multiplied by conversion
$\hat{C}_1$ The value of $C_1$ predicted by an assumed model
$C_2$ An observed value of an intrinsic parameter, which collects all denominator terms of hyperbolic models multiplied by conversion
$\hat{C}_2$ The value of $C_2$ predicted by an assumed model
$C_A$ Concentration of any reactant $A$
$C_{A0}$ Concentration of any reactant $A$ at zero time
$C_L$ Molar concentration of vacant active sites on catalyst
$C_{Ai}$ Concentration of any component $A$ at the $i$th data point, $i = 1, 2, \cdots, N$
$\hat{C}_{Ai}$ Concentration of component $A$ at $i$th data point predicted by a particular model
$\mathbf{C}$ Variance–covariance matrix of observation errors, defined in Eq. (39)
$\mathbf{v}(b_1, b_2)$ Covariance of two parameters, $b_1$ and $b_2$
$D$ Model discrimination criterion defined in Eq. (142)

$D_{ij}$ Quantity related to the divergence $D$ defined in Eq. (143)
$E$ Activation energy
$E'$ Ratio of activation energy to universal gas constant, $E/R$
$E(C_1-\hat{C}_1)$ Expected value of the residual of $C_1$
$F_{\alpha(v_1, v_2)}$ The $100(1 - \alpha)\%$ point of the $F$ statistic with $(v_1, v_2)$ degrees of freedom
$F$ Mass feed rate to reactor
$f'_{iu}$ Partial derivative of nonlinear rate model with respect to the $i$th parameter and evaluated at the $n$th experimental point
$\mathbf{I}$ Matrix of unity diagonal terms and zero off-diagonal terms
$J$ Jacobian of the transformation of the variable $y$ to the variable $y^{(\lambda)}$
$K$ Thermodynamic equilibrium constant for over-all reaction
$K_i$ Adsorption equilibrium constant for species $i$, or product of absorption constant and initial mole fraction of species $i$
$\bar{K}_i$ Adsorption equilibrium constant for species $i$
$K^{(i)}$ Estimate of a parameter after $i$th iteration of nonlinear estimation program
$\mathbf{K}$ Vector of estimated rate and adsorption parameters
$\mathbf{K}^\circ$ Vector of initial estimates of $K$ for use in nonlinear least squares
$k$ Forward reaction-rate constant
$k'$ Reparameterized forward reaction-rate constant
$\bar{k}$ Forward reaction-rate constant
$k_0$ Preexponential factor of forward rate constant
$k_0'$ Reparameterized preexponential factor of forward rate constant
$L$ Total concentration of active sites on catalyst

| | | | |
|---|---|---|---|
| $m$ | Number of rival models | $T$ | Absolute temperature |
| $N$ | Number of experimental observations | $\bar{T}$ | Mean value of absolute temperatures of a particular set of data |
| $n$ | Exponent of denominator of generalized hyperbolic rate equation, such as Eq. (94), or the reciprocal thereof | $t$ | Measure of reaction time |
| | | $t_{1/2}$ | Half-life of a reaction |
| | | $t_{v,\alpha}$ | The $100(1-\alpha/2)\%$ point of the $t$ distribution with $v$ degrees of freedom |
| $\tilde{n}$ | Number of replicated data sets | | |
| $p$ | Number of parameters in a model | $U_i$ | Transformed dependent variable defined by Eq. (125) |
| $p_i$ | Partial pressure of species $i$ in reaction mixture | $v(b_i)$ | Variance of the parameter estimate $b_i$ |
| $q$ | Number of replicated points, used to estimate experimental error | $v(\mathbf{b})$ | Variance–covariance matrix of the vector of parameter estimates $\mathbf{b}$ |
| $R$ | Universal gas constant | $v$ | Designates the response number in Eq. (65) |
| $r$ | Reaction rate | | |
| $r_i$ | Reaction rate predicted from model $i$ | $W$ | Mass of catalyst in reactor |
| | | $X_i$ | The canonical axis $i$, $i = 1, 2, 3, \cdots$ |
| $r_0$ | Initial reaction rate | | |
| $r_s$ | Reaction rate at the stationary point of the reaction-rate surface | $\mathbf{X}$ | A matrix of independent variables, defined for a linear model in Eq. (29) and for nonlinear models in Eq. (43) |
| $r_u^v$ | Reaction rate predicted from response $v$ at the $u$th experimental point | | |
| | | $\mathbf{X}'$ | Transpose of matrix $\mathbf{X}$ |
| | | $\mathbf{X}^{-1}$ | Inverse of the matrix $\mathbf{X}$ |
| $r^{(\lambda)}$ | Transformed reaction rate, where the transformation is defined generally in Eq. (129) | $x$ | Conversion in reactor |
| | | $x_i$ | A generalized independent variable |
| $S$ | Sum of squares of residuals, defined in Eq. (19) | $\hat{x}_i$ | Predicted value of the reactor conversion, for model $i$ |
| $S_c$ | Critical sums of squares contour giving a $100(1-\alpha)\%$ confidence region | $\mathbf{x}$ | A matrix of the independent variables $x_i$ |
| | | $\mathbf{y}$ | A vector of the independent variables $y_i$ |
| $S_{\min}$ | Minimum sum of squares on a sum-of-squares surface, calculated by introducing least-squares parameter estimates into Eq. (19) | $y$ | Dependent variable related to conversion by Eq. (116) |
| | | $y_i$ | Generalized dependent variable |
| $S_r$ | Multiple-response sum of squares defined by Eq. (64) | $\grave{y}$ | Geometric mean of any set of variables $y_i$ |
| $S_\Delta$ | Multiple-response sum of squares defined by Eq. (65) | $\bar{y}$ | Arithmetic mean of any set of variables $y_i$ |
| $S(\lambda)$ | Minimum sum of squares of residuals for a given value of $\lambda$, defined by Eq. (120) | $y_{A0}$ | Initial mole fraction of component $A$ |
| | | $\hat{y}_i$ | Predicted value of the generalized dependent variable $y_i$ |
| $s^2$ | Estimate of experimental error variance | | |
| $s_\theta$ | Standard error of parameter estimate $\hat{\theta}$ | $y^{(\lambda)}$ | Transformation of $y$, given by Eq. (118) |

| | | | |
|---|---|---|---|
| $y_u^v$ | Observed value of the response $v$ at the $u$th experimental point | $\lambda$ | A transforming parameter related to the reaction order |
| $z$ | Dependent variable defined by Eq. (77) | $v$ | Number of degrees of freedom |
| | | $\pi$ | Total pressure |
| $\hat{z}$ | Predicted value of $z$ | $\pi_i$ | Posterior probability associated with model $i$ |
| $Z^{(\lambda)}$ | Normalized transformed dependent variable given by Eq. (119) | $\sigma^2$ | Experimental error variance |
| | | $\sigma_i^2$ | Experimental error variance for $i$th experimental point |
| $\alpha$ | Confidence level of F or $t$ distributions | $\sigma_r^2$ | Variance of the reaction rate, $r$ |
| $\beta_i$ | True value of an $i$th parameter | $\hat{\sigma}_i^2$ | Variance of the predicted rate from model $i$ |
| $\boldsymbol{\beta}$ | Vector of parameter values $\beta_i$ | | |
| $\gamma_i$ | Rate constants of Eq. (7), $i = 1, 2$ | $\chi_1^2(\alpha)$ | The $100(1 - \alpha)\%$ point of the chi-squared distribution with 1 degree of freedom |
| $\Delta$ | Determinant defined in Eq. (65) | | |
| $\varepsilon$ | Experimental error | $\psi$ | A parameter setting the relative contributions of the linearization and steepest descent methods in determining the correction vector $\mathbf{b}$ of Eq. (45) |
| $\hat{\theta}$ | Estimated value of any parameter | | |
| $\Lambda$ | A nonintrinsic parameter defined in Eqs. (77) and (78) | | |

## References

A1. Anscombe, F. J., *Proc. Berkeley Symp. Math. Statist. Probability*, 4*th*, **1**, 19 (1963).
A2. Anscombe, F. J., and Tukey, J. W., *Technometrics* **5**, 141 (1963).
A3. Ayen, R. J., and Peters, M. S., *Ind. Eng. Chem. Proc. Design Develop* **1**, 204 (1962).
B1. Ball, W. E., and Groenweghe, L. C. D., *Ind. Eng. Chem. Fundamentals* **5**, 181 (1966).
B2. Bartlett, M. S., *Biometrics* **3**, 39 (1947).
B3. Beale, E. M. L., *J. Roy. Statist. Soc.*, *Ser. B* **22**, 41 (1960).
B4. Benson, S. W., "The Foundations of Chemical Kinetics." McGraw-Hill, New York, 1961.
B5. Booth, G. W., and Peterson, T. I., IBM SHARE Program "WLNLI" No. 687 (1958).
B6. Boudart, M., and Chembers, R. P., *J. Catalysis* **5**, 517 (1966).
B7. Box, G. E. P., *Biometrics* **10**, 16 (1954).
B8. Box. G. E. P., *Ann. N.Y. Acad. Sci.* **86**, 792 (1960).
B9. Box, G. E. P., Univ. of Wisconsin Dept. Statist. Tech. Rept. No. 111 (1967).
B10. Box, G. E. P., and Coutie, G. A., *Proc. Inst. Elect. Engrs.* (*London*), *Pt. B* **103**, No. 1, Suppl. 100 (1956).
B11. Box G. E. P., and Cox, D. R., *J. Roy. Statist. Soc.*, *Ser. B* **26**, No. 2, 211 (1964).
B12. Box, G. E. P., and Draper, N. R., *Biometrika* **52**, 355 (1965).
B13. Box, G. E. P., and Hill, W. J., *Technometrics* **9**, 57 (1967).
B14. Box, G. E. P., and Hunter, W. G., *Technometrics* **4**, 301 (1962).
B15. Box, G. E. P., and Hunter, W. G., *Proc. IBM Sci. Computing Symp. Statist.*, 113 (1965).
B16. Box, G. E. P., and Lucas, H. L., *Biometrika* **46**, 77 (1959).

B17. Box, G. E. P., and Tidwell, P. W., *Technometrics* **4**, 531 (1964).
B18. Box, G. E. P., and Wilson, H. L., *Biometrika* **46**, 77 (1959).
B19. Box, G. E. P., and Youle, P. V., *Biometrics* **11**, 287 (1955).
C1. Carr, N. L., *Ind. Eng. Chem.* **52**, 391 (1960).
C2. Chou, Chan-Hui, *Ind. Eng. Chem.* **50**, 799 (1958).
C3. Cox, D. R., Proc. Berkeley Symp. Statist. **1**, 105 (1961).
C4. Cox, D. R., *J. Roy. Statist. Soc., Pt. B*, **24**, 406 (1962).
D1. Davies, O. L., ed., "Statistical Methods in Research and Production." Hafner, New York, 1958.
D2. Davies, O. L., ed., "The Design and Analysis of Industrial Experiments." Hafner, New York, 1960.
D3. Draper, N. R., and Hunter, W. G., *Biometrika* **54**, 376 (1967).
D4. Draper, N. R., and Smith, H., "Applied Regression Analysis." Wiley, New York, 1966.
F1. Franckaerts, J. F., and Froment, G. F., *Chem. Eng. Sci.* **19**, 807 (1964).
F2. Franklin, N. L., Pinchbeck, P. H., and Popper, F., *Trans. Inst. Chem. Eng.* **34**, 280 (1956).
F3. Franklin, N. L., Pinchbeck, P. H., and Popper, F., *Trans. Inst. Chem. Eng.* **36**, 259 (1958).
F4. Franks, R. G. E., "Mathematical Modeling in Chemical Engineering," Wiley, New York, 1967.
F5. Frost, A. A., and Pearson, R. G., "Kinetics and Mechanism." Wiley, New York, 1961.
G1. Guttman, I., and Meeter, D. A., Univ. of Wisconsin Dept. Statist. Tech. Rept. No. 37, (1964).
G2. Guttman, I., and Meeter, D. A., Univ. of Wisconsin Dept. Statist. Tech. Rept. No. 38 (1964).
H1. Hartley, H. O., *Technometrics* **3**, 269 (1961).
H2. Hill, W. J., Ph.D. Thesis, Univ. of Wisconsin, Madison, 1966.
H3. Hill, W. J., and Hunter, W. G., *Technometrics* **8**, 571 (1966).
H4. Hill, W. J., and Hunter, W. G., Univ. of Wisconsin Dept. Statist. Tech. Rept. No. 69, (1966).
H5. Hill, W. J., and Mezaki, R., *Am. Inst. Chem. Engrs. J.* (*A.I.Ch.E. J.*) **13**, 611 (1967).
H6. Himmelblau, D. M., Jones, C. R., and Bishoff, K. B., *Ind. Eng. Chem. Fundamentals* **6**, 539 (1967).
H7. Hinshelwood, C. N., and Burk, R. E., *J. Chem. Soc.* **127**, 1105 (1925).
H8. Hunter, W. G., *Ind. Eng. Chem. Fundamentals* **6**, 461 (1967).
H9. Hunter, W. G., and Atkinson, A. C., Univ. of Wisconsin Statist. Tech. Rept. No. 59 (1965).
H10. Hunter, W. G., and Hill, W. J., Univ. of Wisconsin Dept. Statist. Tech. Rept. No. 65 (1966).
H11. Hunter, W. G., and Mezaki, R., *A.I.Ch.E. J.*, **10**, 315 (1964).
H12. Hunter, W. G., and Mezaki, R., *Can. J. Chem. Eng.* **45**, 247 (1967).
H13. Hunter, W. G., and Reiner, A. M., *Technometrics* **7**, 307 (1965).
H14. Hunter, W. G., and Wichern, D. W., Univ. of Wisconsin Dept. Statist. Tech. Rept. No. 33 (1966).
J1. Johnson, R. A., Standel, N. A., and Mezaki, R., *Ind. Eng. Chem. Fundamentals* **7**, 181 (1968).
K1. Kabel, R. L., and Johanson, L. N., *A.I.Ch.E. J.* **8**, 621 (1962).
K2. Kittrell, J. R., Ph.D. Thesis, Univ. of Wisconsin, Madison, 1966.

K3. Kittrell, J. R., and Erjavec, John, *Ind. Eng. Chem. Proc. Design Develop.* **7**, No. 3 321 (1966).
K4. Kittrell, J. R., and Mezaki, R., *Ind. Eng. Chem.* **59**, No. 2, 28 (1967).
K5. Kittrell, J. R., and Mezaki, R., *A.I.Ch.E. J.* **13**, 389 (1967).
K6. Kittrell, J. R., and Watson, C. C., *Chem. Eng. Progr.* **62**, No. 4, 79 (1966).
K7. Kittrell, J. R., Hunter, W. G., and Watson, C. C., *A.I.Ch.E. J.* **11**, 1051 (1965).
K8. Kittrell, J. R. Mezaki, R., and Watson, C. C., *Ind. Eng. Chem.* **57**, No. 12, 19 (1965).
K9. Kittrell, J. R., Mezaki, R., and Watson, C. C., *Brit. Chem. Eng.* **11**, No. 1, 15 (1966).
K10. Kittrell, J. R., Mezaki, R., and Watson, C. C., *Ind. Eng. Chem.* **58**, No. 5, 51 (1966).
K11. Kittrell, J. R., Hunter, W. G., and Watson, C. C., *A.I.Ch.E.J.* **12**, 5 (1966).
K12. Kittrell, J. R., Hunter, W. G., and Mezaki, R., *A.I.Ch.E. J.* **12**, 1014 (1966).
L1. Laible, J. R., Ph.D. Thesis, Univ. of Wisconsin, Madison, 1959.
L2. Laidler, K. J., "Chemical Kinetics," 2nd ed. McGraw-Hill, New York, 1965.
L3. Lapidus, L., and Peterson, T. I., *A.I.Ch.E. J.* **11**, 891 (1965).
L4. Lapidus, L., and Peterson, T. I., *Chem. Eng. Sci.* **21**, 655 (1966).
L5. Levenburg, K., *Quart. Appl. Math.* **2**, 164 (1944).
M1. Marquardt, D. L., *J. Soc. Indust. Appl. Math.* **2**, 431 (1963).
M2. Marquardt, D. L., "Least Squares Estimation of Nonlinear Parameters," IBM SHARE Library Program No. 3094, Exhibit B.
M3. Mathur, G. P., and Thodos, G., *Chem. Eng. Sci.* **21**, 1191 (1966).
M4. Mezaki, R., and Butt, J. B., *Ind. Eng. Chem. Fundamentals* **7**, 120 (1968).
M5. Mezaki, R., and Kittrell, J. R., *Can. J. Chem. Eng.* **44**, No. 5, 285 (1966).
M6. Mezaki, R., and Kittrell, J. R., *A.I.Ch.E. J.* **13**, 176 (1967).
M7. Mezaki, R., and Kittrell, J. R., *Ind. Eng. Chem.* **59**, No. 5, 63 (1967).
M8. Mezaki, R., Kittrell, J. R., and Hill, W. J., *Ind. Eng. Chem.* **59**, No. 1, 93 (1967).
M9. Mickley, H. S., Sherwood, T.K., and Reed, C.E., "Applied Mathematics in Chemical Engineering," 2nd ed. McGraw-Hill, New York ,1957.
P1. Pannetier, G., and Davignon, L., *Bull. Soc. Chim. France* 2131 (1961).
P2. Peterson, T. I., *Chem. Eng. Sci.* **17**, 203 (1962).
P3. Pinchbeck, P. H., *Chem. Eng. Sci.* **6**, 105 (1957).
R1. Rogers, G. B., Lih, ,M.M. and Hougen, O.A., *A.I.Ch.E. J.* **12**, 369 (1966).
R2. Rosenbrock, H. H., and Storey, C., "Computational Techniques for Chemical Engineers." Macmillan (Pergamon), New York, 1966.
R3. Rudd, D. F., and Watson, C. C., "Strategy of Process Engineering," Wiley, New York, 1968.
S1. Shabaker, R. H., Ph.D. Thesis, Univ. of Wisconsin, Madison, 1965.
S2. Spang, H. A., *Soc. Indust. Appl. Math. Rev.* **4**, 343 (1962).
W1. Wei, J., and Prater, C. D., *Advan. Catalysis* **13**, 204 (1962).
W2. Weller, S., *A.I.Ch.E. J.* **2**, 59 (1956).
W3. White, R. R., and Churchill, S. W., *A.I.Ch.E. J.* **5**, 354 (1959).
W4. Williams, E. J., and Kloot, N. H., *Australian J. Appl. Sci.* **4**, 1 (1953).
W5. Williams, E. J., "Regression Analysis." Wiley, New York, 1959).
W6. Wu, S. M., Ermer, D. S., and Hill, W. J., *J. Eng. Ind. Ser. B* **88**, 81 (1966).
Y1. Yang, K. H., and Hougen, O. A., *Chem. Eng. Progr.* **46**, 146, (1950).

# DECOMPOSITION PROCEDURES FOR THE SOLVING OF LARGE SCALE SYSTEMS

### W. P. Ledet* and D. M. Himmelblau

Department of Chemical Engineering
The University of Texas, Austin, Texas

|  |  |
|---|---|
| I. Introduction | 186 |
| II. Information Flow in Process Models | 188 |
|    A. Graphs and Boolean Matrices | 188 |
|    B. Representation of System Equations | 193 |
| III. Finding an Output Set | 196 |
| IV. Partitioning of the System Equations | 198 |
|    A. Criteria for Decomposition Algorithms | 199 |
|    B. How the Irreducible Sets of Equations (Those to be Solved Simultaneously) Correspond to the Maximal Loops in the Adjacency Matrix and are Invariant of the Output Set | 200 |
|    C. Methods of Partitioning | 202 |
| V. Identifying Disjoint Subsystems | 209 |
| VI. Tearing | 211 |
|    A. The Concept of Tearing | 211 |
|    B. Tearing Methods for Large Scale Systems of Equations | 212 |
|    C. Tearing of a System of Process Units | 219 |
| VII. Comparison of Precedence Ordering Techniques | 222 |
| VIII. Two Examples | 226 |
| Appendix A. Computer Program for Precedence Ordering | 237 |
| Appendix B. Reordered Occurrence Matrix of the Hanford N-Reactor System | 252 |
| Nomenclature | 253 |
| References | 253 |

\* Present address: E. I. duPont de Nemours, Orange, Texas.

## I. Introduction

The use in recent years of mathematical models of quite large scale physical processes has confronted the engineer with the task of solving large systems of equations. The system of $n$ real equations represented formally as

$$f_1(x_1, x_2, \ldots x_n) = 0$$
$$f_2(x_1, x_2, \ldots x_n) = 0$$
$$\vdots \qquad \vdots$$
$$f_n(x_1, x_2, \ldots x_n) = 0 \tag{1}$$

in which one or several of the equations may describe one piece of equipment can almost never be solved in closed form. If some of the equations are nonlinear, iterative algorithms of various kinds are employed to obtain a numerical solution of the system equations. These algorithms generally start with an initial vector of all the variables, $x$, and they employ different strategies to modify $x$ so as to converge to a solution by sequential stages. Typical algorithms are Newton's iteration function (T1) which employs first derivatives of the system equations to modify $x$; the method of steepest descent (K2), which also employs first derivatives of the equations to modify $x$ so that

$$\sum_{i=1}^{n} f_i^2(x)$$

is minimized; the method of direct search (H3), which employs a search strategy for modifying $x$, and iteration functions with memory (T1), which employs information from the most recent interations to modify $x$. Many other iterative methods have been described including Rosen (R1), Nagiev (N1), Vela (V1), and Komatsu (K1).

For a process that can be represented by a set of system equations of modest size, say from one to 100, the procedures mentioned above are reasonably straightforward and efficient. However, as the number of equations in the system further increases, these methods run into difficulties because of the large digital computer storage requirements and the excessive amount of computer time needed to solve the entire system of equations simultaneously. Analysis of the system of equations before a solution is attempted can often substantially mitigate these two problems. Experience has demonstrated that a large system of equations that represents a physical process need not be solved simultaneously but can be broken up into subsystems in which only smaller sets of equations need be solved simultaneously. Further, if the system is ordered into a sequence such that each equation or various groups of equations can be solved independently of the remaining equations in the sequence, the solution can proceed serially, and the computer storage needed to effect a solution would correspond roughly to the size of the largest subsystem of equations instead of the entire system. The time for solution would also be substantially reduced. Finally, if the equations within each partitioned subsystem can be ordered so that the number of iterates that need be chosen to obtain a solution of the subsystem is smaller than the number of equations in the subsystem, the computer time needed to solve the entire system may be reduced even further. Therefore, a systematic strategy to identify such subsystems, to order the equations and subsystems in the proper sequence, and to order the equations within the subsystems would have considerable merit. This type of precedence ordering will be called a **decomposition** of the system.

Decomposition leads to a rearrangement of the process equations from their flow chart sequence to a natural sequence based on the **information flow** among the equations. The ultimate goal is to set up an iterative scheme in which each equation is solved for a single variable (by some appropriate root identification method), and where values of unknown variables that must be assumed are checked cyclically. The greatest reduction in the number of iterates that must be assumed, and therefore the greatest reduction in computer storage and time requirements, takes place for those systems of process equations in which the number of variables per equation is small compared to the total number of variables in the system. Clearly, when each of the system equations contains every process variable, no effective decomposition can take place. Fortunately, most models used in the process industries are of such a character that extensive decomposition can be effected.

Decomposition can be applied to models that represent any large scale integrated process whether formed for solving material balances, energy balances, and equilibrium relations in process design, or used for economic evaluation, or for analysis of the control of entire plants. One of the most fruitful applications is to digital executive programs such as PACER (S2), which can be used to design chemical plants in the sense that the parts of the plant can be assembled in the desired sequence and the internal subroutines used in design. Selection of a good precedence order for calculation among the process units can improve materially the operation of the simulation. Another application of decomposition, which has not been attempted yet but which appears promising, is in the optimization of large plants where the model includes many equality constraints that can be decomposed. Industrial scheduling and production planning as well as network analysis are areas of application outside of the process industries.

We first describe in Section II how the information flow takes place in process models, give a compact method of representation of the system of equations, and point out the correspondence between a system of equations and a linear diagraph. In Section III, methods for finding an output set (assignment of a dependent variable to each equation) is presented. Section IV presents procedures for partitioning the system into the smallest irreducible subsystems (groups of equations that must be solved simultaneously). Section V examines the concept of disjoint subsystems and the methods for finding them. Section VI examines the irreducible subsystems and discusses their internal ordering. Various methods of decomposition are discussed and compared in Section VII. Section VIII presents a large scale example of decomposition based on a digital computer program listed in Appendix A.

Certain special problems related to decomposition are also considered. One is whether the process model is determinate, that is, does the model have a solution. Section III indicates how decomposition can help in validating determinacy. Another problem is that of convergence of the iterative strategy. If certain of the equations are too sensitive to the values of the variables being iterated, convergence may not be obtained. Therefore, the decomposition procedure must be constrained so as to choose as iterates only those variables for which the system of equations has suitable sensitivity, as discussed in Section VI.

## II. Information Flow in Process Models

The information flow among the system equations can be represented by information flow diagrams in a fashion roughly analogous to the representation of the process material flows by a process flow sheet. Typical examples of the graphical representation of information flow are organizational charts and signal flow diagrams. Associated with the information flow graphs are Boolean matrices, which have a one to one correspondence with the structure of the graph, but can easily be loaded into a digital computer for analysis. Decomposition of a large system can take place using either the graph or the associated Boolean matrix, but we are concerned primarily with the latter because the decomposition of a large graph by inspection would be impractical. However, decomposition by inspection of an information flow diagram is quite feasible for a small number of system equations say less than 25.

### A. Graphs and Boolean Matrices

We now examine some of the properties of graphs and their associated Boolean matrices (H1, H2), which aid in decomposition. A **graph**, G, is a collection of vertices, $V$ (nodes, points, elements, subsystems) and lines joining the vertices called edges, $E$ (flows, paths, streams). A **digraph** or directed graph is a graph in which the edges are directed (flow may occur in only one direction). A **finite graph** is one which has a finite number of vertices. Associated with each graph is a Boolean matrix R termed the adjacency matrix (associated matrix, relation matrix, structural matrix), which has as many rows and columns as the graph has vertices (the numbering of both the rows and columns correspond to the numbering of the vertices). An element $r_{ij} = 1$ if there is a flow directed along an edge from vertex $j$ to vertex $i$; $r_{ij} = 0$ otherwise. Thus, there is a one to one correspondence between a graph and its adjacency matrix.

Figure 1 illustrates a digraph. The rows of the adjacency matrix correspond to the vertices to which flow is directed. The columns correspond to the vertices from which the flows are directed.

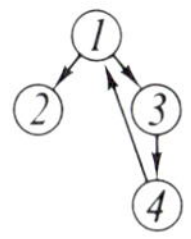

$$\text{Vertex} \begin{array}{c} \\ 1 \\ 2 \\ 3 \\ 4 \end{array} \begin{array}{c} \text{Vertex} \\ 1 \quad 2 \quad 3 \quad 4 \\ \begin{bmatrix} 0 & 0 & 0 & 1 \\ 1 & 0 & 0 & 0 \\ 1 & 0 & 0 & 0 \\ 0 & 0 & 1 & 0 \end{bmatrix} \end{array} = R$$

FIG. 1. Correspondence between a graph and its adjacency matrix R.

The edge of the graph from vertex *1* to vertex *2* indicates a flow of information or perhaps material from vertex *1* to *2* and therefore the element $r_{21} = 1$. Similarly, there is a flow directed from vertex *1* to vertex *3* so that $r_{31} = 1$; also $r_{43} = 1$ and $r_{14} = 1$. All other elements of the matrix are 0 since there are no other flows in the graph. The adjacency matrix thus indicates all of the one step paths (flows between two vertices without an intervening vertex) between vertices. If $r_{ij} = 1$, there is a one-step path from vertex $j$ to vertex $i$. Therefore, all of the structure of the graph is contained in the adjacency matrix.

In the language of symbolic logic the elements $r_{ij}$ of the adjacency matrix are the **propositions**: "There is a direct path from vertex $j$ to vertex $i$" which admit of but two **truth values**: true designated by the symbol 1, and false designated by the symbol 0. Any matrices with elements of which meaningful products and sums are defined can be manipulated by the rules of matrix algebra (B2, P1) to give new matrices that are, in general, meaningful. This is true here if the product of two or more propositions is understood to mean their logical or Boolean product and the sum of a number of propositions is understood to mean their logical sum or Boolean union:

**Boolean multiplication**—If $x, y, z, \ldots$ are propositions, their logical product $(x \cdot y \cdot z \cdot \cdots)$ in any order is the new proposition "*All* of $x, y, z, \ldots$ are true." In the symbolism here adopted, the truth value of $(x \cdot y \cdot z \cdot \cdots)$ is 1 if that of every factor is 1 but it is 0 if that of any factor is 0. Briefly,

$$x \cdot y \cdot z \cdot \cdots = \min\{x, y, z, \ldots\}$$

***Boolean union***—If $x, y, z, \ldots$ are propositions, their Boolean union or logical sum $(x \cup y \cup z \cup \cdots)$ is the new proposition "At least one of $x, y, z, \ldots$ is true." Thus, the truth value of $(x \cup y \cup z \cup \cdots)$ is 0 if that of every addend is 0, but it is 1 if that of any addend is 1. Briefly,

$$z \cup y \cup z \cup \cdots = \max\{x, y, z, \ldots\}$$

Boolean multiplication is distributive with respect to Boolean union; that is,

$$x \cdot (y \cup z) = (x \cdot y) \cup (x \cdot z)$$

wherein the parentheses on the right-hand side are usually omitted, the understanding being that multiplications are to be performed before unions unless parentheses indicate otherwise. In fact, by choosing to indicate true and false by 1 and 0, respectively, Boolean multiplication is conveniently made to obey the rules of ordinary arithmetic.

Boolean union, on the other hand, differs significantly from addition. We have, for example,

$$x \cup x = x$$

Further, Boolean union is distributive with respect to Boolean multiplication; that is,

$$x \cup (y \cdot z) = (x \cup y) \cdot (x \cup z)$$

but here the parentheses may not be removed on the right without ambiguity. There is no meaningful analog of subtraction; consequently the solution of propositional equations may not, as in ordinary algebra, involve the transfer of addends across the equals sign.[1]

Boolean matrix multiplication is carried out according to the usual matrix rules, expressed in terms of the logical product and logical sum as defined above.

Thus, the elements $c_{ij}$ of the matrix

$$C = A \cdot B$$

are found by the formula

$$c_{ij} = \bigcup_{k=1}^{n} a_{ik} \cdot b_{kj} \qquad (2$$

where the operational symbol

$$\bigcup_{k=1}^{n}$$

---

[1] The sign $=$ in propositional equations indicates equality of the truth values of the propositions equated, not necessarily, their identity or equivalence.

indicates the union of the terms typified by the product following it with $k$ taken successively as 1, 2, 3 ... $n$, and $n$ is the common dimension (or order) of the matrices A, B, and C. For example, in the product of the matrix R of Fig. 1 by itself to form $R^2$, the first element in the first row, $c_{11}$, would be calculated as follows:

$$c_{11} = \max[\min\{0, 0\}; \min\{0, 1\}; \min\{0, 1\}; \min\{1, 0\}] = 0$$

and the first element in the fourth row, as follows:

$$c_{41} = \max[\min\{0, 0\}; \min\{0, 1\}; \min\{1, 1\}; \min\{0, 0\}] = 1$$

It will be observed that, in general, $A \cdot B \neq B \cdot A$.

Another operation to be used is the union of rows or columns within a particular matrix. If A is an $n$ by 1 matrix (a column of an $n$ by $n$ matrix), B is an $n$ by 1 matrix (another column), then the $n$ by 1 matrix $C = A \cup B$ is defined by the elements

$$c_i = a_i \cup b_i \text{ for } i = 1, 2, \ldots n \tag{3}$$

For example

$$\begin{bmatrix} 0 \\ 0 \\ 1 \end{bmatrix} \cup \begin{bmatrix} 1 \\ 0 \\ 1 \end{bmatrix} = \begin{bmatrix} 1 \\ 0 \\ 1 \end{bmatrix}$$

A similar definition is applied to the union of the rows.

An important property of the adjacency matrix is that the $i$th power of this matrix gives all the $i$ step paths between vertices. For example, the second and third powers of the adjacency matrix in Fig. 1 are shown in Fig. 2. The

$$R^2 = \begin{array}{c} \\ 1 \\ 2 \\ 3 \\ 4 \end{array} \begin{array}{cccc} 1 & 2 & 3 & 4 \\ \begin{bmatrix} 0 & 0 & 1 & 0 \\ 0 & 0 & 0 & 0 \\ 0 & 0 & 0 & 0 \\ 1 & 0 & 0 & 0 \end{bmatrix} \end{array} \quad R^3 = \begin{array}{c} \\ 1 \\ 2 \\ 3 \\ 4 \end{array} \begin{array}{cccc} 1 & 2 & 3 & 4 \\ \begin{bmatrix} 1 & 0 & 0 & 0 \\ 0 & 0 & 1 & 0 \\ 0 & 0 & 1 & 0 \\ 0 & 0 & 0 & 1 \end{bmatrix} \end{array}$$

FIG. 2. Second and third powers of the adjacency matrix.

rows of $R^2$ and $R^3$ still correspond to the vertices to which the flows are directed, and the columns correspond to the vertices from which the flows are directed. Each nonzero element of the matrix $R^k$ indicates there is a path going through $k$ edges (a $k$ step path) from vertex $j$ to vertex $i$. Figure 2 indicates that a two-step path exists from vertex *1* to vertex *4* and a two-step path exists from vertex *3* to vertex *1*. $R^3$ indicates not only the three-step loop tying together vertices *1*, *3*, and *4* but also the path $3 \to 4 \to 1 \to 2$. A $k$th order **loop** in a graph is defined as a set of $k$ vertices each of which is connected to every other vertex of the loop by a closed path. A $k$th order **maximal loop**, $H$, of a graph $G$, is defined as a set of $k$ vertices in $G$ connected by a closed path such that every other loop in $G$ is either contained in $H$ or has no vertex in common with $H$. In Fig. 3, the loop $1 \to 2 \to 1$ is not maximal, whereas the

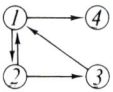

FIG. 3. Maximal and nonmaximal loops.

loop $1 \to 2 \to 3 \to 1$ is a maximal loop. A graph $G$ then may contain several maximal loops of different orders. Because these loops bind together vertices in a common information network, the location of these maximal loops is an important phase in the decomposition of systems.

One method of locating these maximal loops is to compute the reachability matrix, R*(H1), which is the element by element Boolean union of all of the powers of the adjacency matrix up to the $n$th, where $n$ is the number of rows of R. An element of the reachability matrix is defined as

$$r_{ij}^* = \bigcup_{k=1}^{n} r_{ij}^k$$

where $r_{ij}^k$ is the element of the $i$th row and the $j$th column of the $k$th power of the adjacency matrix. The reachability matrix indicates all of the paths of any length between vertices. Figure 4 shows the reachability matrix for the graph of Fig. 1.

$$R^* = \begin{array}{c} \\ 1 \\ 2 \\ 3 \\ 4 \end{array} \begin{array}{c} 1 \ 2 \ 3 \ 4 \\ \begin{bmatrix} 1 & 0 & 1 & 1 \\ 1 & 0 & 0 & 0 \\ 1 & 0 & 1 & 1 \\ 1 & 0 & 1 & 1 \end{bmatrix} \end{array}$$

FIG. 4. Reachability matrix of the graph in Fig. 1.

The maximal loops in the graph can be found from the reachability matrix by finding those sets of vertices that satisfy the following conditions (1) $r_{ij}^* = r_{ji}^* = 1$, where $i$ and $j$ take on all possible combinations of the vertex numbers in the set; (2) no other vertices, not included in the set, satisfy condition (1). The first condition requires that each vertex in the set is reachable by some path from every other vertex in the set. The second condition requires that there is no path from a vertex in the set to a vertex outside the set and back again to a vertex in the set. If a set of vertices meets these two conditions, there must be a closed path that goes through every vertex in the set, which is by definition a loop. The loop found in this manner is maximal because, if there were a larger loop that included it, the second condition would not be satisfied. Other methods for finding maximal loops are discussed in Section IV.

## B. Representation of System Equations

Since the systems of equations to be considered are quite large, it is necessary to use some compact method to represent the information flow among them. A very convenient technique is to relate the system equations to a digraph and its associated Boolean matrix, which represent the structure of the information flow in the system. The Boolean matrix to be used is called the **occurrence matrix** (H1, S3), and is defind as follows: (1) each row of the occurrence matrix corresponds to a system equation, and each column corresponds to a system variable; (2) an element of the matrix, $s_{ij}$, is either a Boolean 1 or 0 according to the rule

$$s_{ij} = \begin{cases} 1, & \text{if variable } j \text{ appears in equation } i \\ 0, & \text{otherwise} \end{cases}$$

This matrix then indicates the occurrence of the dependent variables in each of the system equations. For example, consider the following set of equations written in functional form:

$$\begin{aligned} f_1(x_1, x_2) &= 0 \\ f_2(x_4) &= 0 \\ f_3(x_3, x_6) &= 0 \\ f_4(x_4, x_5) &= 0 \\ f_5(x_1, x_6) &= 0 \\ f_6(x_2, x_3, x_5) &= 0 \end{aligned} \quad (4)$$

Figure 5 is the occurrence matrix corresponding to Eqs. (4).

In decomposing a system of equations, it is necessary to analyze the information flow among the equations concerning the values of the system variables (S3). In order to determine the direction of information flow in the system equations, one must first establish what information each equation is to supply, that is, the identity of the variable whose value is to be obtained from the equation. Further, the system equations together must supply all of the information about the system (the values of all of the variables). The variable for which an equation is to be solved is called its **output variable** and the set of all of the variables assigned to the equations as output variables is called an **output set** (S3, H2). Thus the information that one equation can

$$\text{Equation} \begin{array}{c} \\ f_1 \\ f_2 \\ f_3 \\ f_4 \\ f_5 \\ f_6 \end{array} \overset{\text{Variable}}{\begin{bmatrix} x_1 & x_2 & x_3 & x_4 & x_5 & x_6 \\ 1 & 1 & 0 & 0 & 0 & 0 \\ 0 & 0 & 0 & 1 & 0 & 0 \\ 0 & 0 & 1 & 0 & 0 & 1 \\ 0 & 0 & 0 & 1 & 1 & 0 \\ 1 & 0 & 0 & 0 & 0 & 1 \\ 0 & 1 & 1 & 0 & 1 & 0 \end{bmatrix}}$$

FIG. 5. Occurrence matrix for Eqs. (4).

communicate to the other equations is the value of its output variable. Simultaneously, information flows into an equation through all of the variables it contains except through its output variable. Cast in terms of the occurrence matrix an output set is a set of nonzero elements in the occurrence matrix such that one and only one element appears in each row and simultaneously one and only one element appears in each column. A unique output set may or may not exist for a system of equations, but for the purposes of initiating the decomposition, any output set will suffice. The encircled elements of Fig. 6 designate an output set for the occurrence matrix of Fig. 5. Methods for selecting an output set are discussed in Section III.

$$\text{Equation} \begin{array}{c} \\ f_1 \\ f_2 \\ f_3 \\ f_4 \\ f_5 \\ f_6 \end{array} \overset{\text{Variable}}{\begin{bmatrix} x_1 & x_2 & x_3 & x_4 & x_5 & x_6 \\ \textcircled{1} & 1 & 0 & 0 & 0 & 0 \\ 0 & 0 & 0 & \textcircled{1} & 0 & 0 \\ 0 & 0 & \textcircled{1} & 0 & 0 & 1 \\ 0 & 0 & 0 & 1 & \textcircled{1} & 0 \\ 1 & 0 & 0 & 0 & 0 & \textcircled{1} \\ 0 & \textcircled{1} & 1 & 0 & 1 & 0 \end{bmatrix}}$$

FIG. 6. Occurrence matrix with output set.

Once an output set has been established, the direction of information flow is fixed in the system of equations, and they can be represented either by a linear diagraph or its associated Boolean adjacency matrix. For our purposes it is more convenient to work with the Boolean adjacency matrix, which can be obtained directly from the occurrence matrix and output set as follows. First, assign numbers to the equations that correspond to the rows, and numbers to the variables that correspond to the columns of the occurrence matrix as in Fig. 5. Then pick an output set by the methods described in Section III. For the first equation, and the number of the column containing its output variable. The information flow transmitted by the variable designated by the column number goes to all other equations that have nonzero elements in the column. This flow is identified by entering a nonzero element in the proper row of the first column of the adjacency matrix. Similarly for every other equation, the column in which the output element appears can be located, and the information flow from that equation to all other equations identified. Nonzero entries are made in the columns of the adjacency matrix for each equation (row) having such a flow. For example, in Fig. 6, column 1 of the occurrence matrix contains the output element of Eq. (1), and the nonzero element in column 1, row 5, indicates a flow directed from Eq. (1) to Eq. (5). Therefore, a nonzero entry would be made in the element $r_{51}$ of the adjacency matrix R. The output element of Eq. (2) in Fig. 6 is in column 4, and a flow is indicated from Eq. (2) to Eq. (4). If this procedure is continued for all of the equations, the adjacency matrix of Fig. 7b is obtained.

$$\text{Equation} \begin{array}{c} \\ f_1 \\ f_2 \\ f_3 \\ f_4 \\ f_5 \\ f_6 \end{array} \overset{\text{Variable}}{\begin{bmatrix} x_1 & x_4 & x_3 & x_5 & x_6 & x_2 \\ ① & 0 & 0 & 0 & 0 & 1 \\ 0 & ① & 0 & 0 & 0 & 0 \\ 0 & 0 & ① & 0 & 1 & 0 \\ 0 & 1 & 0 & ① & 0 & 0 \\ 1 & 0 & 0 & 0 & ① & 0 \\ 0 & 0 & 1 & 1 & 0 & ① \end{bmatrix}}$$

(a)

$$\text{Equation} \begin{array}{c} \\ f_1 \\ f_2 \\ f_3 \\ f_4 \\ f_5 \\ f_6 \end{array} \overset{\text{Equation}}{\begin{bmatrix} f_1 & f_2 & f_3 & f_4 & f_5 & f_6^a \\ 0 & 0 & 0 & 0 & 0 & 1 \\ 0 & 0 & 0 & 0 & 0 & 0 \\ 0 & 0 & 0 & 0 & 1 & 0 \\ 0 & 1 & 0 & 0 & 0 & 0 \\ 1 & 0 & 0 & 0 & 0 & 0 \\ 0 & 0 & 1 & 1 & 0 & 0 \end{bmatrix}}$$

(b)

FIG. 7. Occurrence matrix (a) and associated adjacency matrix (b).

To form the adjacency matrix in a simpler way, note that the elements of the first column of the adjacency matrix in Fig. 7b correspond exactly to the elements of the first column of the occurrence matrix in Fig. 6 if the output element is deleted. Similarly, the elements of column 2 of the adjacency matrix correspond exactly to the elements of the fourth column of the occurrence matrix if the output element is deleted. Therefore, if the columns of the occurrence were permuted until all of the output elements appeared on the main diagonal, as in Fig. 7a, the nonzero elements of the occurrence matrix would correspond exactly to those in the adjacency matrix except for the elements appearing on the main diagonal. Thus, after the permutation of columns is completed, the adjacency matrix can be obtained by simply renumbering the columns so that they have the same sequence as the rows and then by deleting the nonzero elements on the main diagonal (which in effect represent self-loops of information flow).

We have shown thus far how the system of equations representing a process can be related to a linear diagraph and its associated Boolean adjacency matrix. In the Section IV, we show how the location of the maximal loops in this adjacency matrix leads to identification of the subsystems of equations that must be solved simultaneously.

## III. Finding an Output Set

The concept of an output set was introduced in Section II. In this section we discuss how to find an output set, and how the procedure that is presented can help in ascertaining system determinacy.

The problem of finding an output set can be viewed as a problem of assigning an output variable to each system equation in such a way that each equation has a single output variable and each variable is assigned to only one equation subject to the constraint that a variable can only be assigned to an equation if it appears in that equation. The problem of finding an output set is a special case of the assignment problem of linear programming (C1) in which there is no objective function to be optimized and all feasible assignments are acceptable.

The algorithm that has turned out to be the most effective in practice for finding an output set is that proposed by Steward (S3). Steward proved that application of the algorithm to a system of equations will lead to one of two outcomes. The first is that an output set will be found. The other is that a subset of equations is found that contains fewer variables than equations, in which case Steward has shown that no output set exists for the system of equations.

Steward's algorithm, presented below, is a logical sequence of steps to assign variables to equations, and is stated in terms of assigning columns to rows. Necessarily, some provision must be made to retain the identification of the variable numbers with the column numbers and the equation numbers with the row numbers. An assignment of a column to a row is indicated by placing the number of the column into the corresponding row of an $n$ by $n$ matrix A. Initially, all of the elements of matrix A are set equal to zero. The algorithm involves two phases, assignment and reassignment. Assignment begins by assigning the first column (variable) of the occurrence matrix to the first row (equation) in the column that has a nonzero element. Then the second column is assigned to the first row in that column that has a nonzero element. This procedure is continued until a column is assigned to a row previously identified with another column. When this occurs reassignment starts; the column that was previously assigned to the row is reassigned to another row that has a nonzero element in the column. If the row to which the column is reassigned also had a previous assignment, the previously assigned column is reassigned in the same manner. This procedure is continued as long as a new column is not reassigned to any row more than once, in order to avoid getting into an infinite loop of reassignments.

Each row and column in which a reassignment has been made is tagged to keep track of the assignments and reassignments. In the computer routine in Appendix A, a column is tagged by placing a 1 in the corresponding column of a 1 by $n$ information matrix C and a row is tagged by placing a 1 in the corresponding row of an $n$ by 1 information matrix T.

If in the reassigment, a column is encountered that cannot be reassigned because it contains no nonzero elements in an untagged row of the occurrence matrix, it is necessary to examine all the columns previously reassigned (tagged columns) to see if any of them can be reassigned to an untagged row. If so, the reassignment continues. If not, the algorithm terminates and an output set cannot be found. The untagged rows correspond to a subset of equations that contain fewer variables than equations.

The reassignment phase continues until a column is reassigned to a row that has not been previously identified with a column. Then the reassigment phase stops and the assignment phase takes up again.

We now consider how Steward's algorithm can help to ascertain whether or not the system of equations describing the process is determinate. It should be noted that if a system of equations having the same number of variables as equations incorporates a subset of equations that contains fewer variables than equations, a unique solution of the system equations is unlikely to exist. We have used the words "is unlikely to" rather than "does not" because there are some special classes of equations that specify more than one variable, and if such an equation is included in the system, the system may have a subset of equations with fewer variables than equations and still be determinate. For example, consider the system of Eq. (5):

$$x^2 + y^4 = 0$$
$$z = 1 \qquad (5)$$
$$2z = 2$$

The second and third equations together contain only one variable and are not independent. The system, however, has a unique solution since the first equation specifies that both $x$ and $y$, if real, must be zero. However, these special classes of equations usually do not appear in models of real physical process, and consequently, if a model is analyzed by Steward's algorithm and no output set can be found, the system of equations is most likely not determinate.

In order to understand why a system of equations that includes a subsystem of equations with fewer variables than equations is not determinate, consider the following cases: Case A—the subsystem of equations contains an equation that is not independent for the remaining equations in the subsystems; Case B—the variables for the subsystem are overspecified, that is, in the subsystem there are more equations than variables. At the same time, in the remaining equations, there are more variables than equations. Case B occurs because a variable was treated as a parameter (constant) in the equations in which it appears, and a parameter was treated as a variable in the equations in which it appears. In Case A, the system equations are not consistent and no solution to the process model exists. In Case B, a solution cannot be obtained either.

Steward's algorithm, thus, can help locate errors in formulating a model of a process by identifying the set of equations that contains improperly specified design variables or parameters. The FORTRAN program listed in Appendix A prints out the equation numbers in the subset that contains fewer variables than equations when an output set cannot be found.

## IV. Partitioning of the System Equations

This section treats the **partitioning** of the system equations into the smallest irreducible subsystems, that is, the smallest groups of equations that must be solved simultaneously. Partitioning represents the first and easiest of the two phases of decomposition; tearing (which is discussed in Section VI) is more difficult.

As an illustration of partitioning consider the following equations written in functional form

$$f_1(x_1, x_2) = 0$$
$$f_2(x_2, x_3) = 0$$
$$f_3(x_1, x_3, x_4) = 0 \qquad (6)$$
$$f_4,(x_4, x_5) = 0$$
$$f_5(x_4, x_5) = 0$$

It can be seen by inspection that equations for functions $f_4$ and $f_5$ contain only variables $x_4$ and $x_5$, and that it is possible to solve these two equations for the values of $x_4$ and $x_5$ independently of the remaining equations. Also it would not be possible to solve either $f_4$ or $f_5$ alone since both contain two variables. Therefore $f_4$ and $f_5$ constitute a subsystem of equations which must be solved simultaneously. After $f_4$ and $f_5$ are solved and the value of $x_4$ is known, the remaining three equations contain only three variables whose values are unknown. Since each remaining equation contains two different unspecified variables, $f_1, f_2$, and $f_3$ must also be solved simultaneously for the values of $x_1$, $x_2$, and $x_3$. Therefore, we can partition the system of Eqs. (6) into two subsystems (which must be solved simultaneously) instead of solving the whole system simultaneously. However, in order to solve the two subsystems of equations independently of one another, the subsystem comprising $f_4$ and $f_5$ must be solved first to obtain a value for $x_4$. If the subsystem comprising $f_1, f_2$, and $f_3$ were to be solved first, a value for $x_4$ would have to be assumed and the subsystem solved again when the correct value of $x_4$ was obtained from the subsystem of $f_4$ and $f_5$. Therefore, an effective partitioning scheme must not only locate the subsystems of equations that must be solved simultaneously, but must also order the subsystems into a sequence such that each subsystem can be solved in its proper sequence.

We first mention certain criteria for partitioning and tearing, and then describe methods of executing the partitioning. Tearing is treated in Section VI.

## A. Criteria for Decomposition Algorithms

Certain desirable features are sought in and certain limitations are imposed on any effective decomposition scheme. These are:

(1) Its logical steps must be appropriate and consistent so that they can be programmed on a digital computer.

(2) Its methods of analysis must be efficient enough so that large systems of equations (on the order of 1000 equations) can be analyzed in a reasonable amount of digital computer time.

(3) It must partition the system into the smallest possible subsystems that can be solved independently in sequence.

(4) It must indicate the precedence order in which the equations or subsystems are to be solved.

(5) It must choose as the output (dependent) variable of each equation a variable that can be solved for explicitly.

(6) It must choose as the variables for iteration those for which the sensitivity of the system is appropriate and consistent with the particular solution technique chosen for the system.

(7) It should order the equations within the irreducible subsystems so that the minimum number of variables need be iterated or specified as system inputs to obtain a complete solution of the system.

A sound decomposition strategy should be applicable to any type of mathematical model of a physical process. Therefore, the set of system equations might include linear or nonlinear equations; algebraic, differential, difference, or integral equations; continuous or discrete variables with the following restrictions:

(1) Each equation must be solvable for at least one of its dependent variables.

(2) The set of $n$ equations together must contain exactly $n$ variables.

(3) Each subset of $i$ equations must be solvable for at least $i$ variables where $i = 1, 2, \ldots n$, as discussed in Section III.

(4) The system of equations must be determinate. Determinate here means that at least one solution of the set of equations exists within the physically allowable range of the variables.

The best decomposition procedure is the one that fits the above criteria and facilitates the most economical solution of large systems of equations. Clearly, not all of the criteria are simultaneously completely compatible; hence any computer routine must necessarily represent certain compromises.

### B. How the Irreducible Sets of Equations (Those to be Solved Simultaneously) Correspond to the Maximal Loops in the Adjacency Matrix and are Invariant of the Output Set

We now demonstrate how the subsystems of equations that must be solved simultaneoulsy correspond to the maximal loops found in the related adjacency matrix. Consider the following set of two equations each containing common variables:

$$f_1(x_1, x_2) = 0 \\ f_2(x_1, x_2) = 0 \qquad (7)$$

Neither equation can be solved independently because both equations contain the same variables. In terms of information flow, regardless of the output set chosen, the first equation must feed information to the second equation and the second to the first, constituting a loop of information flow. For any number of equations, as long as each equation feeds information to the next equation in sequence, and the last equation feeds information to the first equation, the whole system of equations has to be solved simultaneously. If a set of equations comprises part of a larger set of equations, which themselves form a larger loop of information flow, the subset must be solved together with the bigger system. Thus, any set of equations in which each equation is included in a maximal loop of information flow must be solved simultaneously.

Previously we have termed the largest loop of information flow a maximal loop, and indicated that it is not tied into other loops, by definition. One might wonder, because the choice of an output set is not unique, whether the maximal loops of information flow in one adjacency matrix will differ from those of another adjacency matrix, i.e., one formed from a different output set. It is shown in the following paragraphs that the maximal loops will be the same and therefore any output set will suffice for accomplishing the partitioning.

Consider the relationship between output sets. Steward (S3) has shown that all of the output sets of a system of equations can be generated from one output set and the loops of information flow for that set. He showed that the only way another output set could be obtained would be to reassign the output variable of each equation in a loop to the equation in that loop which it feeds. For example, consider the following set of three equations:

$$f_1(x_1, x_3) = 0$$
$$f_2(x_1, x_2) = 0 \qquad (8)$$
$$f_3(x_2, x_3) = 0$$

Suppose $x_1$ is assigned as the output variable of $f_1$, $x_2$ as the output variable of $f_2$, and $x_3$ as the output variable of $f_3$. With this output set, $f_1$ feeds information to $f_2$, $f_2$ feeds information to $f_3$, and $f_3$ feeds information to $f_1$ constituting a loop of information flow.

A new output set can be obtained by reassigning the output variables sequentially in the loop as follows: $x_3$, the output variable of $f_3$, is reassigned to the equation in the loop that $f_3$ feeds, namely $f_1$; then, $x_1$, the former output variable of $f_1$, is reassigned to $f_2$, and finally, $x_2$, the former output variable of $f_2$, is assigned to $f_3$. In the new output set, $x_3$ is the output variable of $f_1$, $x_1$ is the output variable of $f_2$, and $x_2$ is the output variable of $f_3$. With this new output set, $f_1$ feeds information to $f_3$, $f_3$ feeds information to $f_2$, and $f_2$ feeds information to $f_1$, forming a loop of information flow directly opposite to the loop based on the original output set. Steward (S3) also shows that the *only* way to proceed from one output set to another is to reassign the variables sequentially in a loop as described. Intuitively one can see that if a reassignment of an output variable is made in one equation, the former output variable of that equation must be reassigned to another equation, and so on continuing until the equation that contained the first reassigned variable as its output variable had assigned to it a new output variable. This reassignment can only result in a valid output set when the reassignment is carried out sequentially in a maximal or subordinate loop of information flow.

For any maximal loop, which is by definition not connected to a larger loop, it is clear that the information flow into and out of the set of equations in the maximal loop is exactly the same regardless of the output set, because the same variables are assigned to each maximal loop for all output sets. Therefore, no new maximal loops can be formed upon changing the output set. The problem of partitioning, then, is reduced to finding all the maximal loops in an adjacency matrix and ordering these into the proper sequence.

### C. Methods of Partitioning

One method of partitioning the system equations is to compute the maximal loops using powers of the adjacency matrix as discussed in Section II. Certain modifications to the methods of Section II are needed in order to reduce the computation time. The first modification is to obtain the product of the matrices using Boolean unions of rows instead of the multiplication technique previously demonstrated to obtain a power of an adjacency matrix. To show how the Boolean union of rows can replace the standard matrix multiplication, consider the definition of Boolean matrix multiplication, Eq. (2), which can be expanded to

$$c_{ij} = a_{i1} \cdot b_{1j} \cup a_{i2} \cdot b_{2j} \cdots \cup a_{in} \cdot b_{nj} \qquad (9)$$

According to the rules of Boolean operation

$$0 \cdot 1 = 0 \qquad 0 \cup 1 = 1$$
$$1 \cdot 1 = 1 \qquad 1 \cup 1 = 1$$
$$1 \cdot 0 = 0 \qquad 1 \cup 0 = 1$$

Now, if $a_{im} = 0$, the terms with $k = m$ will not contribute to $c_{ij}$ for $j = 1, 2, \ldots n$ (i.e., all of the elements of row $i$ in C). However, if $a_{im} = 1$, every 1 in row $m$ of B will appear in $c_{ij}$ for $j = 1, 2, \ldots n$. Therefore, row $i$ of C is just the Boolean union of the rows of B corresponding to the nonzero entries in row $i$ of matrix A.

This modification for Boolean matrix multiplication permits use of the Boolean union operation (logical OR operation or logical sum) instead of regular multiplication and union operations. The Boolean union operation can be executed much faster on a digital computer. Experience has shown that performing the Boolean union of rows instead of the standard Boolean multiplication of matrices can reduce the computation time by as much as a factor of four.

In order to obtain in a digital computer all of the powers of the adjacency matrix and then compute the reachability matrix by taking the Boolean sum of all of the powers, each power of the adjacency matrix and the reachability matrix would have to be stored in the computer memory. For large systems of equations an unreasonable amount of storage would be required. A second modification to the methods described in Section II is to drastically reduce the number of matrices that must be stored by computing the powers of $(A \cup I)$, where A is the adjacency matrix and I is the identity matrix. Note that the $(A \cup I)$ matrix is just the adjacency matrix in which all of the elements on the main diagonal have been set equal to 1. Steward (S3) has shown that the $n$th power of $(A \cup I)$ is equivalent to the reachability matrix. The square of $(A \cup I)$ is

$$(A \cup I)^2 = [A \cdot (A \cup I)] \cup I \cdot [(A \cup I)] = A^2 \cup A \cup I \qquad (10)$$

and the general relation for the $n$th power of $(A \cup I)$ is

$$(A \cup I)^n = A^n \cup A^{n-1} \cup \cdots \cup A \cup I \qquad (11)$$

Thus, the $n$th power of the matrix $(A \cup I)$ contains all of the nonzero elements of the reachability matrix plus a main diagonal of 1's. In the reachability matrix, only those rows that correspond to equations in a loop have nonzero elements on the main diagonal; however, since the criterion for equation $i$ and equation $j$ to be in the same maximal loop is that $r_{ij}^* = r_{ji}^* = 1$, where $r_{ij}^*$ is an element of the reachability matrix, the extra elements on the main diagonal of $(A \cup I)^n$ will only indicate that each equation is in a maximal loop with itself. Therefore, the maximal loops can be found in the $(A \cup I)^n$ matrix by the same procedure as discussed in Section II. Also, instead of computing each power of the $(A \cup I)$ matrix, some computation time can be saved by multiplying the matrix by itself to obtain the second, fourth, eighth, etc. powers until the power is greater than $n$. Because of the properties of Boolean algebra, any power of the matrix greater than $n$ will contain exactly the same nonzero elements as in the $n$th power, and therefore it is not necessary to compute exactly the $n$th power.

With these modifications to the methods of Section II, only two matrices, the matrix of the current power and the one of the previous power, need be stored in the computer memory at one time, and the number of matrix multiplications is reduced from $n$ multiplications for the methods of Section II to $\log n/\log 2$ multiplications for the modified method.

Actually, only one matrix need be stored if the adjacency matrix is stored initially and thereafter multiplied by itself. Matrix elements are replaced by the resulting product elements as they are computed. The product matrix obtained in this manner for the $k$th power may contain some nonzero elements which correspond to paths longer than $k$ steps instead of strictly $k$ step paths, but this will not affect the final matrix obtained corresponding to the $n$th power, since these paths would eventually be identified in any case. All of the modifications to the methods of Section II mentioned above simplify the calculations needed to obtain the reachability matrix. The procedure for identifying the maximal loops given in Section II remains the same.

In addition to identifying the maximal loops, it is necessary to order them into a sequence such that the equations of each maximal loop feed information only to the equations of maximal loops appearing thereafter in sequence. One scheme of precedence ordering of the maximal loops is accomplished by first forming an adjacency matrix of loops, P, in which the rows and columns correspond to the maximal loops (S3). The elements of the matrix P are either 1 or 0 according to the rule

$$p_{ij} = \begin{cases} 1 & \text{if the output variable of any equation in loop } j \\ & \text{appears in any equation of loop } i \text{ for } i \neq j. \\ 0 & \text{otherwise.} \end{cases}$$

The nonzero elements of a column, $j$, of the matrix P indicate the loops that are fed information by the equations of loop $j$, and the nonzero elements of any row, $i$, indicate the loops that feed information to loop $i$. The ordering procedure for the irreducible sets of equatins (maximal loops) is as follows:

(1) Choose the first column of the matrix P that contains all zeros. This column corresponds to a maximal loop, which is not fed information by any of the other maximal loops; hence the set of equations that make up the maximal loop can be solved independently of the remaining equations in the system. A maximal loop so identified is placed first in the ordering sequence for calculation.

(2) The column of matrix P identified in step 1 and the row corresponding to the same maximal loop are removed from P to obtain a reduced matrix that contains the information flow among the remaining maximal loops.

(3) Steps 1 and 2 are repeated on the reduced matrix and each maximal loop that is isolated from the others as it is detected is placed sequentially in the ordered sequence for solving the subsystems of simultaneous equations. The procedure is continued until all of the columns of the matrix P are removed.

Several other methods of partitioning have been proposed, some of which take less computation time on a digital computer.

Steward (S4) proposed one method of partitioning based on an earlier method of Sargent (S1), a method that involves tracing paths of information flow to find the maximal loops. The algorithm takes advantage of the fact that in many large systems of equations a significant number of the maximal loops contain only one equation and, in addition, in these large systems each equation contains only a small fraction of the total number of system variables. The first phase of Steward's procedure is to remove all of the rows that contain no nonzero elements, and their corresponding columns, from the adjacency matrix to obtain a reduced matrix. These rows correspond to equations that are not fed information from any of the other equations in the system and consequently can be solved independently. Each equation is placed in the precedence scheme in the order in which it is detected. Next there may be some equations that contain, besides their own output variable, only the output variables of equations already placed in the ordering sequence. These equations correspond to rows in the reduced matrix that contain no nonzero elements, and they can be placed successively in the ordering sequence and the appropriate rows and columns removed from the reduced matrix to form an even further reduced matrix. Thereafter, rows with no nonzero elements in the reduced matrix are removed to obtain another reduced matrix, and so on until eventually a reduced matrix is obtained which cannot be further reduced.

The second phase of Steward's method is to trace a path of information flow starting with any equation and going to one of the equations that feeds it. Then the path is traced from this second equation to one of the equations, which in turn feeds it, and so on. Tracing of the path is continued until an equation is encountered twice. When an equation has been encountered twice, a path has been traced around a loop of information flow that includes all of the equations encountered in the path between the first and second times the repeated equation is encountered.

In the adjacency matrix a path is traced by starting with a row, $i$, and tracing a path from that row to the first row, $j$, for which there is a nonzero element in the $j$th column of row $i$. The path is then continued in the same way from row $j$ to the first row that feeds it. When a loop is found, the set of equations in the loop must be solved together and, for the purposes of partitioning, can be considered as one composite equation with the same information for the loop fed to it from the remaining equations in the system and the same information fed by it to the remaining equations in the system. A composite equation can be formed from the adjacency matrix by removing all of the rows and columns of the reduced adjacency matrix that correspond to equations in the loop, and the by adding one row, which is the Boolean union of all of the rows so removed, and one column, which is the Boolean union of all of the columns removed, to the reduced matrix.

When a loop is found by the above procedure, it may or may not be a maximal loop and it may or may not contain the set of equations that should be solved next in sequence. If the loop is maximal and contains the set of equations to be solved next in the sequence, it will not be fed information by any of the other equations of the reduced system, and, in the reduced matrix, the row corresponding to the loop will contain all zero elements. Therefore, when a loop is found and the reduced matrix is formed, the procedure returns to the first phase and removes the rows without nonzero elements. When a row corresponding to a loop is removed from the matrix, the set of equations that correspond to all of the rows that were combined to form the composite row of the loop are placed next in the precedence order. If no rows have all zero elements, the procedure continues with the second phase by tracing a new path starting with any row of the reduced matrix and considering only the loops and equations that remain in the reduced matrix. The procedure continues until all of the equations have been removed from the reduced matrix.

To illustrate how Steward's algorithm is executed, consider the adjacency matrix of Fig. 8a. The first row contains all zero elements and can be removed from the adjacency matrix; the equation represented by row *1* is placed first in the ordering sequence for solving the equtions. The removal of the first row corresponds to a cut shown by the dashed lines in the associated graph in Fig. 8b, and the portion of the graph below the dashed line is the graph of the reduced matrix obtained by removing row *1* and column *1* as in Fig. 9a.

There are no rows with all zero elements in the reduced matrix of Fig. 9a, hence we proceed to the second phase of Steward's algorithm by starting with the first row of the reduced matrix to trace the path $2 \to 3 \to 4 \to 2$. The loop of information flow between vertices *2*, *3*, and *4* is encircled by the dashed line in the graph in Fig. 9b. The rows and columns labeled *2*, *3*, and *4* are next removed from the reduced matrix; and one row, which is the Boolean union of the rows labeled *2*, *3*, and *4*, and one column, which is the Boolean union of the columns labeled *2*, *3*, and *4*, are added to the reduced matrix to obtain the new reduced matrix of Fig. 10a. The added row and column are labeled $L_1$ in Fig. 10a.

There are still no rows with all zero elements in the reduced matrix of Fig. 10a, so we trace the path $L_1 \to 5 \to 6 \to L_1$ starting with the first row of the reduced matrix. The rows and columns labeled $L_1$, 5, and 6 are removed from the matrix, and one row and column, which are the Boolean unions of rows $L_1$, 5, and 6 and columns $L_1$, 5, and 6, respectively, are added to the matrix to obtain the reduced matrix of Fig. 11a. The added row and column are labeled $L_2$ in Fig. 11a. The first row of the reduced matrix contains all zeros. Therefore the loop found is maximal, which indicates that the equations corresponding to rows *2–6* in the adjacency matrix of Fig. 8a then must be solved simultaneously (and subsequently to the equation represented by the first row of that matrix). The row and column labeled $L_2$ in Fig. 11a are removed leaving only row *7*. The equation represented by row *7* must be solved last in the sequence. The final ordering sequence for solving the initial set equations is shown in Table I.

Steward's algorithm has proved to be quite efficient and easily programmed on a digital computer. The FORTRAN program listed in Appendix A employs Steward's algorithm to accomplish the partitioning.

Billingsley (B1) proposed an algorithm for partitioning that also involves tracing paths of information flow in the adjacency matrix. The algorithm begins, as does Steward's algorithm, by removing all of the zero rows from the adjacency matrix until a reduced matrix is obtained containing no rows with all zero elements. Then a path is traced from equation to equation (row to

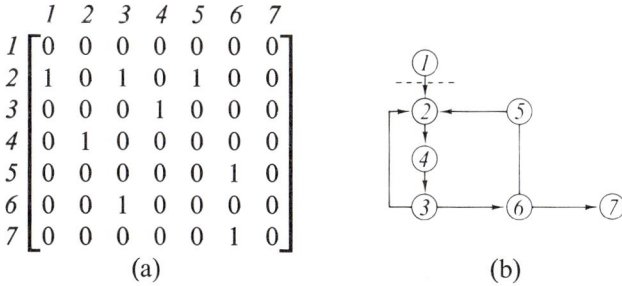

FIG. 8. Adjacency matrix and associated graph.

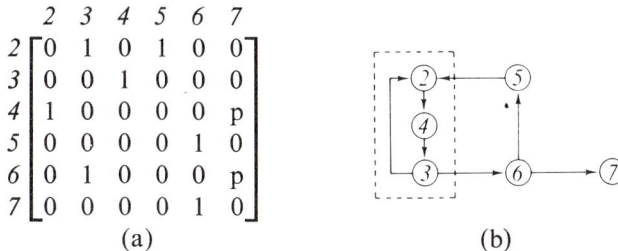

FIG. 9. Reduced matrix and associated graph.

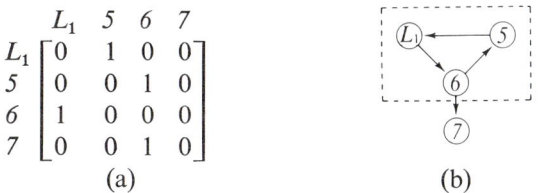

FIG. 10. Reduced matrix with loop included.

FIG. 11. Reduced matrix including maximal loop.

TABLE I

FINAL ORDERING SEQUENCE FOR SOLVING
THE INITIAL SET EQUATIONS

| Order | Equation represented by rows |
|---|---|
| 1 | *1* |
| 2 | 2, 3, 4, 5, 6, simultaneously |
| 3 | 7 |

row in the adjacency matrix) as in Steward's algorithm, except that in the Billingsley algorithm the path does not terminate when an equation is encountered a second time. When an equation is encountered a second time, it is not added to the information flow path. Instead the path tracing continues from the last equation encountered in the path that has a feed from an equation not yet included in the path. A record is kept of all of the equations that are encountered in the path but that do not yet feed the next subsequent equation encountered. These equations are the last equations found in tracing the path around a loop. The path tracing terminates when no equation in the path is fed by an equation not in the path.

Next, for each terminal equation recorded, a new path is traced starting with the terminal equation and going from equation to equation as before. Each of these paths corresponds to a set of equations in which all equations are fed by other equations in the path. Billingsley shows that the shortest of these paths corresponds to a set of equations that must be solved simultaneously (a maximal loop), and, further, this set of equations must be solved next in the ordering sequence since it is not fed by any other equation corresponding to rows of the reduced matrix. The rows and columns corresponding to equations in the shortest path in the reduced adjacency matrix are removed from the matrix and the corresponding set of equations added to the ordering sequence. No rows or columns are added back to the reduced matrix when a loop is found, as in Steward's alogrithm, because every loop found is maximal. The algorithm does proceed to remove rows with all zero elements and the corresponding columns. It terminates when all of the rows are removed from the reduced adjacency matrix.

Billingsley's alogrithm is somewhat less efficient than Steward's algorithm because in the former much longer paths must be traced in the adjacency matrix and each path must be traced twice. However, in small scale systems where the number of loops included within maximal loops is small, Billingsley's method is somewhat faster because of the time required to form the Boolean unions of rows and columns in Steward's method. Both Steward's and Billingsley's methods are significantly faster for computation on a digital computer than the methods of Section II.

## V. Identifying Disjoint Subsystems

In formulating a model of a very large process such as a whole chemical plant, the possibility exists that a subset of the system equations does not contain any variables in common with the remaining equations in the system. Such a subset of equations may physically correspond to a process unit or group of process units that are not connected in any way to the remaining units in the process. If this situation occurs, the subset of equations, which is called a **disjoint** subsystem, can be solved completely independently of the remaining equations in the system. Identification of these disjoint subsystems reduces the dimensionality of the complete system to that of the largest disjoint subsystem.

In order to clarify what a disjoint subsystem is, consider Eqs. (12):

$$\begin{aligned} f_1(x_1, x_3) &= 0 \\ f_2(x_2, x_4) &= 0 \\ f_3(x_2, x_4) &= 0 \\ f_4(x_1) &= 0 \end{aligned} \tag{12}$$

The functions $f_1$ and $f_4$ contain only variables $x_1$ and $x_3$, variables that do not appear in the remaining two equations, and therefore, $f_1$ and $f_4$ constitute a disjoint subsystem ($f_1$ itself is also disjoint). The occurrence matrix of Eq. (12) is shown in Fig. 12. If a column permutation $\pi$ (23) is performed on the

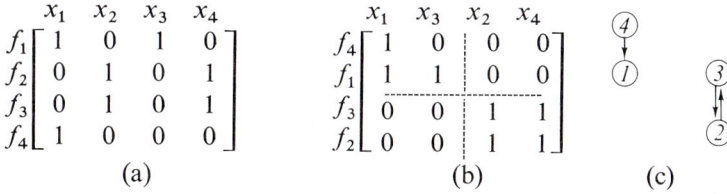

FIG. 12. Occurrence matrices of Eq. (12).

occurrence matrix of Fig. 12a followed by a row permutation $\pi^R(124)$, the occurrence matrix of Fig. 12b is obtained, which is in block diagonal form. The off-diagonal blocks contain all zero elements and each block on the diagonal represents a disjoint subsystem. If an output set was assigned to the system of equations and the corresponding graph drawn, a separate graph would be obtained for each disjoint subsystem with no edges connecting the two graphs (refer to Fig. 12c).

Of course for large systems, it is impractical to ascertain by inspection which row and column permutations will yield a block diagonal form or to draw a graph to assist in the analysis of the system. A systematic procedure for finding disjoint subsystems was proposed by Himmelblau (H1) based on an algorithm of Netter (N2) as follows:

(1) Select the column $k$ of the $n$ by $n$ occurrence matrix that contains the most nonzero entries. For example, in the occurrence matrix of Fig. 12a, choose column $1$. Columns $2$ or $4$ could have been chosen just as well.

(2) Form a new $j$ by $n$ Boolean matrix, $M^{(0)}$ as follows: For each zero entry in column $k$, reproduce the corresponding row as a row in $M^{(0)}$. For example, the second row of the occurrence matrix in Fig. 12a contains a zero in column $k = 1$ and therefore the element $m_{11}^{(0)} = 0$, element $m_{12}^{(0)} = 1$, $m_{13}^{0} = 0$, and $m_{14}^{0} = 1$ comprise the first row of $M^{(0)}$. The second row of $M^{(0)}$ would be exactly like the third row of that occurrence matrix. This process is continued until all of the rows with zero entries in column $k$ have been included as rows of $M^{(0)}$. A final row is added to $M^{(0)}$, whch is the Boolean union of all of the rows in the occurrence matrix which contain nonzero entries in column $k$. For example, rows $1$ amd $4$ of the occurrence matrix contain nonzero elements in column $1$ so that the elements of the last row of $M^{(0)}$ are $m_{31} = 1$, $m_{32} = 0$, $m_{33} = 1$, $m_{34} = 0$. Figure 13a illustrates $M^{(0)}$.

$$\begin{array}{c} \phantom{f_1 \cup f_4} \begin{array}{cccc} x_1 & x_2 & x_3 & x_4 \end{array} \\ \begin{array}{c} f_2 \\ f_3 \\ f_1 \cup f_4 \end{array} \left[ \begin{array}{cccc} 0 & 1 & 0 & 1 \\ 0 & 1 & 0 & 1 \\ 1 & 0 & 1 & 0 \end{array} \right] \\ M^{(0)} \\ (a) \end{array} \qquad \begin{array}{c} \phantom{f_1 \cup f_4} \begin{array}{cccc} x_1 & x_2 & x_3 & x_4 \end{array} \\ \begin{array}{c} f_1 \cup f_4 \\ f_2 \cup f_3 \end{array} \left[ \begin{array}{cccc} 1 & 0 & 1 & 0 \\ 0 & 1 & 0 & 1 \end{array} \right] \\ M^{(1)} \\ (b) \end{array}$$

FIG. 13. $M^{(0)}$ and $M^{(1)}$ matrices of occurrence matrix of Fig. 12.

(3) A new matrix $M^{(1)}$ is formed from $M^{(0)}$ in the same manner that $M^{(0)}$ was formed from the occurrence matrix. The column of $M^{(0)}$ containing the most nonzero elements is identified, and $M^{(1)}$ is made up of rows identical to each row of $M^{(0)}$, which contains a zero in column $k$ and one final row which is the Boolean union of the remaining rows in $M^{(0)}$. Figure 13b illustrates $M^{(1)}$. A record is kept of which rows of the original occurrence matrix have been combined to form each row of $M^{(0)}$, $M^{(1)}$, and so on.

(4) New matrices $M^{(2)}$, $M^{(3)}$, etc. are formed until the matrix $M^{(n)}$ is obtained, which contains exactly one nonzero element in each column. Each row of the matrix $M^{(n)}$ corresponds to a subset of equations in the original system. Since each column of $M^{(n)}$ contains only one nonzero element, each variable of the system appears in only one subset of equations represented by the row in which the nonzero element appears. Therefore, the subset of equations represented by a row of $M^{(n)}$ is a disjoint subsystem. For example, the matrix $M^{(1)}$ illustrated in Fig. 13b has only one nonzero element in each column

Since the first row of $M^{(1)}$ is made of rows *1* and *4* of the original occurrence matrix, $f_1$ and $f_4$ constitute a disjoint subsystem; $f_2$ and $f_3$ comprise the other disjoint subsystem.

The algorithm is easily programmed on a digital computer and is quite efficient in splitting large systems into their separate parts.

## VI. Tearing

Once the complete system of equations has been partitioned into the irreducible subsystems of simultaneous equations, it is desirable to decompose further these irreducible blocks of equations so that their solution can be simplified. The decomposition of the irreducible subsystems is called **tearing**. In the remainder of this section the subsystems of irreducible equations found by partitioning will be referred to as **blocks** to distinguish them from the smaller subsystems of simultaneous equations obtained within a block after the tearing is accomplished.

### A. The Concept of Tearing

All of the methods presented in this section are for tearing one block at a time. Specifically, the concept of tearing a variable consists of removing it from considerations as a variable in one or more of the equations in which it appears. More than one variable can be torn simultaneously. Tearing of a variable is effected by assuming its value as determined, say, by physical principles or an educated guess, followed by reordering of the equations in a block. Because the value of the variable is assumed, it must be checked at some stage in the sequence of calculation for the block, and because the subsequent value rarely agrees with the assumed value, the torn variable becomes an iterated variable.

In tearing, the objective is to wind up with less computation time required to solve the torn system compared with the time required to solve the entire block of equations simultaneously. However, the criteria for evaluating the effectiveness of the tearing are by no means so well defined as those for partitioning, where the objective is clearly to obtain the smallest possible subsystems of irreducible equations. There is no general method for determining the time needed to effect a solution of a set of equations; it is necessary to consider the particular equations involved. Any feasible method of tearing, then, must be based on criteria that are related to the solution time. Some of the more obvious criteria are:

(1) to tear a block so that the number of equations in each subsystem of resulting simultaneous equations in the torn block is mininimized; or

(2) to tear the block of equations so that the minimum number of subsystems of simultaneous equations remains in the torn block; or

(3) to tear a subsystem so that the product of the number of subsystems and the number of equations per subsystem in the torn block is minimized.

In order to illustrate tearing, consider the block of Eqs. (13):

$$
\begin{aligned}
f_1(x_1, x_3) &= 0 \\
f_2(x_1, x_2, x_3) &= 0 \\
f_3(x_2, x_3) &= 0
\end{aligned}
\quad (13)
$$

A tear can be made by removing $x_3$ from $f_1$. With the value of $x_3$ assumed, $f_1$ can be solved for $x_1$. Then $f_2$ and $f_3$ could be solved together for $x_2$ and $x_3$. Finally, the calculated value of $x_3$ could be compared to the assumed value, an iterative sequence of calculations of the values of $x_1$, $x_2$, and $x_3$ executed until the roots of $f_1, f_2$, and $f_2$ are obtained to the desired accuracy. Another equally valid tear would be to remove $x_3$ as a variable in $f_1$ and $f_2$ by assuming a value for $x_3$. Then $x_1$ could be calculated from $f_1$, $x_2$ could be calculated from $f_2$, and $x_3$ calculated from $f_3$. Again the values of the variables can be iterated until a solution is obtained. Whether a variable is torn from one or more than one equation in the block depends upon the particular criterion used for judging the effectiveness of the tear. In general, if a variable is torn from only one equation and the value of the output variable calculated in that equation is relatively insensitive to the value assumed for the torn variable, only a small number of iterations will be needed to obtain a solution. However, when a variable is torn from only one equation, the total number of tears and therefore the total number of variables that are to be iterated, will be more than if the variable were torn from several of the equations in which it appears.

We now discuss two methods of tearing the irreducible blocks.

### B. Tearing Methods for Large Scale Systems of Equations

Steward (S3) proposed an algorithm based on tearing a variable from only one equation at a time and evaluating each tear on the basis of the size of the resulting subsystems of simultaneous equations in the torn system and numerical considerations of the particular equations. Each variable is torn successively from each equation in which it appears and the effectiveness of the tear evaluated.

The algorithm is executed on the adjacency matrix of a block. In order to determine how many subsystems of simultaneous equations will remain after a tear, one must first enumerate all of the loops of information flow in the block and record which equations are included in each loop. The loops are found by tracing a path of information flow starting with any equation and going from that equation to another equation that feeds it information, i.e., opposite to the information flow, and then to an equation that feeds the second equation, and so on exactly as in Steward's algorithm for partitioning (see Section IV). When the path encounters an equation twice, a loop is found that includes all of the equations encountered in the path between the first and second encounters of the repeated equation.

For example, in the adjacency matrix of Fig. 14 we can start with the first row and trace a path from $f_1$ to $f_5$ because there is a nonzero element in row *1* and column *5*. Then the path is traced from $f_5$ back to $f_1$ because of the nonzero element in row *5*, column *1*, yielding the loop $f_1$–$f_5$–$f_1$. After this loop has been found, each path traced from the vertices in this loop will yeild a loop. We can return to the last equation found in the loop, $f_5$, and trace a path from $f_5$ to another equation that feeds it, $f_3$. The path is then continued from $f_3$ to the first equation that feeds it, $f_2$, and from $f_2$ back to $f_3$. Thus the loop $f_3$–$f_2$–$f_3$ has been found. We return to the last equation found, $f_2$, and see that no other equation feeds it, in which case we must return to the equation found just previous to $f_2$, namely $f_3$. Now $f_3$ is fed by $f_4$ and a new path can be traced to obtain a loop $f_3$–$f_4$–$f_5$–$f_3$. Steward continues this procedure until all of the feeds to each equation have been exhausted, but it is not obvious when this situation occurs except if a "tree" is drawn of the paths. For a large block, drawing a tree may not prove to be especially feasible.

Figure 15 is a tree of the paths traced from the adjacency matrix of Fig. 14. The edges drawn from $f_1$ to $f_5$ to $f_1$ comprise the path traced to obtain the $f_1$–$f_5$–$f_1$ loop, and the edges from $f_5$ to $f_3$ to $f_2$ to $f_3$ correspond to the path traced to find the $f_3$–$f_2$–$f_3$ loop above. Each equation in the tree with a line drawn under it is the terminal of a path indicating a loop, and the capital letter identifies the loop. In terms of the tree, Steward's procedure for obtaining the loops is to trace a path starting with any equation until a loop is found, and then start from the last equation in the path upstream that is not repeated to trace another path. If an equation does not have another feed, proceed to the next equation upstream in the path and trace a new path from it. Each edge that leaves an equation and the equations and edges following it in the tree comprise a **branch**. All of the loops have been found when no new path can be traced from any equation in the tree. This method of finding loops does not lend itself to programming on a digital computer very well, but Steward (S3) has proposed another method that is more systematic but too complex to be discussed here. All the loops for small blocks can be found quite readily by forming a tree.

If a variable is torn from an equation, the torn block will retain all of the same information flow as the original block except that there will no longer

$$\begin{array}{c} \phantom{f_1} \quad f_1 \quad f_2 \quad f_3 \quad f_4 \quad f_5 \quad f_6 \\ \begin{array}{c} f_1 \\ f_2 \\ f_3 \\ f_4 \\ f_5 \\ f_6 \end{array} \left[ \begin{array}{cccccc} 0 & 0 & 0 & 0 & 1 & 1 \\ 0 & 0 & 1 & 0 & 0 & 0 \\ 0 & 1 & 0 & 1 & 0 & 0 \\ 0 & 0 & 0 & 0 & 1 & 0 \\ 1 & 0 & 1 & 0 & 0 & 0 \\ 0 & 1 & 0 & 0 & 0 & 0 \end{array} \right] \end{array}$$

FIG. 14. Adjacency matrix of a block.

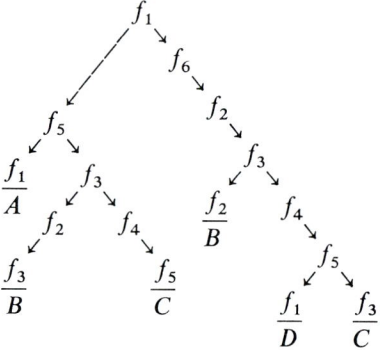

FIG. 15. Tree of paths traced in adjacency matrix of Fig. 14.

be any information flow from the equation for which the torn variable is the output variable to the equation from which the variable is torn. Therefore, the torn block has all of the same loops as the original block except for those that included the path of information flow through the torn element.

In the adjacency matrix a tear is accomplished by removing the nonzero element in the column corresponding to the equation in which the variable is the output and the row corresponding to the equation from which the variable is torn. For example, in Fig. 14 removal of the nonzero element of row *1* and column *5* corresponds to tearing the output variable of $f_5$ from $f_1$. When this tear is made there is no longer any information flow directly from $f_5$ to $f_1$ in the torn block, and loop $A$ $(f_1, f_5)$ is broken; only loops $B$, $C$, and $D$ are retained in the torn block. If the element of row *3* and column *4* is torn instead of the element of the first row and fifth column, loop $D$ is broken, and only loops $A$, $B$, and $C$ remain in this torn system.

All of the elements of the adjacency matrix can be torn in turn and the effectiveness of the tearing of each element evaluated by Steward's criteria. In addition, other tears can be made that correspond to tearing a variable from the equation in which it is the output variable. In order to evaluate these remaining tears, a new output set and adjacency matrix must be found that includes none of the elements of the original output set. The former output elements are located in the adjacency matrix and torn.

Steward's method for choosing the tears breaks the most loops and therefore minimizes the number of loops remaining in the torn system. It is easily seen that if two loops contain two or more equations in common and in the same sequence, the path of information flow between these equations is part of both loops, and if the path is broken by a tear, both loops will be broken, For example, the path from $f_5$ to $f_1$ is contained both in loops $A$ and $D$ of Fig. 15. If the element in the fifth row and first column of the adjacency matrix is removed both loops $A$ and $D$ are broken.

In order to find those tears that will break more than one loop, one must find the paths of information flow that are included as part of more than one loop. This can be done by first forming a new occurrence matrix in which the rows correspond to the loops and the columns to the equations. A nonzero entry is made in a row, $i$, and column, $j$ if equation $j$ is included in loop $i$. Figure 16a shows the occurrence matrix of the equations involved in the loops

$$\begin{array}{c} \begin{array}{cccccc} f_1 & f_2 & f_3 & f_4 & f_5 & f_6 \end{array} \\ \begin{array}{c} A \\ B \\ C \\ D \end{array} \left[ \begin{array}{cccccc} 1 & 0 & 0 & 0 & 1 & 0 \\ 0 & 1 & 1 & 0 & 0 & 0 \\ 0 & 0 & 1 & 1 & 1 & 0 \\ 1 & 1 & 1 & 1 & 1 & 1 \end{array} \right] \end{array} \qquad \begin{array}{c} \begin{array}{cccccc} f_1 & f_2 & f_3 & f_4 & f_5 & f_6 \end{array} \\ \begin{array}{c} A \\ B \\ C \\ D \end{array} \left[ \begin{array}{cccccc} 5 & 0 & 0 & 0 & 1 & 0 \\ 0 & 3 & 2 & 0 & 0 & 0 \\ 0 & 0 & 4 & 5 & 3 & 0 \\ 6 & 3 & 4 & 5 & 1 & 2 \end{array} \right] \end{array}$$

(a)            (b)

FIG. 16. Occurrence matrix of equations in loops.

of the adjacency matrix in Fig. 14. The new matrix contains the information concerning the occurrence of the respective equations in the loops.

If an equation appears in more than one loop, one tear can break all of the loops in which the equation appears if in each of these loops the equation included in both loops has exactly the same equation feeding it in the loop. For example, column 2 of the occurrence matrix in Fig. 16a contains nonzero elements in rows 2 and 4. Upon inspection of the corresponding loops, $B$ and $D$, in the tree of Fig. 15 it is seen that $f_2$ is fed by $f_3$ in both loops ($f_3$ appears immediately after $f_2$ in paths involved in loops $B$ and $D$). Therefore, loops $B$ and $D$ can be broken by tearing the output variable of $f_3$ from the equation $f_2$. Now instead of putting 1's as entries in row $i$ and column $j$ of the occurrence matrix to indicate that equation $j$ appears in loop $i$, we insert instead the number of the equation that feeds equation $j$ in loop $i$ as in Fig. 16b. Then if the same nonzero number appears more than once in the same column, all of the loops represented by the rows in which that number appears can be broken by tearing the output variable of the equation corresponding to the nonzero entry in the column from the equation represented by the column. For example, the two 3's in column 2 of the occurrence matrix of Fig. 16b indicate that loops $B$ and $D$ can be broken by tearing the output variable of Eq. (3) from Eq. (2).

Steward's algorithm is simple and very well suited to hand calculations for smaller systems. However, the computer storage and time requirements needed to tear large systems are prohibitive.

Ledet (L1) proposed a somewhat different algorithm for tearing based on tearing a variable from all the equations except one in a block. He evaluated the effectiveness of the tears in terms of finding the minimum number of tears to reduce the torn block to subsystems of single equations. In this method no initial output set in a block is chosen, and the tearing and assignment of an output set are carried out simultaneously. For example, consider the block of Eqs. (14), which must be solved simultaneously. We could tear $x_3$ from equations $f_1$ and

$$f_1(x_1, x_3) = 0$$
$$f_2(x_1, x_2, x_3) = 0 \qquad (14)$$
$$f_3(x_1, x_2, x_3) = 0$$

$f_2$ and assign $x_1$ as the output variable of $f_1$, $x_2$ as the output variable of $f_2$, and $x_3$ as the output variable of $f_3$. The torn system could be solved by (1) assuming a value of $x_3$; (2) computing the value of $x_1$ in $f_1$; (3) substituting the value of $x_1$ into $f_2$ and computing $x_2$; and (4) finally computing a value for $x_3$ in $f_3$. Then the assumed and calculated values of $x_3$ could be compared, and the value of $x_3$ adjusted by an iterative procedure until the assumed and calculated values coincided.

Another equally valid tear would be to tear $x_2$ from $f_2$ and tear $x_3$ from the equations $f_2$ and $f_3$, and assign $x_3$ as the output variable of $f_1$, $x_1$ as the output variable of $f_2$, and $x_2$ as the output variable of $f_2$. Then the system could be solved by (1) assuming values for $x_2$ and $x_3$; (2) computing the value of $x_1$ in equation $f_2$; (3) computing the value of $x_3$ in $f_1$; and (4) computing the value of $x_2$ in $f_3$. The values of $x_2$ and $x_3$ could then be established by iteration. In both of the suggested tears, the order in which the equations are to be solved for their output variables is determined by the information flow between equations in the torn system. In both cases the torn system results in equations that do not have to be solved simultaneously. But only one variable is torn in the first case as compared with two in the second, hence the first tear is a better tear according to the criteria of evaluation in Ledet's algorithm.

In terms of the occurrence matrix, the best tear is obtained when the rows and columns have been reordered so that each element on the main diagonal is nonzero (the output set of elements lies on the main diagonal) and the number of columns that contain nonzero elements above the main diagonal is minimized. The columns with nonzero elements above the main diagonal represent the torn variables. For example, Fig. 17 illustrates the occurrence

$$\begin{array}{c} & x_1 & x_2 & x_3 \\ f_1 \\ f_2 \\ f_3 \end{array} \begin{bmatrix} 1 & 0 & ① \\ 1 & 1 & ① \\ 1 & 1 & 1 \end{bmatrix}$$

FIG. 17. Occurrence matrix with best tear. Torn variables are encircled.

matrix of the set of Eqs. (14) reordered to correspond to the tearing of $x_3$ from $f_1$ and $f_2$. The encircled elements are the torn elements.

Ledet's algorithm involves two phases. In the first phase an ordering of the rows and columns of the occurrence matrix takes place according to certain optimality criteria. The second phase involves reordering the occurrence matrix to reduce the number of torn variables. The first phase carries out an initial tearing, and the second phase improves the results of the first phase. However, instead of tearing individual variables as in Steward's algorithm, Ledet's method systematically reorders the occurrence matrix as described below.

The optimality criterion given above does not specifically help to reach a decision about which variable to tear. A reordered occurrence matrix is formed one row at a time from the rows of the original occurrence matrix according to the three optimality criteria given below. As each row is placed in the reordered occurrence matrix, it is assigned an output variable (column), the assignment being subject to the restriction that a given column cannot be assigned to more than one row. Because a row may have nonzero elements not assigned as the output variable for that row or any other row, the variables (columns) corresponding to these nonzero elements must be considered as potential iterates. They will become actual iterates if the corresponding column is assigned as an output column for some subsequent row.

The optimal way to choose the rows to form the reordered occurrence matrix is as follows:

(1) Select first the row that has the minimum number of potential iterates.

(2) If two rows are equivalent under criterion 1, select the row that has a nonzero element in the column for which the total (vertically) of nonzero elements is the greatest.

(3) When a row, $i$, is chosen that introduces one or more potential iterates, the next rows are chosen according to the following conditions: (a) Each row contains one and only one nonzero element in a column not yet chosen as an output column (variable), or column which is not yet a potential

iterate. (b) Each row contains one and only one nonzero element in the output column or potential iterate column of all the previous rows up to and including row $i$.

The reason for selection of the rows by criterion 3 is that all of the rows in the set can be reordered simultaneously to permit any of the output and potential iterate columns to become actual interates in the solution of the related equations.

An effective set of iterates can be chosen as follows from the output variables (columns) and potential iterate columns identified by criterion 3 above. Each time a set of rows is selected by criterion 3, a consolidated column is formed of the Boolean sum of all the columns corresponding to the output elements of the set plus one potential iterate column. The consolidated column is substituted for the column corresponding to the potential iterate in the original occurrence matrix. When a row $j$ of the original matrix is placed in the reordered occurrence matrix, a row for which the consolidated column becomes the output variable, the proper output variable (column) for that row must be determined. It can be identified by searching the columns comprising the consolidated column and selecting any one that has a nonzero element in row $j$. This column corresponds to a variable that will become an iterate in solving the system equations.

In choosing rows according to the optimality criteria, certain feasibility conditions must also be met:

(1) Certain variables (columns) may be prohibited from being selected as iterates because, for example, the sensitivity of that equation (row) for the particular variable is too great. Those variables that the user wants to omit as iterates can be specified in the computer program in Appendix A.

(2) Certain variables (columns) may be prohibited from being selected as the output variable for a particular equation (row) because, for example, it is not feasible to solve the equation for that variable. Variables in this category can also be specified in the computer program in Appendix A.

(3) As each row is assigned an output column there must exist at least one possible assignment of the columns that have not yet been assigned as outputs to the rows not yet chosen. This condition can be determined by applying the methods of Section III to the rows and columns not yet placed in the reordered occurrence matrix.

The second phase of Ledet's algorithm is carried out on the reordered occurrence matrix resulting from the first phase of the algorithm described immediately above. Because the second phase is quite complex, it will not be discussed here, but the details can be found in the original reference. Ledet shows that in most cases the first phase of the algorithm will obtain the minimum number of tears, or very nearly the minimum number of tears, and that the extra effort of executing the second phase is seldom justified.

## C. Tearing of a System of Process Units

To this point we have viewed the mathematical representation of a process as just a system of equations and ignored the physical grouping of equations by process units or pieces of equipment. In many instances, the designer or analyst wants to treat each block on his process flow sheet as a single subsystem to and from which information flows by many routes. For example, he may have to use a computer subroutine or a nomograph to relate the unit output(s) to the unit input(s) for a single unit operation such as evaporation or heat exchange. Under these circumstances, each unit corresponds to a node in a graph and there would be several interconnections between pairs of nodes that correspond to material flows. In an adjacency matrix the rows and columns that correspond to individual units and not to individual equations as in the previous discussion.

The problem to be considered now is how to tear effectively a system of such units, units interconnected by material (and probably energy) flows. We assume that the input–output relationships are known for each unit, and that outputs *must* be calculated from the inputs. Each physical flow corresponds to several variables, and the criterion for tearing will be to minimize the number of variables that must be assumed to solve the torn system, i.e., to have the minimum number of variables associated with the total of the torn streams.

The complete process can be represented by a graph in which the vertices represent the process units and the edges represent the material streams. The edges are directed from one unit to another according to the direction in which material flows. Associated with each edge is a number that corresponds to the number of variables (concentrations, temperature, pressure, flow rates, etc.) whose values must be known to completely specify the stream. For each type of process unit (heat exchanger, junction, pump, reactor, etc.), there is a given relationship between the variables of the output streams from a unit and the variables of the input streams to the unit if the design parameters for that unit are given. For example, suppose the process consisted of a reactor followed by a distillation column, and the bottoms stream of the distillation column was recycled to the feed of the reactor. The process could be represented by the graph of Fig. 18, in which vertex *1* is a source of feed for the

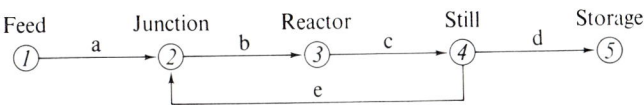

Fig. 18. Graph of reactor process.

reactor, vertex *2* is the junction of the recycle stream and the feed stream, vertex *3* is the reactor, vertex *4* is the distillation column, and vertex *5* is a product storage tank. The lower case letters—a, b, c, d, e—represent the numbers of variables associated with each stream.

Now if all the variables of the feed stream and the design parameters of each unit are given, the variables of each stream could be calculated by assuming the values of the variables in the recycle stream, calculating the variables of the output stream of the junction, calculating the variables of the output stream of the reactor, and finally calculating the variables of the output streams of the distillation column. The calculated values of the variables of the recycle stream could then be compared to the assumed values, and by iteration the proper values of all of the variables determined. In this example, we have torn the lone recycle stream to arrange a precedence order for the solution, and other equally effective tears could be determined by inspection. In more complicated processes the particular streams to tear are not so obvious and some means of optimal tearing is needed.

A process can also be represented by its adjacency matrix, which can be formed without first drawing the graph, by just assigning each unit a number and placing a nonzero entry in column $j$ and row $i$ if there is a stream directed from unit $j$ to unit $i$. Once the adjacency matrix is formed, it can be partitioned into blocks of units that must be solved simultaneously exactly as described in Section IV. It is the tearing of these blocks that is of interest here. In both of the methods of tearing discussed here the objective is to tear a block so that the minimum number of variables will have to be assumed in solving the process equations involved in the block.

Sargent (S1) proposed a method of tearing in which all of the possible orderings of units in a block are considered, and dynamic programming is used to find the optimum order. The objective of the algirithm is to order the units into a sequence such that the minimum number of variables are associated with the output streams of units that feed preceeding units in the sequence. The objective is equivalent to ordering the units so that the maximum number of variables are associated with the feed forward streams.

In general, the algorithm is a systematic procedure for building sequences of units by adding one unit at a time and evaluating each sequence after it is built in terms of the number of variables associated with all of the feed forward streams. Each sequence corresponds to an ordering of the units in a sequence opposite to that in which they are solved. Once all of the possible sequences of every unit in the block have been obtained, the best ordering is chosen according to which sequence has the maximum number of variables associated with feed forward steams.

Sargent also proved the following theorem: "If a sequence of $k$ units is in optimum order, then the subsequence of $(k - 1)$ units obtained by removing the first unit must also be in optimum order." Sargent's algorithm used this principle to avoid storing all of the permutations of orderings of a set of units by retaining instead the optimum permutation.

The algorithm begins by choosing one unit, $i$, of the block and forming a subgroup comprised of unit $i$ and one unit that feeds it material. The number of variables associated with the input stream to the first unit is recorded in terms of the number of feed forward variables associated with the subgroup. Then other subgroups are formed, in the same manner, with unit $i$ and each unit that feeds it. Next every other unit in the block is chosen in turn and subgroups formed in a similar manner. If a subgroup is formed that is just a permutation of the previously formed subgroup, the subgroup that contains the maximum number of variables associated with feed forward streams is retained and the other subgroup discarded. In this manner all of the subgroups of two units are formed. Next the subgroups of three units are formed by adding in turn each unit that feeds one of the pair in the group of two units. Again, if any subgroup is obtained which is a permutation of a previous subgroup, only the one that has the maximum number of variables associated with feed forward streams is retained. Subgroups of four, five, six, etc. units are formed in the same manner until each subgroup contains all of the units in the block. The optimum sequence is the one that has the maximum number of variables associated with feed forward streams.

Sargent's algorithm is easily programmed on a digital computer especially if a list processing language is used; however, the storage requirements are quite large. Sargent indicates that 1144 words of storage are needed to process a block of eight units, and that the storage requirements approximately double for each additional unit. Thus 16 units would take roughly 250,000 words of storage if an extended extrapolation is made.

Lee (L2) proposed a different algorithm for tearing, which also obtains the tears for the minimum number of variables associated with the torn streams. The procedure is completely different from that of Sargent's in that the loops of material flow are determined first and tears made so that all the loops are broken. The algorithm requires a previous knowledge of the loops of material flow and which units and streams are included in each loop. The loops can be found by forming the adjacency matrix for the process, and determining the loops in the matrix by Steward's method as described in Section IV.

Lee's algorithm makes use of an occurrence matrix in which the rows correspond to the loops and the columns correspond to the streams. A nonzero entry is placed in column $j$ and row $i$ if stream $j$ is included in loop $i$. The first step of the algorithm is to obtain a set of independent columns. The independent columns are obtained by removing a column, $k$, from the occurrence matrix if (1) there is another column, $j$, which corresponds to a stream with less variables than the stream represented by column $k$; and (2) column $j$

contains nonzero entries in all of the rows in which column $k$ also contains nonzero entries. Columns are removed according to these criteria until no more such columns exist in the matrix. The remaining columns then are independent. The columns removed correspond to the poorest possible tears. Tearing one of the streams corresponding to an independent column breaks at least all of the same loops as the removed column and a smaller number of variables would be assumed in solving the process equations.

In the residual occurrence matrix comprised solely of independent columns, if a row contains only one nonzero element, the stream corresponding to the column in which the nonzero element appears is the only stream that can be torn to break the loop corresponding to that row. The next step of the algorithm is (1) to remove the columns in which a lone nonzero element in a row appears and (2) to remove all of the rows that have nonzero elements in that column. Each column removed in this step corresponds to tearing the stream represented by the column, and all of the rows removed represent loops that are broken by the tear.

After all of the rows that contain only one nonzero element are removed from the matrix, the columns that are strictly contained in a set are removed. A column, $k$, is strictly contained in a set of columns if (1) at least one column in the set contains a nonzero element in each row in which column $k$ also has nonzero entries, and if (2) together, the columns of the set have less variables associated with the streams they represent than does the stream represented by column $k$. The columns representing streams with the largest number of variables that are strictly contained in the set are removed first, after which the reduced matrix is checked for rows with only one nonzero element. Whenever a row is obtained with only one nonzero element, it is removed and a new tear is performed as before.

Once all of the rows have been removed from the reduced occurrence matrix, the tears corresponding to the minimum number of variables that must be assumed in the solution of the process material and energy balances have been found.

Lee does not give a systematic method of ordering the units after tearing except by inspection of the graph with the torn streams removed. The method is satisfactory for tearing small systems by inspection but cannot be programmed without supplementation by a systematic method of ordering.

## VII. Comparison of Precedence Ordering Techniques

Various methods of carrying out the different phases of decomposition are compared in this section. Only one method of finding an output set has been reported in the literature, namely that of Steward (see Section III)

Steward's method is quite efficient and very easily programmed for a digital computer. Subroutine OUTSET in the FORTRAN program listed in Appendix A executes Steward's algorithm for finding the output set. Table II lists the execution times on a Control Data Corporation 6600 computer to obtain the output sets of several problems of increasing scale. No attempt has been made to relate the computation times versus the number of equations by a function because the times will not only vary with the number of equations but also with the number and location of the nonzero elements in the occurrence matrix. However, since the primary operation performed in the algorithm is searching of the rows and columns of the occurrence matrix, since the number of searches performed is proportional to the number of equations in the system, and since the length of each search is limited by the number of equations, the times should be roughly proportional to the square of the number of equations.

The three methods of partitioning discussed in Section IV were programmed for a digital computer, and Table III compares the execution times on the CDC 6600 computer to partition the same problems as in Table II. It was found that Steward's algorithm was the most efficient for large systems and Billingsley's method the most efficient for small systems. From Table III it can be seen that Steward's algorithm was executed in less time than Billingsley's only for two examples; however, those examples contained the greatest number of loops of information flow within the maximal loops. It can be expected that in most large scale problems the number of loops within maximal loops will be large and therefore Steward's algorithm will usually be more efficient. Subroutine STEWARD in the FORTRAN program listed in Appendix A executes Steward's algorithm.

Himmelblau's algorithm discussed in Section V for finding the disjoint subsystems is easily programmed on a digital computer and quite efficient. Because the algorithm for finding the disjoint subsystems was so straightforward and efficient, no other methods have been considered here. A computer program to find the disjoint subsystems in a process was not included in Appendix A because the analysis on disjoint subsystems should be carried out prior to the more detailed decomposition of partitioning and tearing. Table IV lists execution times for several problems by a program prepared by the authors and available from them.

The two methods for tearing systems of equations described in Section VI,B could not be compared directly because no computer program was available to execute Steward's algorithm. However, some qualitative observations can be made with respect to the computer storage requirements and computation time, which would be required for executing Steward's algorithm on a digital computer. First of all, the storage requirement for storing the numbers of all of the equations that appear in each loop would be quite large.

## TABLE II

### Time to Obtain an Output Set by Steward's Algorithm

| Number of equations | Time (sec) |
|---|---|
| 12 | 0.017 |
| 24 | 0.088 |
| 59 | 0.423 |
| 71 | 0.588 |
| 95 | 1.116 |
| 112 | 1.920 |
| 126 | 1.272 |
| 134 | 1.846 |
| 143 | 1.491 |
| 148 | 1.616 |

## TABLE III

### Computation Times for Partitioning

| Number of equations | Time (sec) Steward | Billingsley | Boolean Powers |
|---|---|---|---|
| 12 | 0.029 | 0.022 | 0.059 |
| 17 | 0.059 | 0.083 | 0.116 |
| 18 | 0.087 | 0.055 | 0.162 |
| 24 | 0.181 | 0.103 | 0.231 |
| 42 | 0.694 | 0.315 | 0.745 |
| 59 | 1.540 | 0.649 | 1.457 |
| 71 | 2.533 | 1.180 | 2.273 |
| 95 | 5.823 | 2.537 | 4.163 |
| 113 | 9.555 | 3.455 | 5.798 |
| 126 | 11.959 | 13.093 | 36.530 |

## TABLE IV

### Computation Times to Find Disjoint Subsystems

| Number of equations | Time (sec) |
|---|---|
| 42 | 0.231 |
| 59 | 0.673 |
| 71 | 1.210 |
| 95 | 2.598 |
| 113 | 4.009 |

Second, the adjacency matrix would require $n^2$ words of storage where $n$ is the number of equations in the system. In finding the loops by either the method described here or by the other method proposed by Steward (S3), all of the loops are found more than once, which tends to reduce the efficiency of the procedure. Finally, since no exact criteria was given by Steward for evaluating the effectiveness of each tear, all possible tears must be performed and the best tear chosen by inspection of all the tears. Steward's algorithm, however, is simpler than Ledet's algorithm, and therefore better suited for decomposition of small systems by hand.

In Ledet's algorithm, if both the first and second phases are carried out to obtain the absolute minimum number of tears, the computer storage required is quite large and the amount of computation time will be unreasonable. However, if only the first phase is executed, the number of iterates will be drastically reduced and in many cases the minimum number of tears will be found anyway. The computer storage required to perform the tearing with only the first phase is quite reasonable and the computer execution time is short compared to the time required to solve the system. To be specific, the storage required for the computer program listed in Appendix A is approximately $n^2 + 11,000$ words, where $n$ is the number of equations in the system. Table V lists the execution times on the CDC 6600 computer to accomplish the tearing for the same problems as in Table II. The unusual distribution of times was caused by the quite different structure of information flow in each problem.

Ledet's algorithm has the advantage that a realistic tearing can be accomplished because the output variables chosen by the algorithm can be controlled to correspond with the accustomed information flow for the solution of each

TABLE V

EXECUTION TIMES FOR TEARING BY THE FIRST PHASE OF LEDET'S ALGORITHM

| Number of equations | Time (sec) |
|---|---|
| 12 | 0.026 |
| 24 | 0.351 |
| 59 | 0.894 |
| 71 | 0.893 |
| 95 | 2.183 |
| 112 | 30.074 |
| 126 | 31.455 |
| 139 | 88.593 |
| 143 | 5.800 |
| 148 | 3.675 |

equation, and the choice of variables as iterates can be limited to those variables for which it has been determined that the solution technique will converge.

In tearing of the material and energy flows between units as described in Section VI,C, the method proposed by Sargent is more effective for digital computation. Unfortunately, though, the storage requirements are high, and because of the combinatorial approach in the ordering of units, the time required for tearing would be substantial. However, the advantages of Sargent's algorithm are (1) that it obtains the ordering with absolute minimum number of variables associated with the recycle streams, and (2) that the order of solving the units is designated. It should also be pointed out that in most processes the number of units in a process will be considerably less than the number of equations that mathematically describe the units, so that Sargent's algorithm may prove adequate.

In Lee's algorithm the amount of storage required for accomplishing the tearing would be small compared to Sargent's, but the storage required initially to find the loops would be almost as large as that required in Sargent's algorithm.

## VIII. Two Examples

To illustrate a complete decomposition for a process represented by a large number of equations is not feasible in this review because of the practical difficulties of printing the resulting matrices. Also, the problem statements would probably be longer than the review itself. Consequently, decomposition of two modest-sized processes is illustrated in this section, one a process of sufficient scale to be quite impracticable to decompose by inspection. The first example model has been taken from "Analog Simulation of the Hanford N-Reactor Plant" (S5), a documentation readily available from the Clearinghouse for Technical and Scientific Information as well as AEC depositories.

An analog simulation model of the Hanford N-reactor and its principal coolant systems had been used to study the dynamic characteristics of the plant; to determine the effects of plant, operator, and control system malfunctions; and to study various operational procedures and proposed design changes. Figure 19 shows the process flow diagram. The basic subsystems modeled were the reactor, a pump in the primary loop, and a steam generator and a dump condenser in the secondary loop. Each pump, steam generator, or dump condenser could be augmented to represent any desired number of parallel units by repetition of the model.

The size and complexity of the N-reactor plant and the limited amount of computing equipment that was available necessitated a judicious use of simplifying assumptions. For instance, primary coolant temperature transport lags were lumped into two groups, one each for the hot and cold loop legs thermodynamic effects in the secondary system condensate headers and surge

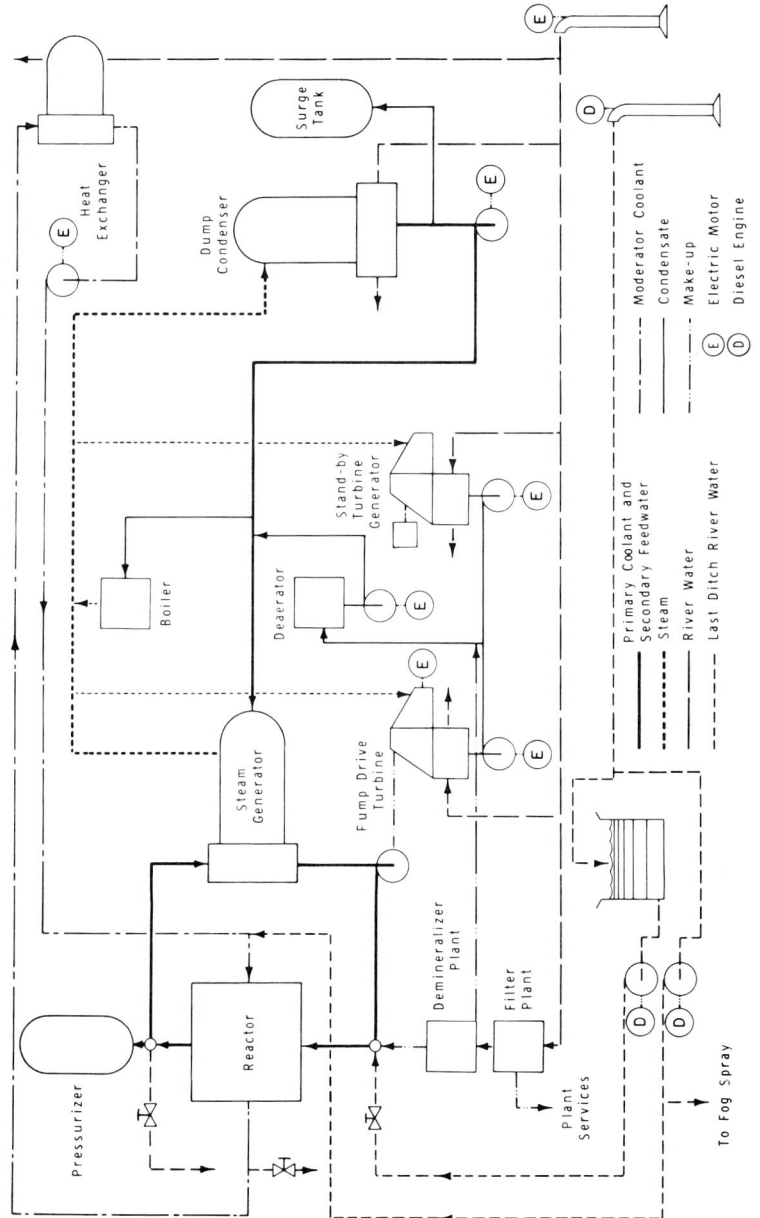

Fig. 19. Flow diagram for the Hanford N-reactor.

tanks were neglected; and the reactor was considered to act as a lumped subsystem. For many types of studies, these assumptions resulted in only second-order errors.

To provide this first example, a system of 112 equations were taken from the original reference comprising the following subsystems:

*Primary Loop:*

(1) Reactor (power level model with scram and power setback trip circuits).

(2) Primary coolant hydraulic model (one primary pump).

(3) Primary coolant thermal model with lumped transport and mixing lags (the transport lag was varied as a function of the flow rate).

(4) Steam generator primary side (one steam generator model with bypass valve flow and thermal characteristics).

(5) Automatic primary coolant flow control (both total and loop flow controllers).

(6) Automatic primary coolant temperature control with inlet and outlet temperature measurement lags and associated control devices.

*Secondary Loop:*

(1) Steam generator secondary side.

(2) Main steam header.

(3) Dump condenser and butterfly control valve.

(4) Surge tank (including auxiliary standpipe volumes).

(5) Condenser header and pumping system.

(6) Steam generator level control system.

(7) Main header pressure control system.

(8) Surge tank level and inventory for normal fill, emergency fill, and emergency spill systems.

(9) Condensate/main header differential pressure control system.

Of the 112 equations, about 12% were ordinary differential equations, 75% were algebraic equations, 10% were integral equations, and 3% were transfer functions. About 40% were nonlinear equations. The detailed list of equations and variables is far too long to be repeated here; the equations take fourteen pages to list in the original reference. The interested reader will find the derivation of the equations carefully described and the equations themselves clearly arranged by subsystem in the original reference.

In order to decompose the system equations successfully, it was necessary to prepare carefully the occurrence matrix (of equations and variables) for introduction into the computer program listed in Appendix A. The major difficulty that faces any analyst is distinguishing between system variables and parameters included in the equations. Actually, several sets of system variables could have been chosen depending upon whether one wished to study the design or simulation of the N-reactor plant. It was found that assigning an output set to the equations by hand helped immensely in isolating the parameters in the equations and also in detecting redundant equations, which were presented in the article for clarity. Groups of three engineers were assigned the task of decomposing the system by inspection, but this did not prove possible.

The recorded occurrence matrix after partitioning and tearing obtained from the computer program listed in Appendix A for the N-reactor problem is shown in Appendix B. Fifty-two seconds were required to execute the program; thirty manhours were required to set up the original occurrence matrix. The zeros in the matrix have been replaced by dots so that the nonzero elements can be more easily discerned. A box has been drawn around each set of rows and columns that represent an irreducible subsystem. A box included within another box indicates a subsystem of equations that must be solved simultaneously within a larger set of simultaneous equations. The column labeled EQN. in Appendix B is a list of the equations corresponding to the original sequentially numbered rows of the occurrence matrix. The second column, labeled VAR., is a list of the output variables corresponding to the adjacent equation in the column labeled EQN. Because the output set elements have been arranged to lie on the main diagonal, each column represents the variable on the main diagonal. The variable can be identified by proceeding across a row from the main diagonal to the column labeled VAR.

The matrix in Appendix B shows that the system was partitioned into three main blocks and nineteen subsidiary blocks, each comprising a single equation that could be solved for one variable in the system. The largest block in the system includes (a) the equations of the primary coolant loop between the steam generator and the reactor, and (b) the equations of the secondary coolant loop between the steam generator and the dump condenser. The two smaller blocks correspond to the equations of an injection system in the primary coolant loop (not shown in Fig. 19). Each loop indicated by a box in the occurrence matrix in Appendix B has one iterate and the last column (right-hand column) in the box corresponds to the iterated variable. Note that there are no nonzero elements above the main diagonal except in the columns corresponding to iterates.

The total number of tears in the entire system was 14. No further reductions in the number of tears could be found by inspection of the matrix.

As a second example, consider the classic multiple effect evaporator with forward feed, illustrated in Fig. 20. The problem is as follows:

A feed-forward triple-effect evaporator is to concentrate 5% NaOH to 50% NaOH. Feed enters at 60°F. Overall heat transfer coefficients are: $U_I = 800$, $U_{II} = 500$, $U_{III} = 300$ Btu/(hr)(ft$^2$)(°F). Steam entering steam chest I is at 125 psia, saturated (344°F) while the vacuum on effect number III is 1 psia (102°F). Ten tons/hr of 50% solution are to be produced. Find the heating areas if they are all equal.

A graph of the flows indicates no recycle streams exist, but the system cannot

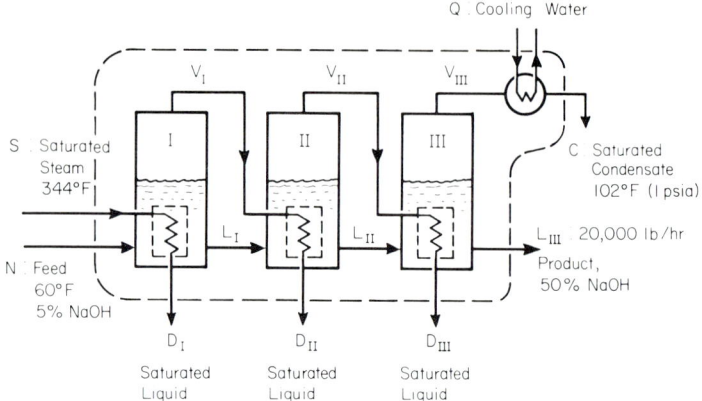

FIG. 20. Triple effect evaporator.

be decomposed successfully by a linear sequence of calculations because many of the system equations are nonlinear.

The steady state material and energy balances for the evaporator are listed in Table VI and VII, and the notation in Table VIII. Table IX lists the enthalpy relationships for the various streams as well as the boiling point versus pressure and concentration relationships in functional form for NaOH solutions and pure water. The list of unknown variables and the numbers assigned to each is given in Table X. At this stage in the analysis there are 25 equations and 27 unknown variables. Another pair of equations comes from the problem statement in which the following is given

$$A_\mathrm{I} = A_\mathrm{II} = A_\mathrm{III}$$

to give a total of 27 equations in 27 unknowns.

Table XI lists the equation numbers and the variables that appear in each equation. The numbers in parentheses are the assigned numbers of each variable. Only five of the material balances are independent, and those selected are designated as Eqs. 1–5 in Table XI. Altogether there are six independent material balances, but because the overall balance for NaOH can be solved directly for $F$, $F$ is removed from the role of a variable and the number of independent material balances can be reduced to five. The five selected were

(1) NaOH balance, effect II
(2) NaOH balance, effect III
(3) Total balance, effect I
(4) Total balance, effect II
(5) Total balance, effect III

The total balances were selected in lieu of the component water balances because fewer variables were involved in the equations. The energy balances of Table VII form Eqs 6–12, in the same order, in Table XI. Eqs. 13–18 are the enthalpy relationships for the NaOH solution in each effect and the water vapor in each effect in the same order as in Table IX. Equations 19–21 are the boiling point relationships for each effect while Eqs. 22 and 23 are the condensate–temperature relations for effects II and III, respectively. Equations 24 and 25 are the statements that the areas should be equal. The last two equations are the enthalpy versus pressure relationships for the condensate from the steam chests of effects II and III, respectively. Values of certain of the variables as specified in the problem statement, such as the steam temperature, the vapor temperature from the last evaporator, etc., have been substituted directly into the system equations.

The occurrence matrix for the evaporator problem was formed and analyzed using the computer program listed in Appendix A. Figure 21 is the

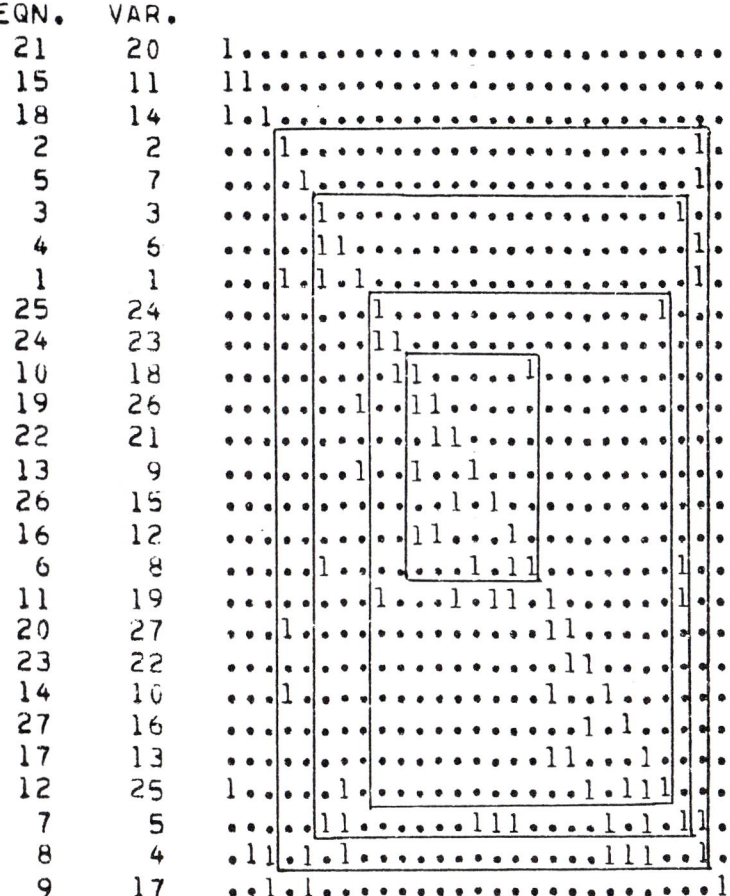

FIG. 21. Reordered occurrence matrix for the triple effect evaporator system illustrated in Fig. 20.

## TABLE VI

### Material Balances

| | |
|---|---|
| NaOH | |
| Effect I | $(0.05)(F) = \omega_I^b L_I$ |
| Effect II | $\omega_I^b L_I = \omega_{II}^b L_{II}$ |
| Effect III | $\omega_{II}^b L_{II} = (0.50)(20,000)$ |
| Over-all | $(0.05)(F) = (0.50)(20,000)$ |
| Water | |
| Effect I | $(0.95)(F) = \omega_I^w L_I + V_I$ |
| Effect II | $\omega_I^w L_I = \omega_{II}^w L_{II} + V_{II}$ |
| Effect III | $\omega_{II} L_{II}^w = (0.50)(20,000) + V_{III}$ |
| Over-all | $(0.95)(F) = (0.50)(20,000 + V_I)$ $+ V_{II} + V_{III}$ |
| Total | |
| Effect I | $F = L_I + V_I$ |
| Effect II | $L_I = L_{II} + V_{II}$ |
| Effect III | $L_{II} = L_{III} + V_{III}$ |
| Over-all | $F = L_{III} + V_I + V_{II} + V_{III}$ |

## TABLE VII

### Energy Balances

Effect I
$(1191.0)S + (26)(F) = (315.6)S + H_{L_I}L_I + H_{V_I}V_I$

Effect II
$H_{V_I}V_I + H_{L_I}L_I = H_{D_{II}}V_I + H_{L_{II}}L_{II} + H_{V_{II}}V_{II}$

Effect III
$H_{V_{II}}V_{II} + H_{L_{II}}L_{II} = H_{D_{III}}V_{II} + H_{L_{III}}(20,000) + H_{V_{III}}V_{III}$

Condenser
$H_{V_{III}}V_{III} = Q + (68)(V_{III})$

Steam chest I
$S(1191.0 - 315.6) = (800)(A_I)(344 - T_I)$

Steam chest II
$V_I(H_{V_I} - H_{D_{II}}) = (500)(A_{II})(T_{D_{II}} - T_{II})$

Steam chest III
$V_{II}(H_{V_{II}} - H_{D_{III}}) = (300)(A_{III})(T_{D_{III}} - T_{III})$

## TABLE VIII

### Notation

| | |
|---|---|
| $A$ | Area of heat transfer in steam chest |
| $F$ | Mass flow rate of feed |
| $H$ | Enthalpy per unit mass |
| $L$ | Mass flow rate of solution |
| $Q$ | Condenser duty (rate of energy removed) |
| $P$ | Pressure |
| $S$ | Mass flow rate of steam |
| $T$ | Temperature |
| $V$ | Mass flow rate of vapor |

*Greek*

| | |
|---|---|
| $\omega^b$ | Mass fraction NaOH in solution $L$ |
| $\omega^w$ | Mass fraction water in solution $L$ |

*Subscripts*

| | |
|---|---|
| I, II, III | Effect number |
| $D$ | Condensate |
| $L$ | Liquid |
| $V$ | Vapor |

## TABLE IX

### Physical Property Relationships

NaOH solution enthalpies
$H_{L_I} = f_I(T_I, \omega_I^b)$
$H_{L_{II}} = f_{II}(T_{II}, \omega_{II}^b)$   Available in the form of charts
$H_{L_{III}} = f_{III}(T_{III}, \omega_{III}^b)$

Water vapor enthalpies
$H_{V_I} = g_I(T_I, P_I)$
$H_{V_{II}} = g_{II}(T_{II}, P_{II})$   Superheated steam from the steam tables
$H_{V_{III}} = g_{III}(T_{III}, 1 \text{ psia})$

NaOH solution boiling points
$T_I = k_I(P_I, \omega_I^b)$
$T_{II} = k_{II}(P_{II}, \omega_{II}^b)$   Duhring charts and steam tables
$T_{III} = k_{III}(1 \text{ psia}, 0.5)$

Condensate temperatures
$T_{D_{II}} = g_I(P_I)$   Saturated steam from the steam tables
$T_{D_{III}} = g_{II}(P_{II})$

Condensate enthalpies
$H_{D_{II}} = h_{II}(T_{D_{II}})$   Saturated steam from the steam tables
$H_{D_{III}} = h_{III}(T_{D_{III}})$

TABLE X

Unknown System Variables and Assigned Numbers

| Variable number | Variable | Variable number | Variable |
|---|---|---|---|
| 1  | $\omega_I^b$   | 14 | $H_{V_{III}}$ |
| 2  | $\omega_{II}^b$ | 15 | $H_{D_{II}}$ |
| 3  | $L_I$          | 16 | $H_{D_{III}}$ |
| 4  | $L_{II}$       | 17 | $Q$ |
| 5  | $V_I$          | 18 | $T_I$ |
| 6  | $V_{II}$       | 19 | $T_{II}$ |
| 7  | $V_{III}$      | 20 | $T_{III}$ |
| 8  | $S$            | 21 | $T_{D_{II}}$ |
| 9  | $H_{L_I}$      | 22 | $T_{D_{III}}$ |
| 10 | $H_{L_{II}}$   | 23 | $A_I$ |
| 11 | $H_{L_{III}}$  | 24 | $A_{II}$ |
| 12 | $H_{V_I}$      | 25 | $A_{III}$ |
| 13 | $H_{V_{II}}$   | 26 | $P_I$ |
|    |                | 27 | $P_{II}$ |

reordered occurrence matrix obtained by the program. In order to obtain this sequence of equations it was necessary to specify that none of the enthalpies could be chosen as output variables of Eqs. 6–9, 11, 12, and only the enthalpies could be chosen as output variables in Eqs. 13–18 and Eqs. 26 and 27 (by prohibiting all the other variables from being selected). Specifying these prohibited variables was necessary because of the difficulty of determining temperatures, pressures, or concentrations from the steam tables or Duhring plots when the enthalpy is given. If these prohibited variables were not specified, another precedence ordering of the system equations was obtained in which an enthalpy was assigned as the output variable of one of the energy balances. In attempting to solve the system in the designated precedence order in this latter case, the value of the enthalpy calculated in the energy balance was not within the range of values given in the steam tables unless the assumed values of the iterates were quite close to the correct values.

The classical procedure to solve the triple effect evaporator problem is as follows:

(1) Assume values for $V_I$, $V_{II}$, $V_{III}$. Solve the material balances in sequence for $L_I$, $L_{II}$, $\omega_I^b$, $\omega_{II}^b$.

(2) Assume values for $T_I$, $T_{II}$. Find the values of $T_{D_{II}}$ and $T_{D_{III}}$ from the Duhring charts.

(3) Solve the energy balances simultaneously for $S$, $V_I$, $V_{II}$. Use the values of $V_I$, $V_{II}$, and the material balances to check the assumptions for step 1; iterate, if necessary.

## TABLE XI

### Equations and Unknown Variables for Triple Effect Evaporator

| Equation number | Variables[a] |
|---|---|
| 1 | $\omega_I^b(1)$, $L_I(3)$, $\omega_{II}^b(2)$, $L_{II}(4)$ |
| 2 | $\omega_{II}^b(2)$, $L_{II}(4)$ |
| 3 | $L_I(3)$, $V_I(5)$ |
| 4 | $L_I(3)$, $L_{II}(4)$, $V_{II}(6)$ |
| 5 | $L_{II}(4)$, $V_{III}(7)$ |
| 6 | $S(8)$, $H_{L_I}(9)$, $L_I(3)$, $H_{V_I}(12)$, $V_I(5)$ |
| 7 | $H_{V_I}(12)$, $V_I(5)$, $H_{L_I}(9)$, $H_{D_{II}}(15)$, $H_{L_{II}}(10)$, $L_{II}(4)$, $H_{V_{II}}(13)$, $V_{II}(6)$, $L_I(3)$ |
| 8 | $H_{V_{II}}(13)$, $V_{II}(6)$, $H_{L_{II}}(10)$, $L_{II}(4)$, $H_{D_{III}}(16)$, $V_{II}(6)$, $H_{L_{III}}(11)$, $H_{V_{III}}(14)$, $V_{III}(7)$ |
| 9 | $H_{V_{III}}(14)$, $V_{III}(7)$, $Q(17)$ |
| 10 | $S(8)$, $A_I(23)$, $T_I(18)$ |
| 11 | $V_I(5)$, $H_{V_I}(12)$, $H_{D_{II}}(15)$, $A_{II}(24)$, $T_{D_{II}}(21)$, $T_{II}(19)$ |
| 12 | $V_{II}(6)$, $H_{V_{II}}(13)$, $H_{D_{III}}(16)$, $A_{III}(25)$, $T_{D_{III}}(22)$, $T_{III}(20)$ |
| 13 | $H_{L_I}(9)$, $T_I(18)$, $\omega_I^b(1)$ |
| 14 | $H_{L_{II}}(10)$, $T_{II}(19)$, $\omega_{II}^b(2)$ |
| 15 | $H_{L_{III}}(11)$, $T_{III}(20)$ |
| 16 | $H_{V_I}(12)$, $T_I(18)$, $P_I(26)$ |
| 17 | $H_{V_{II}}(13)$, $T_{II}(19)$, $P_{II}(27)$ |
| 18 | $H_{V_{III}}(14)$, $T_{III}(20)$ |
| 19 | $T_I(18)$, $P_I(26)$, $\omega_I^b(1)$ |
| 20 | $T_{II}(19)$, $P_{II}(27)$, $\omega_{II}^b(2)$ |
| 21 | $T_{III}(20)$ |
| 22 | $T_{D_{II}}(21)$, $P_I(26)$ |
| 23 | $T_{D_{III}}(22)$, $P_{II}(27)$ |
| 24 | $A_I(23)$, $A_{II}(24)$ |
| 25 | $A_{II}(24)$, $A_{III}(25)$ |
| 26 | $H_{D_{II}}(15)$, $T_{D_{II}}(21)$ |
| 27 | $H_{D_{III}}(16)$, $T_{D_{III}}(22)$ |

[a] The numbers in parentheses are the assigned variable numbers.

(4) Solve the steam chest equations in sequence for $A_I$, $A_{II}$, $A_{III}$.

(5) Check the values of $A_I$, $A_{II}$, $A_{III}$ to see if they are equal; if not, start with step 1 or step 2 again.

(6) Calculate $Q$ from the condenser equation.

The classical method of solution is effective and the iterates are quite insensitive to the initial assumptions, if properly chosen as described in books on unit operations. No claim is made that the precedence scheme of Fig. 21 is either better or quicker than the classical method. However, the procedure can be applied automatically to other more complicated problems for which calculational schemes have not yet been developed. For large scale systems it is quite difficult to develop precedence schemes by inspection alone, and the algorithms presented here should be quite helpful in determining the best precedence schemes.

ACKNOWLEDGMENT

This work was supported by the National Science Foundation under grant NSF-GK-2849.

## Appendix A. Computer Program for Precedence Ordering

```
PROGRAM MATRIX(OUTPUT,INPUT)
C****
C****        PROGRAM MATRIX FOR DECOMPOSITION OF LARGE SCALE SYSTEMS
C****
C****        THE OBJECTIVE OF THE PROGRAM IS TO ARRANGE IN PRECEDENCE
C****        ORDER FOR SOLUTION A SYSTEM OF N EQUATIONS AND N VARIABLES.
C****
C****                          INPUT DATA CARDS
C****
C****   GROUPS OF    FIELD                    DESCRIPTION OF DATA
C****   DATA CARDS   COLUMNS    FORMAT             IN FIELD
C****   _____   _____    _____        _____
C****
C****       1         1-3         I3        NUMBER OF EQUATIONS IN THE SYSTEM.
C****
C****       2        1-3,5-7,   20(I3,     THE NONZERO ELEMENTS OF THE OCCUR-
C****                9-11, ..    1X)       RENCE MATRIX ARE GIVEN IN TERMS OF ORDER
C****                                      PAIRS ALTERNATING BETWEEN EQUATION NUMBER
C****                                      AND VARIABLE NUMBER.  THE ORIGINAL AS-
C****                                      SIGNMENT OF EQUATION AND VARIABLE NUMBERS
C****                                      ARE ARBITRARY. ENTER IN COLUMNS 1-3 AN EQ-
C****                                      UATION NUMBER.  ENTER IN COLUMNS 5-7
C****                                      ONE OF THE VARIABLE NUMBERS IN THE
C****                                      SELECTED EQUATION.  IN COLUMNS 9-11
C****                                      ENTER AN EQUATION NUMBER (THE SAME EQ-
C****                                      UATION OR ANOTHER) AND IN COLUMNS 13-
C****                                      15 ENTER THE NUMBER OF ONE OF THE VAR-
C****                                      IABLES IN THE SPECIFIED EQUATION.
C****                                      CONTINUE INTRODUCING EQUATION NUMBERS
C****                                      AND VARIABLE NUMBERS UNTIL ALL THE NONZERO
C****                                      ELEMENTS HAVE BEEN ACCOUNTED FOR.  EACH CARD
C****                                      MUST BE FILLED COMPLETELY, EXCEPT THE
C****                                      LAST CARD WHICH MUST HAVE TWO BLANK
C****                                      FIELDS, OR THE READING WILL STOP.
C****                                      NOTE THERE IS A SPACE BETWEEN FIELDS.
C****                                      EVERY NUMBER MUST BE RIGHT JUSTIFIED
C****                                      IN THE FIELD.  THE LAST CARD MUST CONTAIN AT
C****                                      LEAST TWO BLANK FIELDS FOLLOWING THE
C****                                      DATA.  THIS MAY NECESSITATE ADDING A BLANK
C****                                      CARD AT THE END OF THE DATA.
C****
C****       3        1-3,5-7,   20(I3,     THESE DATA CARDS READ IN THOSE VARIABLES
C****                9-11, ..    1X)       WHICH ARE PROHIBITED FROM BEING ASSIGN-
C****                                      ED AS OUTPUT VARIABLES IN A PARTICULAR
C****                                      EQUATION.  THE DATA MUST BE PRESENTED
C****                                      IN PAIRS AS ON DATA CARDS IN GROUP NO. 2.
C****                                      ENTER IN THE FIRST FIELD OF THE PAIR AN
C****                                      EQUATION NUMBER AND IN THE SECOND FIELD
C****                                      THE NUMBER OF A VARIABLE IN THE EQUATION
C****                                      WHICH CANNOT BE ASSIGNED AS THE OUTPUT
C****                                      VARIABLE OF THAT EQUATION.  REPEAT FOR
C****                                      ADDITIONAL PROHIBITED VARIABLES.  EVERY
C****                                      FIELD MUST BE FILLED ON EACH CARD EXCEPT
C****                                      THE LAST DATA CARD WHICH MUST HAVE AT
C****                                      LEAST TWO BLANK FIELDS FOLLOWING THE
C****                                      DATA.  THIS MAY NECESSITATE ADDING A BLANK
C****                                      CARD AT THE END OF THE DATA.
C****
C****       4        1-3,5-7,   20(I3,     THESE DATA CARDS READ IN ANY VARIABLES
C****                9-11, ..    1X)       WHICH ARE PROHIBITED FROM BEING CHOSEN
C****                                      AS ITERATES.  IN THE FIRST FIELD ENTER
C****                                      A VARIABLE NUMBER WHICH SHOULD NOT BE
C****                                      CHOSEN AS AN ITERATE.  IN THE SECOND
C****                                      FIELD ENTER ANOTHER VARIABLE NUMBER WHICH
C****                                      SHOULD NOT BE AN ITERATE.  AT LEAST ONE
C****                                      OF THESE CARDS IS REQUIRED, EVEN IF NO
C****                                      VARIABLES ARE TO BE EXCLUDED AS ITERATES.
C****                                      THIS MAY NECESSITATE ADDING A BLANK CARD
C****                                      AS DATA.
```

```
C****                         PROGRAM OUTPUT
C****                         ───────────────
C**** THE OUTPUT FROM THIS PROGRAM IS THE FINAL REORDERED OCCURRENCE
C**** MATRIX OF THE SYSTEM.
C****
C**** FORM OF THE OUTPUT  WILL APPEAR AS IN THE FOLLOWING SAMPLE
C****                        OUTPUT MATRIX (X,X)
C****                  EQNS.      VARS.
C****                    3          5        1.....1
C****                    2          1        .1.....
C****                    1          2        1.1...1
C****                    7          3        .1.1.. 
C****                    6          4        11111..
C****                    4          6        1..1.1.
C****                    5          7        ...1..1
C****
C**** THE PERIODS IN THE OCCURRENCE MATRIX INDICATE ZEROS FOR EASE OF
C**** READING.  THE MATRIX OUTPUT SET IS ON THE MAIN DIAGONAL, AND THE ROWS
C**** ARE IN THE FINAL PRECEDENCE ORDER.  THE ELEMENTS IN A MATRIX ROW ARE
C**** THOSE VARIABLES ASSOCIATED WITH THE EQUATION GIVEN UNDER THE
C**** COLUMN HEADED EQNS.  THE COLUMN HEADED VARS. IDENTIFIES THE OUTPUT
C**** VARIABLE FOR THAT EQUATION.  BY TRANSPOSING THE VECTOR OF NUMBERS
C**** UNDER THE COLUMN VARS. AND PLACING IT ACROSS THE TOP OF THE MATRIX(X,X)
C**** YOU CAN IDENTIFY EACH OF THE VARIABLES BY NUMBER.  EACH COLUMN WITH
C**** NONZERO ELEMENTS ABOVE THE MAIN DIAGONAL REPRESENTS AN ITERATED
C**** VARIABLE.
C****
C**** IF THERE ARE MORE THAN 112 EQUATIONS IN THE SYSTEM, ONLY 112 COL-
C**** UMNS OF THE MATRIX WILL BE PRINTED ON EACH PAGE.  THEN THE MATRIX
C**** SHOULD BE PUT TOGETHER AS FOLLOWS.
C****       OUTPUT MATRIX (1,1)   OUTPUT MATRIX (1,2) ... ETC.
C****
C**** IF THE PROGRAM CANNOT FIND AN OUTPUT SET, A LIST OF EQUATION
C**** NUMBERS IS GIVEN WHICH COMPRISES THE SUBSET IN WHICH THE EQUATIONS
C**** ARE GREATER IN NUMBER THAN THE NUMBER OF VARIABLES APPEARING IN
C**** ALL OF THE EQUATIONS OF THE SUBSET.  THE PROBLEM MUST BE REVIEWED
C**** TO MAKE IT DETERMINATE.  (IF THE NUMBER OF EQUATIONS IS LESS THAN
C**** THE NUMBER OF VARIABLES, THE PROGRAM WILL PROCESS THE EQUATIONS.)
C****
      DIMENSION K(162,162),IND(162),JND(162),ISTOR(665),IOUTST(162),INZ(
     110),JNZ(20)
      COMMON IND,JND,N,NPO,K,IOUTST
      COMMON /A/ ISTOR, IL
      COMMON /L/ LOOP(162,11)
      COMMON /T/ LI(20), LO(20)
      COMMON /N/ KSTOR(665),IKJ
      DIMENSION IIND(162), JJND(162)
C**** THE MAIN PROGRAM MATRIX ONLY READS THE OCCURRENCE MATRIX AND CALLS
C**** SUBROUTINES TO DO THE DECOMPOSITION.
C**** THE OCCURRENCE MATRIX IS STORED IN MATRIX K.  THE MATRIX LOOP IS
C**** USED TO STORE THE NON OUTPUT ELEMENTS OF THE EQUATIONS.  THE VEC-
C**** TOR ISTOR IS USED TO STORE THE NONZERO ELEMENTS OF THE OCCURRENCE
C**** MATRIX.  N IS THE NUMBER OF EQUATIONS.  IND AND JND ARE USED TO
C**** KEEP THE CORRESPONDENCE BETWEEN THE EQUATION NUMBERS AND THE ROW
C**** NUMBERS AND THE VARIABLE NUMBERS AND THE COLUMN NUMBERS RESPECT-
C**** IVELY.  THE VECTOR IOUTST IS USED TO STORE THE OUTPUT SET.
    1 READ 5,N
C**** THE NUMBER OF EQUATIONS IS READ HERE.
      NPO = N +1
    5 FORMAT(I3,I17)
      NPT = N + 2
      DO 10 I = 1,N
      IND(I) = I
      IOUTST(I) = 0
      JND(I) = I
      DO 10 J = 1,N
   10 K(I,J) = 0
      IL = 0
C**** THE NONZERO ELEMENTS OF THE OCCURRENCE MATRIX ARE READ HERE.
   40 READ 50 ,(INZ(L),JNZ(L),L= 1,10)
   50 FORMAT(20(I3,1X))
      DO 60 L = 1,10
      IF(INZ(L).EQ.0) GO TO 70
```

# DECOMPOSITION OF LARGE SCALE SYSTEMS 239

```
           I = INZ(L)
           J = JNZ(L)
           K(I,J) = 1
           IL = IL +1
C**** THE NONZERO ELEMENTS ARE STORED IN THE VECTOR ISTOR HERE.
           ISTOR(IL) = 1000*I + J
        60 CONTINUE
           GO TO 40
C**** THE NONZERO ELEMENTS ARE ORDERED FOR EASY REFERENCE.
        70 DO 80 I = 1,IL
           DO 80 J=I,IL
           IF(ISTOR(I).LE.ISTOR(J)) GO TO 80
           KDUM = ISTOR(I)
           ISTOR(I) = ISTOR(J)
           ISTOR(J) = KDUM
        80 CONTINUE
           DO 81 I = 1,N
           DO 81 J = 1,11
        81 LOOP(I,J) = 0
           IKJ = 0
C**** THE NON OUTPUT ELEMENTS OF THE OCCURRENCE MATRIX ARE READ HERE.
        82 READ 50,(INZ(L),JNZ(L),L = 1,10)
           DO 83 L = 1,10
           IF(INZ(L).EQ.0) GO TO 85
           I = INZ(L)
           IKJ = IKJ +1
           KSTOR(IKJ) = I*1000 + JNZ(L)
C**** THE NUMBER OF NON OUTPUT ELEMENTS IN EACH EQUATION IS COMPUTED
C**** AND STORED IN THE 11TH COLUMN OF LOOP.
           LOOP(I,11) = LOOP(I,11) +1
           IA = LOOP(I,11)
        83 LOOP(I,IA) = JNZ(L)
           GO TO 82
        85 CONTINUE
           DO 87 I = 1,IKJ
           DO 87 J = 1,IKJ
           IF(KSTOR(I).LE.KSTOR(J)) GO TO 87
           KDUM = KSTOR(I)
           KSTOR(I) = KSTOR(J)
           KSTOR(J)    = KDUM
        87 CONTINUE
C**** THE NON OUTPUT ELEMENTS ARE REMOVED BEFORE FINDING AN OUTPUT SET.
           DO 91 I = 1,N
           IE = LOOP(I,11)
           IF(IE.EQ.0) GO TO 91
           DO 92 L = 1,IE
           J = LOOP(I,L)
        92 K(I,J) = 0
        91 CONTINUE
C**** SUBROUTINE OUTSET FINDS THE OUTPUT SET, IF NONE CAN BE FOUND FLAG
C**** IS SET = 1 AND THE PROGRAM TERMINATES.
           CALL OUTSET(1,N,FLAG,1)
           IF(FLAG.NE.0) GO TO 999
C**** THE NON OUTPUT ELEMENTS ARE RESTORED IN THE OCCURRENCE MATRIX BE-
C**** FORE PARTITIONING.
       115 DO 117 I = 1,N
           IA = LOOP(I,11)
           IF(IA.EQ.0) GO TO 117
           DO 116 L = 1,IA
           JK= LOOP(I,L)
           DO 118 J = 1,N
           IF(JND(J).EQ.JK) GO TO 116
       118 CONTINUE
       116 K(I,J) = 1
       117 CONTINUE
           MIN = 1
           MAX = N
           INDEX = 0
C**** THE SUM OF THE NONZERO ELEMENTS IN THE ROWS AND COLUMNS ARE COMP-
C**** UTED AND STORED IN THE N+1 COLUMN AND ROW OF K RESPECTIVELY BY
C**** SUMRC.
       120 CALL SUMRC(MIN,MAX)
C**** SUBROUTINE STEWRD PARTITIONS THE SYSTEM BY STEWARD S ALGORITHM.
           CALL STEWRD(MIN,MAX)
```

```
C****  SUBROUTINE PACK RESTORES THE ORIGINAL NONZERO ELEMENTS OF THE OC-
C****  CURRENCE MATRIX.
       CALL PACK(MIN,MAX,MIN,MAX,1)
       DO 130 J = 1,N
  130  K(N+3,J) = 0
C****  THE VARIABLES WHICH CANNOT BE TAKEN AS ITERATES ARE READ HERE.
  135  READ 50,(JNZ(L),L=1,20)
       DO 150 L = 1,20
       IF(JNZ(L).EQ.0) GO TO 155
       DO 140 J = 1,N
       IF(JND(J).EQ.JNZ(L)) GO TO 145
  140  CONTINUE
  145  K(N+3,J) = 1
  150  CONTINUE
  155  CONTINUE
       DO 200 I = MIN,MAX
       IA = LOOP(I,11)
       IF(IA.EQ.0) GO TO 200
       DO 190 L = 1,IA
       JK= LOOP(I,L)
       DO 180 J = 1,N
       IF(JND(J).EQ.JK) GO TO 190
  180  CONTINUE
  190  K(I,J) = 0
  200  CONTINUE
  250  MAX = 0
C****  THE PARTITIONS ARE INDICATED IN THE N+2 COLUMN OF K AND THE FOL-
C****  LOWING SECTION LOCATES ONE PARTITION AT A TIME FOR TEARING.
       DO 300 I = 2,N
       IF ( K(I,NPT).GE.1000) GO TO 300
       IF(K(I,NPT).NE.K(I-1,NPT)) MIN = I
       IF(K(I,NPT) .EQ.K(I-1,NPT)) MAX = I
       IF((K(I-1,NPT).EQ.K(I,NPT)).AND.(K(I+1,NPT).NE.K(I,NPT)))GO TO 350
  300  CONTINUE
  350  IF((INDEX.EQ.0).AND.(MIN.EQ.0)) MIN = 1
       ICMP = MAX - MIN
       IF(MAX.EQ.0.OR.MAX.GE.159) GO TO 500
       IF(ICMP.GT.1) GO TO 400
       K(MIN,NPT) = 1000 + MIN
       K(MAX,NPT) = 2000 + MAX
       GO TO 250
  400  CONTINUE
       INDEX = 1
C****  THE ROWS AND COLUMNS OF K ARE SUMMED AS BEFORE.
       CALL SUMRC(MIN,MAX)
C****  SUBROUTINE TEAR EXECUTES THE INITIAL PHASE OF LEDET S ALGORITHM.
       CALL TEAR(MIN,MAX,L)
C****  THIS SECTION REORDERS THE ROWS WITH THE CHECK EQUATIONS APPEARING
C****  AS SOON AS POSSIBLE.
       DO 990 I = 1,L
       ILL = LI(I)
       IUU = LO(I)
       IF(IUU.LT.ILL) GO TO 990
       DO 980 J = ILL,IUU
       IF(K(NP3,J).EQ.1) GO TO 980
       DO 970 IK = MIN,MAX
  970  K(IK,MAX-I+1) = K(IK,MAX-I+1) .OR.K(IK,J)
  980  CONTINUE
  990  CONTINUE
       LI(L+1) = MAX-L+1
       DO 1400 IK = 1,L
       I = L -IK +1
       MXL = MAX -I
       DO 1100 IJ = MIN,MXL
       J = MXL - IJ + MIN
       IF(K(MAX-I+1,J).EQ.1) GO TO 1110
 1100  CONTINUE
 1110  IF(J.LE.MAX-L) GO TO 1180
       JDM = J -1
       DO 1120 II = MIN,JDM
       IJ = JDM -II + MIN
       IF(K(MAX-I+1,IJ).EQ.1) GO TO 1130
 1120  CONTINUE
 1130  LO(I) = LO(MAX-J+1)
```

```
      IAM = MAX -J
      IPO = I +1
      DO 1150 IAC = IPO, IAM
      IF(LO(I).GE. LO(IAC)) GO TO 1150
      LO(I) = LO(IAC)
 1150 CONTINUE
      IF(IJ.GT.MAX-L) GO TO 1400
      IF(IJ+1.GT.LO(I)) LO(I) = IJ +1
      GO TO 1400
 1180 IF(J.GE.LI(I+1)) GO TO 1200
      LO(I) = J +1
      GO TO 1400
 1200 DO 1220 IA = 2,L
      IF(J.GE.LI(IA-1).AND.J.LT.LI(IA)) GO TO 1230
 1220 CONTINUE
      IA = L +1
 1230 LO(I) = LO(IA-1)
      IF(IA-1.EQ.I+1) GO TO 1250
      IPO = I +1
      IAM = IA -2
      DO 1240 IAC = IPO , IAM
      IF(LO(I).GE.LO(IAC)) GO TO 1240
      LO(I) = LO(IAC)
 1240 CONTINUE
 1250 IF(J+1.GT.LO(I)) LO(I) = J +1
 1400 CONTINUE
      ID = MIN -1
      MXL = MAX -L+1
      DO 1600 I = MIN,MXL
      DO 1450 IJ = 1,L
      J = L-IJ +1
      IF(I.NE.LO(J)) GO TO 1450
      ID = ID +1
      IIND(ID) = IND(MAX-J+1)
      JJND(ID) = JND(MAX-J+1)
 1450 CONTINUE
      IF(I.EQ.MXL) GO TO 1600
      ID = ID +1
      IIND(ID) = IND(I)
      JJND(ID) = JND(I)
 1600 CONTINUE
      DO 1650 I = MIN,MAX
      IND(I) = IIND(I)
 1650 JND(I) =JJND(I)
      CALL PACK(MIN,MAX,MIN,MAX,2)
      GO TO 250
  500 CONTINUE
      CALL PACK(1,N,1,N,2)
C**** AFTER ALL OF THE TEARING IS DONE, SUBROUTINE PRNT PRINTS THE RE-
C**** ORDERED OCCURRENCE MATRIX.
      CALL PRNT(1,N,1)
  999 CONTINUE
C**** RETURN TO READ ANOTHER OCCURRENCE MATRIX.
      GO TO 1
      END
C****
C****
      SUBROUTINE PACK(MINR,MAXR,MINC,MAXC,KASE)
C**** THIS SUBROUTINE REPLACES THE ORIGINAL ELEMENTS IN ANY PART OF
C**** MATRIX K.  MINR AND MAXR ARE THE FIRST AND LAST ROWS OF THE POR-
C**** TION TO BE REPLACED, AND MINC AND MAXC ARE THE FIRST AND LAST COL-
C**** UMNS OF THE PORTION TO BE REPLACED.  KASE INDICATES WHETHER TO RE-
C**** PACK ACCORDING TO THE OUTPUT SET  IN OUTST OR ACCORDING TO THE
C**** ORDER IN IND AND JND.
      DIMENSION K(162,162),IND(162),JND(162),ISTOR(665),IOUTST(162),INZ(
     110),JNZ(10)
      COMMON /A/ ISTOR, IL
      COMMON IND,JND,N,NPO,K,IOUTST
      COMMON /N/ KSTOR(665),IKJ
      COMMON /L/ LOOP(162,11)
      IRW(I) = ISTOR(I)/1000
      ICL(I) = ISTOR(I) -ISTOR(I)/1000*1000
      DO 100 I = MINR,MAXR
      IF(KASE.EQ.2) GO TO 50
```

```
              JI = IND(I)
              JND(I) = IOUTST(JI)
           50 DO 100 J = MINC,MAXC
          100 K(I,J) = 0
              DO 350 IK = MINR,MAXR
              DO 300 L = 1,IL
              IR = IRW(L)
              IF(IR - IND(IK)) 300,250,350
          250 JC = ICL(L)
              DO 275 JK = MINC,MAXC
              IF(JC.EQ.JND(JK)) GO TO 280
          275 CONTINUE
              GO TO 300
          280 K(IK,JK) = 1
          300 CONTINUE
          350 CONTINUE
              DO 400 I = MINR,MAXR
              DO 400 J = 1,11
          400 LOOP(I,J) = 0
              IF(IKJ.EQ.0) GO TO 600
              DO 500 I = MINR,MAXR
              IB = 0
              DO 450 J = 1,IKJ
              IA = KSTOR(J)/1000
              IF ( IA-IND(I)) 450,430,500
          430 IB = IB +1
              LOOP(I,11) = IB
              LOOP(I,IB) = KSTOR(J) - KSTOR(J)/1000*1000
          450 CONTINUE
          500 CONTINUE
          600 CONTINUE
              RETURN
              END
C****
C****
              SUBROUTINE PRNT(MIN,MAX, IOP)
C****    THIS SUBROUTINE PRINTS THE OCCURRENCE MATRIX.
              DIMENSION K(162,162),IND(162),JND(162)
              COMMON IND,JND,N,NPO,K
              NA = MAX -MIN  + 1
              NN = NA/112-(1/(1 + NA- NA/112*112)) +1
              DO 14 MMK = 1 , NN
              ID = 1
              PRINT 6
            6 FORMAT(1H1)
              IF(IOP)1,1,3
            1 PRINT 2,ID,MMK
            2 FORMAT( 44X,  *INPUT MATRIX (*,I2,*,*,I2,*)*)
              PRINT 16
           16 FORMAT(                 2X,*EQN.*,2X,*VAR.*)
              GO TO 5
            3 PRINT 4,ID,MMK
            4 FORMAT(44X,   *OUTPUT MATRIX (*,I2,*,*,I2,*)*)
              PRINT 16
            5 CONTINUE
              JLOW =112*(MMK - 1) + MIN
              IF(MMK - NN) 11 , 12 , 12
           12 JUP = MAX
              GO TO 13
           11 JUP = JLOW + 111
           13 DO 14 I = 1 , N
              DO 20 J = JLOW,JUP
              IF(K(I,J).EQ.1) GO TO 19
              K(I,J) = 1H.
              GO TO 20
           19 K(I,J) = 1H1
           20 CONTINUE
           15 FORMAT(2X,I3,3X,I3,3X,112A1)
              PRINT 15,             IND(I),JND(I),(K(I,J),J=JLOW,JUP)
           14 CONTINUE
              DO 100 I = MIN,MAX
              DO 100 J = MIN,MAX
              IF(K(I,J).EQ.1H1) GO TO 50
              K(I,J) = 0
```

# DECOMPOSITION OF LARGE SCALE SYSTEMS 243

```
         GO TO 100
      50 K(I,J) = 1
     100 CONTINUE
         RETURN
         END
C****
C****
         SUBROUTINE TEAR (MIN,MAX,L)
C**** THIS SUBROUTINE PERFORMS THE TEARING ON A PARTITION BY LEDET S
C**** ALGORITHM.
         DIMENSION K(162,162),IND(162),JND(162),ISTOR(665),IOUTST(162),INZ(
        1 10),JNZ(20),KSET(162)
         COMMON IND,JND,N,NPO,K,IOUTST
         COMMON /T/ LI(20),LO(20)
         COMMON /L/ LOOP(162,11)
         NPT = N +2
         NP3 = N +3
         NP4 = N +4
         ID = MIN -1
         L = 0
C**** THE ELEMENTS OF THE N+3 COLUMN OF K ARE SET EQUAL TO THE NUMBER
C**** OF NEW VARIABLES IN EACH ROW.
         DO 10 I = MIN,MAX
         K(NP4,I) = 0
         K(I,NP3) = 0
         DO 10 J = MIN,MAX
         K(NP4,I) = K(NP4,I) + K(J,I)
      10 K(I,NP3) = K(I,NP3) + K(I,J)
         DO 12 I = MIN,MAX
      12 KSET(I) = K(I,NP3)
         IK = 0
C**** IF ALL OF THE EQUATIONS HAVE BEEN REORDERED, STOP.
       1 IF(ID.GE.MAX-L) GO TO 900
         IDP = ID +1
C**** THE ELEMENTS OF THE N+4 COLUMN OF K ARE SET EQUAL TO THE NUMBER
C**** OF VARIABLES IN EACH ROW WHICH ARE INCLUDED IN THE LAST OPEN LOOP
         DO 20 I = ID,MAX
         KJD = 0
         IU = LOOP(I,11)
         IF(IU.EQ.0) GO TO 19
         DO 18 ILA = 1,IU
         MXL = MAX -L+IK -1
         DO 16 JLA = ID,MXL
         IF(JND(JLA).EQ.LOOP(I,ILA)) GO TO 17
      16 CONTINUE
         GO TO 18
      17 KJD = KJD +1
      18 CONTINUE
         IF(KSET(I).EQ.1.AND.KJD.EQ.1)KJD = 2
      19 CONTINUE
      20 K(I,NP4) = KSET( I) + KJD
         KSUM = 1
         IK = 1
      25 IF(ID.GE.MAX-L) GO TO 900
         INDEX = 1
         IDP = ID +1
C**** THE ROWS ARE SEARCHED HERE TO FIND THE ROW WHICH CONTAINS IK VAR-
C**** IABLES.
         DO 700 I = IDP, MAX
         IF(K(I,NP3).NE.IK) GO TO 700
C**** THIS BRANCH ALLOWS FOR CHOOSING THE EQUATION WITH THE MINIMUM
C**** NUMBER OF NONZERO ELEMENTS IN THE LAST LOOP.
         KOMPAR = 0
         MBL = MAX -L
         DO 29 IAC = I,MAX
         IF(K(IAC,NP3).NE.IK) GO TO 29
         ISUM = 0
         DO 26 JAC = IDP,MBL
      26 ISUM = ISUM + K(IAC,JAC)*K(NP4,JAC)
         IF(ISUM.LE.KOMPAR) GO TO 29
         KOMPAR = ISUM
         CALL ROWXC(IAC,I)
      29 CONTINUE
         IF(K(I,NP4) -KSUM) 700,35,30
```

```
      30 INDEX = 2
         GO TO 700
C**** ROW I IS CHECKED TO SEE IF IT HAS ANY NON OUTPUT VARIABLES WHICH
C**** HAVE NOT YET BEEN SPECIFIED.  IF ANY EXIST THE SEARCH CONTINUES.
      35 DO 45 IL = 1,10
         IF(LOOP(I,IL).EQ.0) GO TO 45
         DO 40 JL = 1,ID
         IF(JND(JL).EQ.LOOP(I,IL)) GO TO 45
      40 CONTINUE
         DO 42 JL = 1,L
         IF(JND(MAX-JL+1).EQ.LOOP(I,IL)) GO TO 45
      42 CONTINUE
         GO TO 700
      45 CONTINUE
         IF(IK.LT.2) GO TO 90
C**** THIS BRANCH IS TAKEN IF A NEW LOOP IS OPENED BY TAKING ROW I.
         IKM = IK -1
         IDP = ID +1
         MXA = MAX -L
         KOUNT = 0
         DO 50 IA = IDP , MXA
         IF(K(I,IA).EQ.1.AND.K(NP3,IA).NE.1) KOUNT = KOUNT +1
      50 CONTINUE
         IF(KOUNT.LT.IKM) GO TO 700
         DO 53 IA = IDP ,MAX
      53 KSET(IA) = K(IA,NP3)
C**** THE COLUMNS OF THE ITERATES ARE PLACED AT THE END OF THE BLOCK
C**** AND THE COLUMN OF THE OUTPUT VARIABLE PLACED NEXT IN THE SEQUENCE
C**** THE COLUMNS REMOVED ARE SUBTRACTED FROM THE N+3 COLUMN.
         DO 60 IA = IDP , MXA
         JB = MXA +IDP -IA
         IF(K(I,JB).NE.1.OR.K(NP3,JB).EQ.1) GO TO 60
         L = L +1
         CALL COLXC(JB,MAX-L+1)
         LI(L) = ID +1
         LO(L) = 0
         DO 55 IB = IDP, MAX
      55 K(IB,NP3) = K(IB,NP3) - K(IB,MAX-L+1)
      60 CONTINUE
         IF(KOUNT.GT.IKM) GO TO 75
         LO(L) = ID +1
         DO 70 JA = IDP, MXA
         IF(K(I,JA).NE.1) GO TO 70
C**** THE BOOLEAN SUM OF ALL OF THE COLUMNS WHICH CAN BE ITERATES IS
C**** COMPUTED AND PLACED IN THE PROPER COLUMN.
         DO 65 IA = IDP,MAX
      65 K(IA,NP3) = K(IA,NP3) - K(IA,JA)
         GO TO 85
      70 CONTINUE
      75 JA = MAX -L+1
         L = L -1
         LO(L) = ID     +1
         DO 80 IA = ID,MAX
      80 K(IA,MAX-L+1) = K(IA,MAX-L+1) .OR.K(IA,JA)
      85 CALL COLXC(ID +1,JA)
         GO TO 145
C**** THIS BRANCH IS TAKEN IF EQUATION I HAS ONLY ONE NEW VARIABLE.
      90 MXA = MAX -L
         IDP = ID +1
         DO 100 JA = IDP,MXA
         IF(K(I,JA).NE.1) GO TO 100
C**** THE COLUMN IS SUBTRACTED FROM THE N+3 COLUMN.
         DO 95 IA = IDP,MAX
      95 K(IA,NP3) = K(IA,NP3) -K(IA,JA)
         GO TO 110
     100 CONTINUE
C**** THE COLUMN IS PLACED NEXT IN THE SEQUENCE.
     110 CALL COLXC(ID+1,JA)
         IF(KSUM.NE.2.OR.K(NP3,ID+1).EQ.1) GO TO 145
         IF(LO(L).EQ.0) GO TO 125
         ILL = LI(L)
         DO 120 IA = ILL,ID
         IF(K(IA,NP4).GT.2) GO TO 145
```

```
      120 CONTINUE
          GO TO 135
      125 ILL = LI(L)
          DO 130 IA = ILL , ID
          IF(K(IA,NP4).GT.2) GO TO 145
      130 CONTINUE
      135 LO(L) = ID +1
C**** IF THE VARIABLE CAN BE AN ITERATE THE BOOLEAN UNION OF THE COLUMN
C**** AND THE LAST LOOP COLUMN IS PLACED IN THE LAST LOOP COLUMN.
          DO 140 IA = IDP,MAX
      140 K(IA,MAX-L+1) = K(IA,MAX-L+1) .OR.K(IA,ID+1)
      145 ID = ID +1
C**** ROW I IS PLACED NEXT IN THE SEQUENCE.
          CALL ROWXC(ID,I)
          KDUM = KSET (ID)
          KSET(ID) = KSET(I)
          KSET(I) = KDUM
          IA = 1
C**** THE REMAINING EQUATIONS ARE CHECKED TO SEE IF AN OUTPUT SET CAN
C**** BE ASSIGNED.
      150 CALL OUTSET(ID+1,MAX,FLAG,2)
          IF(FLAG.EQ.0) GO TO 250
C**** IF AN OUTPUT SET COULD NOT BE FOUND AND THERE IS SOME EQUATION
C**** WHICH CONTAINED MORE THAN ONE ITERATE THE ITERATES OF THAT EQUA-
C**** TION ARE REARRANGED AND ANOTHER CHECK MADE FOR AN OUTPUT SET.
          LM = L -1
          IF(IA.GE.L) GO TO 260
          DO 240 IB = IA,LM
          IF(LI(IB).NE.LI(IB+1)) GO TO 240
          IDP = ID +1
          DO 160 IC = IDP , MAX
          IF(K(IC,MAX-IB+1).EQ.1) GO TO 240
      160 CONTINUE
          IC = LI(IB)
          IF(K(NP3,IC).EQ.1) GO TO 240
          CALL COLXC(IC,MAX-IB+1)
          IC = IB +1
      165 IF(LI(IC).NE.LI(IC+1).OR.IC.EQ.L) GO TO 170
          IC = IC +1
          GO TO 165
      170 IL = LI(IC)
          IU = LO(IC)
          DO 185 JC = IL , IU
          IF(K(NP3,JC).EQ.1) GO TO 185
          DO 180 IE = MIN,MAX
      180 K(IE,MAX -IC +1) = K(IE,MAX-IC+1).OR.K(IE,JC)
      185 CONTINUE
          IA = IC
          GO TO 150
      240 CONTINUE
C**** THE ITERATES COULD NOT BE REARRANGED AND EQUATION I CANNOT BE PUT
C**** NEXT IN THE SEQUENCE.
          GO TO 260
C**** IF EQUATION I CAN BE PUT NEXT IN THE SEQUENCE, RETURN TO SEARCH
C**** ROWS AGAIN.
      250 IF(IK.GT.1) 1, 255
      255 KSUM = 1
          GO TO 25
C**** THIS BRANCH IS TAKEN WHEN AN OUTPUT SET COULD NOT BE FOUND AND
C**** THE SYSTEM IS RESTORED TO THE WAY IT WAS BEFORE EQUATION I WAS
C**** CHOSEN.
      260 IF(IK.GT.1) GO TO 290
          IF(KSUM.NE.2.OR.K(NP3,ID).EQ.1) GO TO 280
          IF(LO(L).NE.ID) GO TO 280
          CALL PACK(MIN,MAX,MAX-L+1,MAX-L+1,2)
          KD = ID
      265 KD = KD -1
          IF(K(NP3,KD).EQ.1) GO TO 265
          LO(L) = 0
          IF(KD.LT.LI(L)) GO TO 280
          LO(L) = KD
          IL = LI(L)
          IU = LO(L)
          DO 275 JA = IL,IU
```

```
      IF(K(NP3,JA).EQ.1) GO TO 275
      DO 270 IA = MIN,MAX
  270 K(IA,MAX-L+1) = K(IA,MAX-L+1) .OR.K(IA,JA)
  275 CONTINUE
  280 ID = ID -1
      DO 285 IA = ID ,MAX
  285 K(IA,NP3) = K(IA,NP3) + K(IA ,ID+1)
      GO TO 700
  290 JA = ID
      IKM = IK
      IDM = ID -1
  300 DO 310 IB = IDM,MAX
  310 K(IB,NP3) = K(IB,NP3) + K(IB,JA)
      IKM = IKM -1
      IF(IKM.LE.0) GO TO 320
      JA = MAX-L+IKM
      CALL PACK(MIN,MAX,JA,JA,2)
      GO TO 300
  320 L = L -IK +1
      ID = ID -1
  700 CONTINUE
C**** NO MORE EQUATIONS CAN BE FOUND AT THIS POINT WHICH CONTAIN IK NEW
C**** VARIABLES AND KSUM VARIABLES IN THE LAST OPEN LOOP.  IF INDEX = 2
C**** THERE ARE SOME ROWS LEFT WITH IK NEW VARIABLES BUT THEY HAVE MORE
C**** THAN KSUM VARIABLES IN THE LAST OPEN LOOP.
      IF(INDEX.NE.2) GO TO 710
      KSUM = KSUM +1
      IF(KSUM.GT.(MAX-MIN+1)) 800,25
  710 IF(IK.GT.(MAX-MIN+1)) GO TO 800
      KSUM = 1
      IK = IK +1
      GO TO 25
  800 PRINT 810 , IND(MIN)
  810 FORMAT(* NO TEAR COULD BE FOUND FOR THE BLOCK BEGINNING WITH EQUAT
     1ION*,I3,* WITH THE NON ITERATES SPECIFIED. * )
  900 CONTINUE
C**** THE LAST OUTPUT SET ASSIGNMENT IS OBTAINED FROM THE N+1 COLUMN
C**** AND THE CHECK EQUATIONS ARE ORDERED PROPERLY.
      DO 915 I = 1,L
      II = MAX -I+1
      IF(K(II,NPO).GE.10000) K(II,NPO) = K(II,NPO) - 10000
      IF(K(II,NPO).GE.1000 ) K(II,NPO) = K(II,NPO) - 1000
  915 CONTINUE
      DO 916 I = 1,L
      II = MAX -I+1
      DO 913 J = 1,L
      JJ = MAX-J+1
      IF(K(JJ,NPO).NE.II) GO TO 913
      CALL ROWXC(II,JJ)
      GO TO 916
  913 CONTINUE
  916 CONTINUE
      DO 910 I = MIN,MAX
  910 K(I,NPT) = I
      LI(L+1) = MAX +1
      DO 950 I = 1,L
      CALL REORD (I,MIN,MAX)
  950 CONTINUE
      K(MIN,NPT) = 1000 +MIN
      K(MAX,NPT) = 2000 + MAX
      CALL PACK(MIN,MAX,MIN,MAX,2)
 1010 RETURN
      END
C****
C****
      SUBROUTINE REORD(IL,MIN,MAX)
C**** THIS SUBROUTINE REORDERS LOOP IL SO THAT THE PROPER ITERATE IS
C**** PLACED IN THE LOOP COLUMN AND THE OUTPUT VARIABLES ARE REASSIGNED
C**** PROPERLY.
      DIMENSION K(162,162),IND(162),JND(162),ISTOR(665),IOUTST(162),INZ(
     1 10),JNZ(20),KSET(162)
      COMMON IND,JND,N,NPO,K,IOUTST
      COMMON /T/ LI(20),LO(20)
      DIMENSION IIND(162),JJND(162)
```

```
      NP3 = N +3
      INDL = LI(IL)
      JNDL = LI(IL)   -1
      JLL = LI(IL)
      JUU = LO(IL)
      IF(LO(IL).EQ.0) GO TO 900
      DO 20 J = JLL,JUU
      IF(K(MAX-IL+1,J).EQ.1.AND.K(NP3,J).NE.1) GO TO 30
   20 CONTINUE
      GO TO 900
   30 JJND(MAX-IL+1) = JND(J)
      IIND(JLL) = IND(J)
      JM = J -1
      ITER = J
   40 IF(JM.LT.JLL) GO TO 100
      DO 50 JK = JLL,JM
      IF(K(J,JK).EQ.1) GO TO 60
   50 CONTINUE
      GO TO 100
   60 JNDL = JNDL +1
      INDL = INDL +1
      IIND(INDL) = IND(JK)
      JJND(JNDL) = JND(JK)
      JM = JK -1
      J = JK
      GO TO 40
  100 JNDL = JNDL +1
      JJND(JNDL) = JND(MAX-IL+1)
      JND(MAX-IL+1) = JJND(MAX-IL+1)
      DO 150 I = JLL,JUU
      K(JLL,I) = K(NP3,I)
      K(NP3,I) = 0
      IF(I.EQ.ITER) GO TO 150
      KOMPAR = JND(I)
      DO 120 J = JLL,JNDL
      IF(KOMPAR.EQ.JJND(J)) GO TO 150
  120 CONTINUE
      JNDL = JNDL +1
      JJND(JNDL) = KOMPAR
      IIND(JNDL) = IND(I)
  150 CONTINUE
      DO 175 I = JLL,JUU
      IND(I) = IIND(I)
      IF(K(JLL,I).NE.1) GO TO 175
      KOMPAR = JND(I)
      DO 160 J = JLL,JUU
      IF(JJND(J).EQ.KOMPAR) GO TO 170
  160 CONTINUE
  170 K(NP3,J) = 1
  175 JND(I) = JJND(I)
  900 RETURN
      END
C****
C****
      SUBROUTINE OUTSET(MIN,MAX,FLAG,KASE)
C**** THIS SUBROUTINE FINDS AN OUTPUT SET FOR ROWS AND COLUMNS FROM MIN
C**** NO MAX.  FLAG INDICATES WHETHER AN OUTPUT SET COULD BE FOUND.
C**** KASE IS AN INDICATION TO THE SUBROUTINE WHETHER THE OUTPUT SET
C**** SHOULD BE PLACED ON THE MAIN DIAGONAL.
      DIMENSION K(162,162),IND(162),JND(162),IOUTST(162)
      COMMON IND,JND,N,NPO,K,IOUTST
  303 FORMAT(///I2//)
      FLAG = 0.0
      DO 50 I =MIN,MAX
      K(NPO,I) = 0
   50 K(I,NPO) = 0
  100 CONTINUE
C**** A PHI IS PLACED IN A ROW OR A COLUMN TO INDICATE AN ASSIGNMENT OF
C**** A COLUMN TO A ROW.
C**** CHOOSE ANY COLUMN IN WHICH NO PHI EXIxTS -- COL. J.
      DO 150 J=MIN,MAX
      IF(K(NPO,J).LE.1000)200,150
  150 CONTINUE
      IL = 150
```

```
      GO TO 1500
C**** IN COL. J FIND A NONZERO ELEMENT IN AN UNTAGGED ROW.
C**** FIRST LOOK FOR A ROW WITH NO PHI
  200 DO 250 I = MIN,MAX
      IF(          K(I,NPO) .LT.1000 .AND. K(I,J)              .GT.0)500,250
  250 CONTINUE
C**** IF NO NONZERO ELEMENT COULD BE FOUND IN A ROW WITHOUT A PHI, THEN
C**** LOOK FOR ONE WITH A PHI ALREADY IN IT.
      DO 275 I =MIN,MAX
      IF(          K(I,NPO) .LT.10000.AND. K(I,J)              .GT.0)500,275
  275 CONTINUE
C**** IF NO NONZERO ELEMENT IN AN UNTAGGED COLUMN COULD BE FOUND TAG
C**** THIS COLUMN.
      IF(K(NPO,J).GE.10000)400,300
  300 K(NPO,J) = K(NPO,J) +10000
C**** FIND A TAGGED COLUMN CONTAINING A NONZERO ELEMENT IN AN UNTAGGED
C**** ROW.
C**** FIRST LOOK FOR A ROW WITH NO PHI.
  400 DO 450 J = MIN,MAX
      IF(K(NPO,J).GE.10000)425,450
  425 DO 440 I=MIN,MAX
      IF(K(I,NPO).LT.1000 .AND.K(I,J).GT.0)1000,440
  440 CONTINUE
  450 CONTINUE
C**** IF NO  NONZERO ELEMENT COULD BE FOUND IN A ROW WITHOUT A PHI, THEN
C**** LOOK FOR ONE WITH A PHI ALREADY IN IT.
      DO 475 J=MIN,MAX
      IF(K(NPO,J).GE.10000)455,475
  455 DO 445 I =MIN,MAX
      IF(K(I,NPO).LT.10000.AND.K(I,J).GT.0)1000,445
  445 CONTINUE
  475 CONTINUE
C**** IF NONE COULD BE FOUND GO TO STEP 1200.
      GO TO 1200
C**** FIND ANOTHER NONZERO ELEMENT IN COL. J WHICH IS IN AN UNTAGGED ROW
C**** AND PUT PHI IN IT AND ERASE OLD PHI.
 1000 LY = K(NPO,J) -11000
      IF(K(LY,NPO).LT.10000)1025,1050
 1025 K(LY,NPO) = 0
      GO TO 1075
 1050 K(LY,NPO) = 10000
C**** TAG COLUMN J
 1075 K(NPO,J) = 11000 + I
C**** SAVE OLD POSITION OF PHI IN ROW I, IF ANY.
      KDUM = K(I,NPO)
C**** PUT NEW POSITION OF PHI IN ROW I AND TAG IT.
      K(I,NPO) = 11000 +J
      GO TO 900
C**** TAG COLUMN J
  500 K(NPO,J) = 11000 +I
C**** SAVE OLD POSITION OF PHI.
      KDUM = K(I,NPO)
C**** PUT NEW POSITION OF PHI IN ROW I AND TAG IT.
      K(I,NPO) = 11000 +J
C**** IF ANOTHER PHI EXISTS IN ROW I ERASE IT.
  900 IF(KDUM.GT.10000)KDUM = KDUM - 10000
      IF (KDUM.GT.1000)KDUM=KDUM -1000
      IF(KDUM.GT.0)600,800
C**** CONSIDER THE COLUMN IN WHICH THE PHI HAS JUST BEEN ERASED.
  600 J= KDUM
      KDUM = 0
      IF(K(NPO,J).GT.10000)625,650
  625 K(NPO,J) = 10000
      GO TO 200
  650 K(NPO,J) = 0
      GO TO 200
  800 KDUM = 0
C**** REMOVE ALL TAGS ON ROWS AND COLUMNS.
      DO 120 I=MIN,MAX
      IF(K(NPO,I).GE.10000)105,115
  105 K(NPO,I) = K(NPO,I) -10000
  115 IF(K(I,NPO).GE.10000)117,120
```

```
      117 K(I,NPO) = K(I,NPO) -10000
      120 CONTINUE
      700 DO 750 I=MIN,MAX
          IF(K(I,NPO).LT.1000.OR.K(NPO,I).LT.1000)100,750
      750 CONTINUE
C**** THE OUTPUT SET ELEMENTS ARE PLACED IN IOUTST AND THE COLUMNS PER-
C**** MUTED TO PUT THE OUTPUT SET ON THE MAIN DIAGONAL.
     1100 IF(KASE.EQ.2) GO TO 70
          PRINT 1111
     1111 FORMAT (1H1)
          PRINT 1120
     1120 FORMAT (10X,//34H AN OUTPUT SET HAS BEEN OBTAINED.      )
          DO 1150 I =MIN,MAX
          IF(K(I,NPO).GT.10000)K(I,NPO) = K(I,NPO) -10000
          IF(K(I,NPO).GT.1000)K(I,NPO) = K(I,NPO) -1000
          IO = K(I,NPO)
          IOUTST(I) = JND(IO)
     1175 FORMAT( 14X, I3, 14X, I3)
     1150 CONTINUE
          DO 30 I=MIN,MAX
          DO 10 J=MIN,MAX
          IF(JND(J).EQ.IOUTST(I))20,10
       10 CONTINUE
       20 CALL COLXC(I,J)
       30 CONTINUE
       70 RETURN
C**** IF AN OUTPUT SET COULD NOT BE FOUND THE EQUATIONS WITH FEWER VAR-
C**** IABLES THAN EQUATIONS ARE PRINTED OUT IF KASE = 1.
     1200 IF(KASE.EQ.2) GO TO 1260
          PRINT 1220
     1220 FORMAT(10X,//35H AN OUTPUT SET COULD NOT BE FOUND.       /10X,70H TH
         1E FOLLOWING EQUATIONS CONTAIN FEWER VARIABLES THAN EQUATIONS.
         2  /10X,9H EQUATIONS         )
          DO 1250 I=MIN,MAX
          IF(K(I,NPO).LT.10000)1225,1250
     1225 PRINT 1230,I
     1230 FORMAT (14X,I3)
     1250 CONTINUE
     1260  FLAG = 1.0
          RETURN
     1500 PRINT 1550,IL
     1550 FORMAT(10X,45H AN ILEGAL BRANCH WAS ATTEMPTED AT STATMENT ,I4)
          FLAG = 1.0
          RETURN
          END
C****
C****
          SUBROUTINE SUMRC(MIN,MAX)
C**** THIS SUBROUTINE SUMS EACH ROW AND COLUMN AND PLACES THE RESULT IN
C**** THE N +1 COLUMN AND ROW RESPECTIVELY.
          DIMENSION K(162,162),IND(162),JND(162)
          COMMON IND,JND,N,NPO,K
          DO 100 I=MIN,MAX
          K(I,NPO)= 0
          DO 100 J=MIN,MAX
      100 K(I,NPO) = K(I,NPO) + K(I,J)
          DO 200 J=MIN,MAX
          K(NPO,J) = 0
          DO 200 I = MIN,MAX
      200 K(NPO,J) = K(NPO,J) + K(I,J)
          RETURN
          END
C****
C****
          SUBROUTINE COLXC(JD,JSAVE)
C**** THIS SUBROUTINE EXCHANGES THE ELEMENTS OF COLUMN JD AND JSAVE IN
C**** MATRIX K.
          DIMENSION K(162,162),IND(162),JND(162)
          COMMON IND,JND,N,NPO,K
          JJDUM = JND(JD)
          JND(JD) = JND(JSAVE)
          JND(JSAVE) = JJDUM
          NP4 = N +4
```

```
      DO 850 I = 1,NP4
      JDUM = K(I,JD)
      K(I,JD) = K(I,JSAVE)
  850 K(I,JSAVE) = JDUM
      RETURN
      END
C****
C****
      SUBROUTINE ROWXC(ID,ISAVE)
C**** THIS SUBROUTINE EXCHANGES THE ELEMENTS OF ROW ID AND ISAVE OF K
C**** AND THE CORRESPONDING ROWS OF LOOP.
      DIMENSION K(162,162),IND(162),JND(162)
      COMMON IND,JND,N,NPO,K
      COMMON /L/ LOOP(162,11)
      IDUM = IND(ID)
      IND(ID) = IND(ISAVE)
      IND(ISAVE) = IDUM
      NP4 = N +4
      DO 515 JM = 1,NP4
      KDUM = K(ID,JM)
      K(ID,JM) = K(ISAVE , JM)
  515 K(ISAVE,JM) = KDUM
      DO 514 JM = 1,11
      LDUM = LOOP(ID,JM)
      LOOP(ID,JM) = LOOP(ISAVE,JM)
  514 LOOP(ISAVE,JM) = LDUM
      RETURN
      END
C****
C****
      SUBROUTINE STEWRD(MIN,MAX)
C**** THIS SUBROUTINE EXECUTES STEWARD S ALGORITHM FOR PARTITIONING FOR
C**** THE SUBMATRIX OF ROWS AND COLUMNS OF K FROM MIN TO MAX.
      DIMENSION K(162,162),IND(162),JND(162) , LST(162)
      COMMON IND,JND,N,NPO,K
C**** THE N+1 COLUMN CONTAINS THE SUM OF EACH ROW.  THE DIAGONAL ELE-
C**** MENTS (OUTPUT SET ELEMENTS) ARE REMOVED.
      ID = MIN
      INDEX = 0
      LSTL = 0
      NPT = N+ 2
      DO 10 I = MIN,MAX
      K(I,NPO) = K(I,NPO) -K(I,I)
      K(I,NPT) = 0
   10 K(I,I) = 0
C**** THE N+1 ROW IS SEARCHED FOR COLUMNS WHICH CONTAIN NO NONZERO ELE-
C**** MENTS AND HAS NOT YET BEEN REMOVED FROM K.
    1 DO 50 I = MIN,MAX
      IF((K(I,NPO).EQ.0).AND.(K(I,NPT).LT.1000 )) GO TO 60
   50 CONTINUE
      INDEX = 1
      GO TO 65
C**** IF A ROW HAS ALL ZERO ELEMENTS, IT IS REMOVED AND INDICATED NEXT
C**** IN THE SEQUENCE FOR SOLVING THE EQUATIONS.
   60 K(I,NPT) = 10000 + ID
      DO 62 MN = MIN,MAX
      K(MN,I) = 0
      IF(K(MN,NPT).NE.(1000 + I)) GO TO 62
      K(MN,NPT) = 10000 + ID
   62 CONTINUE
      CALL SUMRC(MIN,MAX)
      ID = ID + 1
   65 CONTINUE
C**** THE ROWS ARE CHECKED HERE TO DETERMINE IF ALL ROWS HAVE BEEN AS-
C**** SIGNED AN OUTPUT COLUMN.
      DO 70 IJ = MIN,MAX
      IF(K(IJ,NPT).LT.1000)GO TO 75
   70 CONTINUE
      GO TO 500
   75 IF(INDEX.EQ.0) GO TO 1
      INDEX = 0
C**** ADD ROW IJ TO THE LIST.
      LSTL = LSTL + 1
```

```
      LST(LSTL) = IJ
C**** FORM A LIST OF ROWS EACH OF WHICH FEEDS THE ONE IMMEDIATELY ABOVE
C**** IT IN THE LIST.  WHEN ANY ROW IS ENCOUNTERED TWICE BRANCH TO 200.
C**** THIS IS THE PATH TRACING TECHNIQUE.
   80 DO 100 J = MIN,MAX
      IF((K(IJ,J).EQ.1).AND.(K(J,NPT).LT.1000 )) GO TO 110
  100 CONTINUE
      GO TO 1000
  110 IF(LSTL.LT.1)GO TO 175
      DO 150 I=1,LSTL
      IF(LST(I).EQ.J) GO TO 200
  150 CONTINUE
  175 LSTL = LSTL +1
      LST(LSTL) =J
      IJ = J
      GO TO 80
C**** TAG EACH COLUMN IN THE LIST BETWEEN THE ELEMENT OF THE REPEATED
C**** ROW AND THE END OF THE LIST AS A LOOP.
  200 MNAI = LST(I)
      DO 220 MN = I,LSTL
      MNA = LST(MN)
      IF (MN .EQ.I) GO TO 205
      K(MNA,NPT) = 1000 +MNAI
  205 DO 208 L = MIN,MAX
      IF(K(L,NPT).NE.(1000 + MNA)) GO TO 208
      K(L,NPT) = 1000 + MNAI
  208 CONTINUE
  220 CONTINUE
      I = I + 1
C**** FORM ONE ROW AND ONE COLUMN TO REPRESENT THE LOOP.  THE ROW IS THE
C**** BOOLEAN UNION OF ALL ROWS IN THE LOOP AND THE COLUMN IS THE BOO-
C**** LEAN UNION OF ALL THE COLUMNS IN THE LOOP.
      DO 217 MN = I,LSTL
      MNA = LST(MN)
      DO 217 J = MIN,MAX
      K(MNAI,J) = K(MNAI,J).OR.K(MNA,J)
      K(J,MNAI) = K(J,MNAI).OR.K(J,MNA)
      K(MNA,J) = 0
  217 K(J,MNA) = 0
      K(MNAI,MNAI) = 0
      CALL SUMRC(MIN,MAX)
      DO 250 I= MIN,MAX
  250 LST(I) = 0
      LSTL = 0
      GO TO 65
C**** THE TAGS ARE REMOVED FROM THE ROWS AND COLUMNS HERE.
  500 DO 600 I = MIN,MAX
      IF(K(I,NPT).GT.10000)K(I,NPT) = K(I,NPT) -10000
      IF(K(I,NPT).GT.1000 )K(I,NPT) = K(I,NPT) -1000
      K(I,NPO) = K(I,NPT)
      K(I,1) = IND(I)
  600 CONTINUE
      ID = ID - 1
      IN = MIN
C**** GROUP THE ROWS AND COLUMNS OF EACH LOOP TOGETHER AND ORDER THEM IN
C**** THE CORRECT PRECEDENCE ORDER.
      DO 750 I = MIN,ID
      DO 700 J = MIN,MAX
      IF(K(J,NPT).NE.I) GO TO 700
      IF(IN.EQ.J) GO TO 690
      CALL ROWXC(IN,J)
  690 IN = IN +1
  700 CONTINUE
  750 CONTINUE
      GO TO 1500
 1000 PRINT 1001
 1001 FORMAT (38H AN ILLEGAL BRANCH WAS ATTEMPTED             )
 1500 RETURN
      END
```

# Appendix B. Reordered Occurrence Matrix of the Hanford N-Reactor System

## Nomenclature[2]

$a_{ij}$ An element of matrix A
A   Adjacency matrix
A   Boolean matrix in general
$b_{ij}$ An element of matrix B
B   Boolean matrix in general
$c_{ij}$ An element of matrix C
C   Information matrix
C   Boolean matrix in general
E   A set of edges in a graph
f   Function in general
G   A digraph
H   A digraph
i   Index for a row of a matrix
j   Index for a column of a matrix
M   Reduced occurrence matrix used in finding disjoint subsystems
n   The number of equations or variables in a system
P   Adjacency matrix of information flow among the loops in a system of equations
$r_{ij}$ An element of matrix R
$r_{ij}^k$ Element of the $k$th power of matrix R
R   Adjacency matrix
$r_{ij}^*$ Element of R*
R*  Reachability matrix
$s_{ij}$ An element of matrix S
S   Occurrence matrix
T   Information matrix
x   System variable
y   System variable
z   System variable

## References

B1. Billingsley, D. S., *Chem. Eng. Sci.* **22**, 719 (1967).
B2. Bôcher, M., "Introduction to Higher Algebra." Macmillan, New York, 1919.
C1. Churchman, C. W., Ackoff, R. L., and Arnoff, E. L., "Operations Research." Wiley, New York, 1961.
H1. Himmelblau, D. M., *Chem. Eng. Sci.* **21**, 719 (1966).
H2. Himmelblau, D. M., and Bishoff, K. B., "Process Analysis and Simulation." Wiley, New York, 1967.
H3. Hooke, R., and Jeeves, T. A., *J. A. C. M.* **8**, 212 (1961).
K1. Komatsu, S., *Ind. Eng. Chem.* **62**, No. 2, 36 (1968).
K2. Kunz, K. S., "Numerical Analysis." McGraw-Hill, New York, 1957.
L1. Ledet, W. P., Ph.D. Thesis. Univ. of Texas, 1968.
L2. Lee, W., and Rudd, D. F., *Am. Inst. Chem. Engrs. J.* **12**, 1184 (1966).
N1. Nagiev, M. F., *Chem. Eng. Progr. (CEP)* **53**, No. 6, 297 (1957).
N2. Netter, Z., *IRE Trans. Circuit Theory*, **CT-8**, 77 (1961).
P1. Perry, J. H., Chilton, C. H., and Kirkpatrick, S. D., "Chemical Engineer's Handbook," 4th ed., Section 2. McGraw-Hill, New York, 1963.
R1. Rosen, E. M., *CEP* **58**, No. 10, 69 (1062).
S1. Sargent, W. H., and Westerberg, A. W., *Trans. Inst. Chem. Engrs. (London)* **42**, T190 (1964).
S2. Shannon, P. J., Johnson, A. I., Crowe, C. M., Hoffman, T. U., Hamielec, A. E., and Woods, D. R., *CEP* **62**, No. 6 (1966).
S3. Steward, D. V., *Soc. Ind. Appl. Math. Rev.* **4**, 321 (1962).
S4. Steward, D. V., *J. Soc. Appl. Math. Numerical Anal.* **B2**, 345 (1965).

[2] The notation used in the second example of Section VIII (multiple effect evaporator) is given in Table VIII.

S5. Swanson, C. D., Conghran, K. D., Dionne, P. J., and Thieme, G. G., *AEC Doc.* HW-82589 (1964).
T1. Traub, J. F., "Iterative Methods for the Solution of Equations." Prentice–Hall, Englewood Cliffs, New Jersey, 1964.
V1. Vela, M. A., *Petrol. Refiner* **40**, 189, No. 6 (1961).

# THE FORMATION OF BUBBLES AND DROPS

R. Kumar and N. R. Kuloor*

Department of Chemical Engineering, Indian Institute of Science
Bangalore, India

| | |
|---|---|
| I. Introduction | 256 |
|   A. Film Method | 256 |
|   B. Rupture of Bulk Fluid | 257 |
|   C. Dispersion through Submerged Orifices | 257 |
| II. Measurement of Bubble Volume | 257 |
|   A. Direct Methods for Evaluating Bubble Volume | 258 |
|   B. Indirect Methods for Evaluating Bubble Volume | 260 |
| III. Influence of Various Factors on Bubble Size | 265 |
|   A. Experimental Set-up | 265 |
|   B. Factors Influencing Bubble Size | 266 |
|   C. Effect of Orifice Characteristics | 267 |
|   D. Effect of Chamber Volume | 268 |
|   E. Effect of Submergence | 270 |
|   F. Effect of Surface Tension of the Liquid and the Wetting Properties of the Orifice | 271 |
|   G. Effect of Liquid Viscosity | 272 |
|   H. Effect of Liquid Density | 273 |
|   I. Effect of Gas Properties | 274 |
|   J. Effect of Gas-Flow Rates | 276 |
|   K. Effect of Continuous-Phase Velocity | 277 |
|   L. Conclusions | 277 |
| IV. Bubble Formation under Constant Flow Conditions | 277 |
|   A. Bubble Formation in Inviscid Liquids | 278 |
|   B. Bubble Formation in Highly Viscous Liquids | 289 |
|   C. Model of Ramakrishnan, Kumar, and Kuloor | 295 |
| V. Bubble Formation under Constant Pressure Conditions | 304 |
|   A. Correlation of Hayes, Hardy, and Holland | 304 |
|   B. Davidson and Schuler's Model | 307 |
|   C. Model of Satyanarayan, Kumar, and Kuloor | 310 |
| VI. Bubble Formation in Non-Newtonian Fluids | 316 |
| VII. Bubble Formation in Gas Fluidized Beds | 318 |
|   A. Introduction | 318 |

* Deceased February 5, 1970.

|        |                                                                    |     |
| ------ | ------------------------------------------------------------------ | --- |
|        | B. Model of Davidson and Harrison                                  | 319 |
|        | C. Model of Kumar and Kuloor                                       | 320 |
| VIII.  | The Effect of Orifice Geometry on Bubble Size                      | 321 |
| IX.    | The Influence of Orifice Orientation on Bubble Formation           | 324 |
|        | A. Introduction                                                    | 324 |
|        | B. Models of Kumar, Kuloor, and Co-workers                         | 325 |
| X.     | The Influence of Continuous Phase Velocity on Bubble Size          | 332 |
| XI.    | Formation of Bubbles at Multiple Orificed Plates                   | 333 |
| XII.   | Formation of Drops                                                 | 334 |
|        | A. Influence of Variables on Drop Size                             | 334 |
|        | B. Hayworth and Treybal Model                                      | 335 |
|        | C. Null and Johnson Model                                          | 337 |
|        | D. Models of Kumar, Kuloor, and Co-workers                         | 337 |
|        | E. Comparison of Various Models                                    | 340 |
| XIII.  | Drop Formation in Non-Newtonian Fluids                             | 343 |
|        | A. Drop Formation in Power-Law Fluids                              | 343 |
|        | B. Power-Law Drops in Newtonian Fluid                              | 345 |
|        | C. Power-Law Drops in Power-Law Fluids                             | 346 |
| XIV.   | Drop Formation from Vertically Oriented Orifices                   | 346 |
| XV.    | Spraying and Atomization                                           | 347 |
|        | A. Liquid-Liquid Systems                                           | 347 |
|        | B. Liquid-Gas Systems                                              | 348 |
| XVI.   | Unified Model for Drop and Bubble Formation                        | 350 |
| XVII.  | Bubble and Drop Size in Stirred Vessels                            | 354 |
| XVIII. | Bubble Formation under Intermediate Conditions and at Sintered Disks | 356 |
|        | A. Bubble Formation under Intermediate Conditions                  | 356 |
|        | B. Bubble Formation at Sintered Disks                              | 359 |
| XIX.   | Concluding Remarks                                                 | 362 |
|        | Nomenclature                                                       | 363 |
|        | References                                                         | 365 |

## I. Introduction

Two immiscible fluids, in contact with each other, share a common surface, called the "interface." Operations involving transfer of matter or of heat across an interface are very common in chemical industry. In such operations a large interfacial area per unit volume is necessary if the desired transfer is to be obtained rapidly in equipment of finite size. Three common methods of providing a high ratio of interfacial area to volume are now discussed.

### A. Film Method

Here, either both the fluids are spread over an existing surface or one of them is spread and the other is present in bulk. Usually, the walls of a column are utilized for this purpose by wetting them with one of the fluids. Wetted-wall columns and rotating film reactors (K11) make use of this principle to bring about gas-liquid or liquid-liquid contact. Two liquids may be contacted cocurrently by allowing them to flow in films down opposite sides of a vertical wire mesh.

The drawback of these techniques is the requirement of artificial surfaces for the stabilization of the films created.

### B. Rupture of Bulk Fluid

In this method a large bulk of one of the fluids is broken into smaller particles inside a continuous phase of the other fluid. This can be caused either by an artificial stirring action or by the instability of the big globule itself. A typical example of this method is the rupture of bubbles during their ascent in a stirred-tank reactor.

### C. Dispersion through Submerged Orifices

The method of dispersion through submerged nozzles, slots, or holes is the simplest and hence the most common. It permits equipment of extremely simple design and leads to reasonably large interfacial areas.

As the dispersed phase moves, the bubbles (gas in liquid) or drops (liquid in liquid or gas) may either coalesce or rupture, depending on the conditions existing. Sometimes packings or stirrers are introduced to facilitate these phenomena. When the dispersed-phase particles arrive at the top of the column, they coalesce and form a bulk interface with the continuous phase, across which transfer processes can continue.

Some industrial operations involving bubble and drop formation are extraction, direct contact heat exchange, distillation, absorption, sparger reactors, spray drying and atomization, fluidization, nucleate boiling, air lifts, and flotation.

In all these operations involving bubbles and drops, three stages have to be studied, viz. (i) the formation of bubbles or drops, (ii) the movement of bubbles or drops through the continuous phase and possible coalescence therein, and (iii) the formation of the interface. This review is an attempt to understand the first of the three aspects of the study, especially in the case of submerged orifices.

## II. Measurement of Bubble Volume

In any study on bubble formation, the measurement of bubble volume is of primary importance. The bubbles formed at a nozzle ascend through the liquid column and rupture at the surface of the liquid. During their ascent, the bubbles may coalesce to form bigger bubbles, or may rupture, giving rise to smaller bubbles. In an actual operation, both these processes go on inside the liquid column. Thus, the bubble size in the middle of the column is not the same as that at the tip of the nozzle. Though approximate evaluation of bubble volumes at any position in the column can be made, the methods discussed in this section are meant mainly for evaluating the bubble volume at the nozzle-tip.

### A. DIRECT METHODS FOR EVALUATING BUBBLE VOLUME

Methods which yield the bubble size directly make use of photography in one form or another.

#### 1. *Cinephotography*

A number of investigators (D8, H7, K18, L2) have employed cinephotography to obtain qualitative information about the phenomenon of bubble formation. Speeds varying from 80 frames per second (D8) to 4000 frames per second (H7) have been used. This method can also be used to obtain quantitative information. The photographs taken are projected, and their size is measured. As the size of the image on the film depends upon the distance of the nozzle from the lens, it is necessary to have a graduated scale or other object of known size placed by the side of the nozzle at the same distance from the lens as the nozzle, and visible in the photograph. Thus, when the photograph is projected, a contour of the bubble is obtained from which the bubble volume can be calculated by assuming symmetry about the vertical axis. Sometimes, photographs are taken simultaneously from two or three mutually perpendicular directions to obtain bubble contours from different angles. This is particularly advantageous when highly distorted bubbles are examined.

If the reduction ratio is independently known in terms of the distance between the object and the camera, then the bubble size can be known by measuring the size of image by means of an ocular micrometer.

Alternatively, the camera is kept at a fixed distance, and a scale is first photographed. The ratio of the image size to the actual size gives the reduction ratio. If the camera is then positioned at the same distance and the bubble photographed, the actual dimension of the bubble can be obtained from the image size by dividing it by the reduction ratio.

The photographic method for evaluating the bubble volume suffers from the disadvantage that it does not yield the volume directly but only gives the contour of the bubble in a single plane. The lighting used is purely a matter of trial and error. Assumptions have to be made regarding the symmetry of the bubble about a fixed axis. Thus, the bubble volume obtained may sometimes be seriously in error, particularly when the bubbles are highly distorted. Another disadvantage of this method is that it is very laborious and a number of frames must be examined to find the proper dimension of the bubble.

There are, however, a few points which make this method an attractive one.

It can be used for measuring bubble sizes not only at the nozzle tip but also at any position in the liquid column. Thus, even coalesced and ruptured bubbles can be examined by this method.

It examines each bubble individually and does not depend on the average of many bubbles. Thus, if subsequent bubbles are of different sizes as sometimes happens when secondary bubbles are formed, this method is the only choice.

It is the only method which can give the rate of growth of the bubble with time; the intermediate stages of bubble formation are accurately recorded.

It is the best method that can be used when the surface area of the bubble is required.

## 2. *X-Ray Cinephotography*

The cinephotographic method discussed above can be used only when the fluid in which the bubbles are formed is transparent. If the fluid is opaque, like some non-Newtonian fluids or fluidized beds, x-ray cinephotography has to be used. Rowe *et al.* (R10, R11) have used this technique for studying gas bubbles in fluidized beds. The column in which the bubbles are formed is placed between the x-ray tube and the cine camera (normally 35 mm). Photographs up to 50 frames per second have been obtained by Rowe and Partridge (R10) with an exposure time per frame of the order of 0.01 sec.

This method has recently been employed by Botterill *et al.* (B6) for studying the behavior of bubble chains in fluidized beds. Most of the discussion given for the cinephotographic method is valid for the x-ray cinephotographic method also.

## 3. *γ-Ray Absorption*

This method gives the variation in voidage at any position in the column. It consists essentially of passing a $\gamma$-ray beam through the column and measuring the intensity of the emerging beam by its count rate. By calibration, the count rate is related to the voidage and hence the voidage corresponding to any count rate is obtained. Bloore and Botterill (B5) found the bubble diameter by scanning the face of the column and determining the average projections of the voidage contours. The breadth of the injected disturbance was considered to be the bubble diameter. The volume of the bubble could be determined by assuming it to be spherical at the nozzle tip.

Baumgarten and Pigford (B2) have employed the $\gamma$-ray method for their study of density fluctuations in fluidized beds. The method is laborious and time-consuming, and yields only approximate values based on a large number of bubbles. Because of this, the x-ray cinephotographic method is to be preferred for the study of the behavior of bubbles in fluidized beds.

## B. INDIRECT METHODS FOR EVALUATING BUBBLE VOLUME

In these methods the volumetric flow corrected to the nozzle tip, $Q$, and the frequency of bubble formation, $f$, are directly measured. The bubble volume is then calculated. These methods have a number of limitations.

First, they cannot be used for the evaluation of volumes of individual bubbles. Only an average bubble volume is obtained. Bubble formation is generally a cyclic phenomenon, and for a definite flow rate in a particular system, the frequency and the bubble volume are time-independent. However, there are situations where each bubble is followed by smaller secondary bubbles. In such cases, the above methods cannot yield reliable values and photographic methods have to be resorted to.

Second, these methods do not give any information about the shape and thereby the surface area of the bubble. Although Quigley *et al.* (Q1) have found from photographic studies that the area evaluated assuming sphericity of bubbles is not very different from that actually measured, this evidently cannot be true for large and highly distorted bubbles since a sphere has the minimum area and any distortion in shape tends to increase it.

Even with the above-mentioned limitations, the indirect methods constitute the simplest way of evaluating bubble volumes from single nozzles, and hence are most extensively used. As these methods involve a knowledge of the two quantities $Q$ and $f$, the ways of measuring each of them are separately discussed below.

### 1. *Measurement of $Q$, the Volumetric Flow Rate*

The volumetric flow rate is measured either by cumulative devices which collect the bubbles for a definite period of time or by rate measuring devices.

*a. Cumulative Devices.* These devices have been generally employed for low gas flow rates or when the gas is required to be collected for gas analysis, etc. Various arrangements ranging from inverted burettes to soap-film flow meters have been used. For higher flow rates, wet gas meters which yield cumulative flows have been used. Some of the devices employed particularly for low flow rates are given below.

Reservoir method: This method makes use of the displacement principle. Brine or any other saturated solution in which a gas has low solubility is used as the liquid. Gas from the column is collected in a burette from which the displaced liquid flows to a reservoir. As the gas collection proceeds, the gas is collected under increasing pressure conditions, thereby changing the flow rate as well as the frequency of bubble formation. In order to collect gas under atmospheric conditions, the levels of the liquid in the burette and the reservoir must always be kept equal. This requires manual adjustments.

Kumar and Kuloor's device (K15): This device avoids the necessity of continuous control, and it can collect gas samples without disturbing the flow conditions in the main equipment as shown in Fig. 1. The device consists of a graduated burette (1) for collecting the bubbles and a levelling reservoir (2). The stopcock (3) at the top of (1) permits the levelling liquid to fill the burette, and thus prepare it for bubble collection. The reservoir (2) is divided into two chambers by an overflow tube (4). The outer chamber is connected to this tube by the stopcock (5). The burette and the reservoir are connected through a rubber tube. The capacity of the reservoir is about one and a half times that of the burette.

When the bubbles are to be collected, the burette is filled with the liquid by opening the stopcocks (3) and (5) and raising the reservoir (2). Then, these

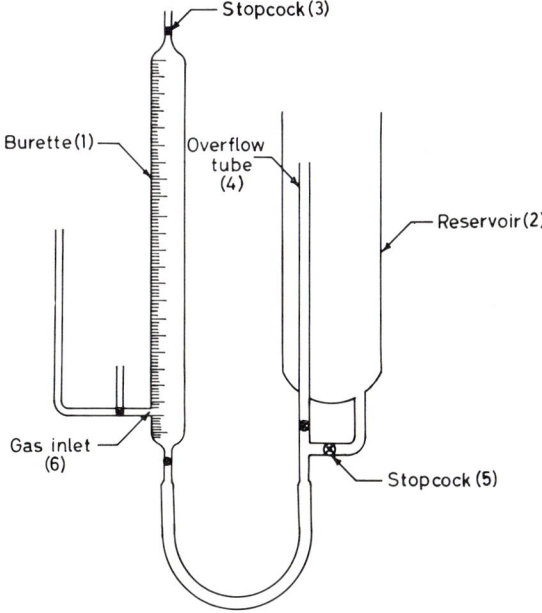

FIG. 1. Kumar and Kuloor's device for collecting gas samples.

are closed and the reservoir is placed in such a way that the top of the overflow tube (4) and the gas inlet (6) are at the same level. The bubbles are then collected under atmospheric pressure. The bubbles are collected for a known interval of time and the volume collected is found from the reading of the burette when the levels of the liquid in the burette and the reservoir are equalized.

This device has been successfully employed for measuring bubble volumes when the chemical reaction between a gas and a liquid is being studied (V2). The conditions under which the gas is measured are converted to the conditions of pressure, temperature, etc. existing at the nozzle tip.

Soap-film flow meter: A modified form of the soap film flow meter used by Krishnamurthi, Kumar, Datta, and Kuloor (K10) for collecting bubbles at atmospheric pressures is shown in Fig. 2. This device, which makes use of the movement of a soap film in a burette, can also be employed to calibrate low rate gas flow meters. Gas enters the apparatus from the top. A soap film is formed at the tip of the calibrated burette by raising a metallic loop dipped in soap solution. A water-seal is used to prevent the gas from escaping through

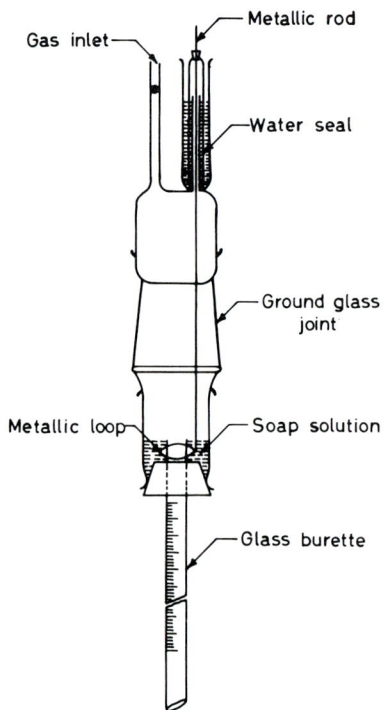

FIG. 2. Modified form of soap film flow meter.

the opening provided for the rod supporting the metallic loop. As the bubbles go on collecting in the burette, the soap film moves along its length, permitting the measurement of the volume of the collected gas at any time.

The advantages of this device are the ease of handling, the absence of continuous manual control and the negligible error due to mass transfer.

Wet gas meter: Wet gas meters can be employed at high flow rates.

*b. Rate Measuring Devices.* Rotameters and orifice meters are extensively used as rate measuring devices, although any other kind of flow meter can be used with equal advantage. The flow can be measured at any position in the apparatus but must be converted to the conditions existing at the nozzle tip.

## 2. Measurement of f, the Frequency of Bubble Formation

When the frequency of bubble formation is very low ($<200$ bubbles per minute), the bubbles can be visually counted without the aid of any instrument. When the frequency is higher than this, other methods have to be employed.

The simplest instrument that finds extensive use is the strobotac which illuminates the object at varying frequencies. When the frequency of the strobotac flash coincides with that of bubble formation, the image of the bubble appears to be stationary. This method is evidently useful only when the system is transparent.

Various other methods devised for measuring frequency are all based on the fact that the bubble creates a disturbance in a continuously occurring phenomenon. This distrubance can be utilized as an obstruction to the passage of light, x-rays, or $\gamma$-rays; or as a disturbance in the noise pattern or in the flow of current. Following are some of the methods which make use of the above ideas.

*a. Photoelectric Cell Counter.* When a light beam passes just above the nozzle and through the column on to a photoelectric cell, a voltage which depends on the intensity of the emergent beam is developed. If a bubble is formed at the nozzle tip it obstructs the light beam in its passage and the voltage at the cell drops suddenly. The cyclic formation of bubbles causes corresponding fluctuations in the voltage. The rate of voltage fluctuations directly gives the value of the frequency of bubble formation. This method has been successfully employed by Sullivan *et al.* (S17) and Lamont and Scott (L1).

In order to carry out the counting by mechanical devices, the count rate is divided by a factor of two or four and the output is amplified.

*b. Cinephotographic Method.* High-speed photography can also be used for measuring the frequency. This method is particularly useful when direct readings of bubble volume are required. The camera is operated at a known speed and the film is projected at a much lower speed. Knowing the camera speed, the projector speed, and the number of bubbles per unit time at the projected nozzle, the frequency of bubble formation can be directly evaluated. This method has been used by Venugopal, Kumar, and Kuloor (V2) during their studies in liquid-phase oxidation of acetaldehyde to acetic acid.

All the above methods which make use of the passage of light through the liquid-gas mixture are applicable only when the liquid transmits light. For opaque systems like fluidized beds and some non-Newtonian fluids, methods based on different principles have to be employed.

*c. Pressure Pulse Technique.* When a bubble detaches itself from the nozzle tip, it generates a pressure fluctuation which can be picked up by a suitably situated microphone. The pulse frequency can be measured in comparison with a standard variable-frequency generator. Two groups of investigators have employed this method for the measurement of bubble frequency.

Calderbank (C1) employed a crystal microphone located in the gas supply line near the nozzle tip, which was connected to an oscilloscope through a preamplifier. The photographic comparison of this signal with a constant-frequency (60 cps in this case) test signal, yielded the frequency of bubble formation.

Newman and Lerner (N2) have used an arrangement where the signal picked up by a microphone attached to the flat surface below the orifice plate is amplified and fed to a loud speaker. The amplified bubble signal is then fed to one pair of fixed contacts of a double-pole double-throw switch of which the other pair of fixed contacts is connected to an audio-frequency generator. The movable contacts of the switch are connected to the vertical and ground terminals of an oscilloscope. This arrangement permits the observation of either the bubble signal or the sine wave as a function of the internal linear time-base of the oscilloscope.

The signal, amplified to a good sound level at the loudspeaker, is fed to the vertical input of the oscilloscope. A stationary trace is obtained on the oscilloscope. The frequency of the sine wave of the oscilloscope is varied until a single trace is obtained. This frequency is then equal to the frequency of bubble formation.

*d. Capacitance Technique.* The capacitance of a system changes when a bubble is formed between its two probes. The frequency of capacitance change corresponds to the frequency of bubble formation.

Calderbank (C1) in studying gas-liquid contacting on plates, placed one platinum probe ($\sim 0.6$ cm in length) in the path of the bubble stream and another in the undisturbed liquid, both probes being connected in series with a 4.5 V dry cell. A scaler with a readily adjustable discrimination was used in conjunction with the above arrangement. Each bubble passing the probe produced a pulse, every sixty-fourth of which was recorded on a magnetic counter with interpolation by a scaler.

This arrangement could be used with distilled water, without the use of any electrolyte.

A similar technique, different in detail, has been employed by Harrison and Leung (H3) for studying bubble formation in fluidized beds.

Other methods based on $\gamma$-ray absorption or x-ray cinematography may be used for special cases by employing the principles already explained. However, for most of the ordinary contacting devices, the methods enumerated above are adequate.

## III. Influence of Various Factors on Bubble Size

### A. Experimental Set-up

Having determined the bubble size, the next step is to examine the several factors that influence it. Before doing this, it is important to bear in mind a general set-up of the equipment commonly used in the formation and study of bubbles. The bubble studies normally conducted and reported in literature are concerned with single nozzles, even though multiple nozzles are commonly employed in industry. This is done to facilitate an understanding of the bubble formation phenomenon at a single nozzle, which can then be extended to multinozzled systems. Although the equipment employed varies from investigator to investigator, a typical set-up is that used by the present authors and their colleagues and described below.

The set-up can be divided into three sections as shown in Fig. 3: the gas supply, the bubble forming section, and the gas metering section.

The supply of gas under constant pressure can be obtained either by the constant-level tank arrangement or by a compressor or cylinder through a surge tank. If mass transfer is to be avoided during bubble formation, then the gas can be passed through a bubbler containing the same liquid as that in which the bubble formation is to be studied.

The bubble-forming section consists of a control valve (needle valve type), a chamber, and the liquid column. It is customary to have flat glass plates as two sides of the column, to assist photography and visual observation of the phenomenon. A manometer is also provided to evaluate the pressure difference across the nozzle.

The bubbles formed break at the surface, where an inverted funnel having an outlet at the top collects them and conveys the gas to the measuring devices which have already been discussed in Section II.

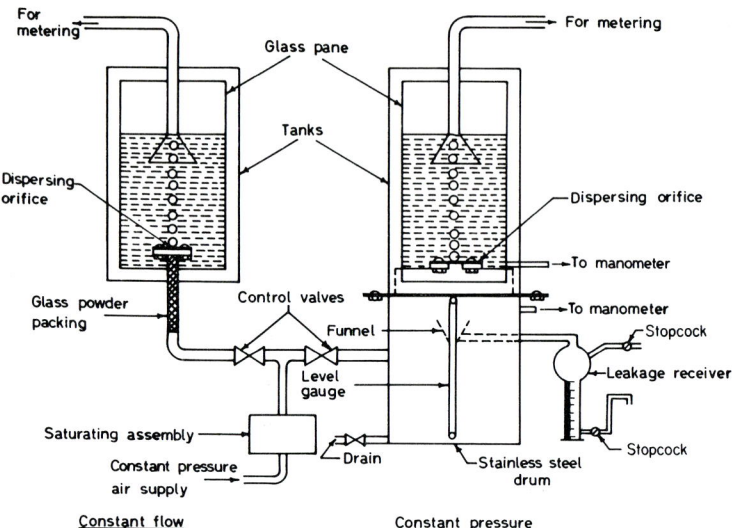

Fig. 3. Equipment for bubble formation under constant flow and constant pressure conditions.

The measuring devices like rotameters, orifice meters, etc. can also be introduced into the equipment prior to the bubble forming device. Keeping the above general set-up in mind, we can now proceed to examine the factors which influence the bubble size.

B. Factors Influencing Bubble Size

These factors can be broadly classified as equipment variables, system variables, and operating variables.

1. *Equipment Variables*

These variables are directly decided by the bubble forming device, whose major parts are the chamber, the nozzle, and the liquid column. The important factors are enumerated below:

*a.* diameter, geometry, orientation, nature and material of construction of the nozzle,

*b.* chamber volume,

*c.* arrangement of slots for multiholed plate systems.

Most of these factors remain unstudied, though the influence of the diameter of the nozzle and chamber volume have been reported in the literature (B3, D4, D5, H7, P1).

## 2. System Variables

These factors are associated with the liquid-gas combination chosen for investigation. The most important system variables are:
  a. surface tension $\gamma$
  b. density of the liquid $\rho_1$, viscosity of the liquid $\mu_1$
  c. density of the gas $\rho_g$ and viscosity of the gas $\mu_g$
  d. contact angle $\theta$
  e. velocity of sound in the gas $c$.

The considerable discrepancies in literature regarding the effects of many of the above variables will be discussed later.

## 3. Operating Variables

These are the factors which can be varied by the investigator at will. They include
  a. volumetric flow rate of the gas
  b. velocity of continuous phase
  c. head of the liquid in the column ("submergence")
  d. pressure drop across the nozzle
  e. temperature of the system.

The pressure drop has been included here for want of a better place. The temperature of the system does not directly influence the bubble size, but does so indirectly by varying the physical properties of the gas-liquid system. Hence it can be omitted from the analysis. Even the other variables are not entirely independent. For example, the pressure drop across the nozzle is a function of the flow-rate, chamber volume, etc.

Normally, the various factors do not influence the bubble size to the same extent over the entire range of the other variables. Thus surface tension, which is one of the most important factors determining the bubble size at vanishingly small flow rates, assumes much less importance at higher flow rates. Similarly, the viscosity of the liquid is much more important at higher flow rates than at the lower ones.

The known information regarding the influence of various factors is given below. Most of the investigators have tried to study the effect of one variable at a time, but often minor variations of other variables also occur alongside. For example, variation of viscosity by the use of glycerol solution varies the surface tension by a few dynes per centimeter.

### C. Effect of Orifice Characteristics

The observations made below are applicable mainly to wetted nozzles.

At low gas flow rates, the effect of orifice diameter is such that the volume of the bubble is directly proportional to it (B3, D4, V1). At higher flow rates also the bubble volume has been reported to be a strong function of the orifice diameter. Thus Leibson *et al.* (L2) and Davidson and Amick (D5) find that the bubble volume is proportional to the 3/2 power of the orifice diameter, whereas Quigley *et al.* (Q1) find the power to be 2.01. The bubble size is found to be independent of the orifice diameter by Silberman (S12) in a study of bubble formation in jets. In the region of single bubble formation, the data of Ramakrishnan, Kumar, and Kuloor (R1) show that the bubble volumes for various nozzle diameters tend to join in a single curve at high flow rates.

Apart from the size of the orifice, the nature of the orifice is of considerable importance. Thus the use of an orifice and a tube of the same diameter yields different bubble sizes. The orifice plate is found to yield a higher bubble volume than the corresponding tube.

The thickness of orifice plate may also affect the bubble size. This effect has been reported to become significant (H4) if the thickness is equal to or greater than 100 times the value of orifice diameter. A similar observation has also been made by Hughes *et al.* (H7).

One of the major variables associated with the orifice is the geometry of the opening. Though most of the investigators have used circular orifices, slots of various shapes are used in the industry. In bubble-cap columns, the slots are not only rectangular but also vertically oriented and for gas–liquid contacting, porous unglazed cylinders are used, where the pores are generally not circular .Thus the slot geometry is seen to be an important factor that requires detailed study. A systematic approach to this has been made by Krishnamurthi, Kumar, and Datta (K5, K7). They have compared the bubble volume obtained by using alternatively, a standard circular orifice of arbitrary diameter and two sets of orifices of other geometries like triangular, square, etc., chosen to have either their perimeters or their areas equal to that of the standard one. Their work, which was confined to low flow rates ($<0.05$ $cm^3/sec$), indicated that the bubble volumes obtained from the circular orifice did not correspond exactly to those from the noncircular orifices whether compared on an equal-perimeter or equal-area basis. They were closest for orifices of equal area. On extending this work to higher flow rates (up to 200 $cm^3/sec$), Ramakrishnan, Kumar, and Kuloor (R1) found that an orifice having a noncircular geometry gives bubble volumes equal to those obtained from a circular orifice of the same area. Thus any theory for the formation of bubbles from circular orifices can also be extended to situations where noncircular orifices are involved.

D. Effect of Chamber Volume

Spells and Bakowski (S15) were the first to recognize the importance of the chamber volume as a variable. They observed that the inclusion of a

35 litre reservoir between the air supply and the bubble slot influenced the bubble formation phenomenon. However they did not come to any definite conclusions and remarked that their findings needed further confirmation.

This was done by Hughes et al. (H7). They found that for small chamber volumes and small flow rates, the bubble volume is virtually independent of the chamber volume. Similar is the case when the chamber volume is large and the flow rates are normal. At very small flow rates and large chamber volumes, the bubbles normally form in doublets, and triplets and their size cannot be definitely determined. These conclusions have been verified by Davidson and Amick (D5) who varied the chamber volume in their equipment from 4 cm$^3$ to 4000 cm$^3$.

The chamber volume can be varied by varying either the diameter or the height of the chamber. It is of interest to know whether the geometrical proportions of the chamber have any influence on the bubble volumes obtained. Hayes et al. (H4) studied this aspect of the problem and found that the effect of the diameter of the gas chamber is insignificant in the formation of the bubbles, provided that the ratio of internal diameter of the gas chamber to that of the orifice is equal to or greater than 4.5. When the diameter of the gas chamber approaches that of the orifice, the bubble formation occurs as if at zero gas chamber volume. The bubble formation is further found to be independent of the chamber volume, provided that the latter is larger than about 800 cm$^3$.

When the chamber volume is between 0 and 800 cm$^3$, resonance effects are important because then the velocity of sound plays an important role. Very little work has been done regarding the resonance effects.

The influence of chamber volume is closely related to that of the pressure difference, $\Delta P$, across the orifice. It should, however, be defined as the volume between the orifice and that point in the gas stream where the pressure drop is very large. Normally, this point occurs at the control valve, but does not necessarily have to. Davidson and Schuler (D8) have used a sintered plate just before the orifice, even though a large gas chamber exists below it. Similarly, Krishnamurthi, Kumar, and Kuloor (K9) have obtained high pressure drops near the orifice by packing the chamber with fine glass powder. When the pressure drop across the orifice is large and the pressure variations occurring during the formation of bubble are small compared to the total pressure drop, the gas flow rate does not change during bubble formation and the bubbles are said to form under constant flow conditions. This is true also when the chamber volume is zero. Thus, under constant flow conditions, zero chamber volume corresponds to a high pressure drop across the orifice.

Similarly, if conditions are such that on the air-supply side of the orifice the pressure is maintained constant during bubble formation, as the bubble size increases, the pressure inside it decreases, resulting in a higher flow rate to the bubble in the constant pressure arrangement. Therefore, constant flow conditions are not maintained and the rate of bubble-growth varies. These are the conditions which prevail when large chamber volumes ($>800$ cm$^3$) prevail.

Davidson and Schuler (D8) observed the above two situations of constant flow and constant pressure on the basis of pressure drop. In industry, constant pressure conditions are normally employed.

E. Effect of Submergence

Most of the investigators studying the formation of bubbles at horizontal orifices have chosen the orifice submergence as one of the variables (D4, D5, H4, H7, S15). It is generally agreed that this variable does not influence the bubble volume at the tip. This fact has been verified by the work of Datta et al. (D4) in a study of bubble formation from capillary tips (under essentially constant flow conditions), where the hydrostatic head was varied from 213.3 cm to 60.95 cm for an orifice diameter of 0.141 cm and a frequency range of 1–125 bubbles per minute. The negligible influence of submergence on the bubble volume has also been reported by Hughes et al. (H7), who varied the submergence from 20.07 cm to 2.032 cm for orifice diameters of 0.107 cm and 0.1864 cm; and by Davidson and Amick (D5), who varied the submergence from 25.4 cm to 2.54 cm. The region of submergence where the bubble volume is affected is when the submergence is low (S15) and less than approximately twice the bubble diameter (H4).

In apparent contradiction to these results are the data reported by Padmavathy, Kumar, and Kuloor (P1), who varied the submergence from 163 cm to 66 cm using an orifice diameter of 0.29 cm and found that the bubble volume changed by more than 2.5 times. Figure 4 shows a graph plotted from the data of the above authors where the bubble volume is seen to decrease with increasing submergence. The discrepancy is mainly due to the fact that all the earlier investigators had employed either constant flow or constant pressure conditions, whereas Padmavathy, Kumar, and Kuloor (P1) worked in the intermediate region by having a chamber volume of 60 cm$^3$. Though Hughes et al. (H7) also had a chamber volume of 60.63 cm$^3$, the maximum submergence studied by them was of the order of 20 cm$^3$. It is seen from the graph that at higher values of submergence the bubble volume is very sensitive to variations in submergence. At the chamber volume of 66 cm$^3$ itself the bubble volume hardly varies with submergence, and at still lower values of the chamber volume, the effect of submergence is negligible. Thus the study of Hughes et al. (H7) was conducted in too low a range of submergence to detect any appreciable change in the bubble volume.

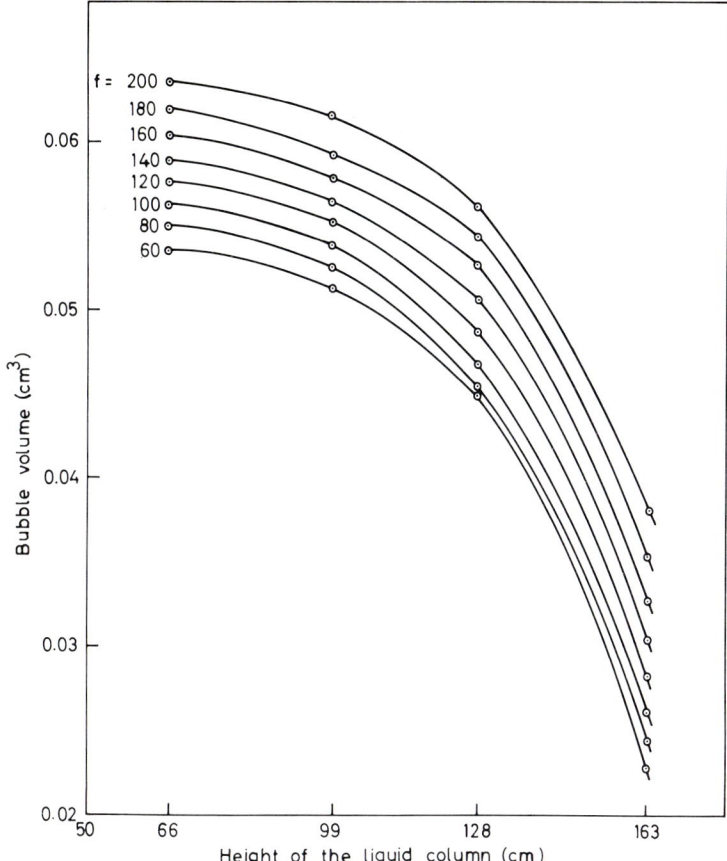

FIG. 4. Effect of orifice submergence on bubble volume in intermediate region.

The submergence, therefore, has no influence on the bubble volume under constant flow and constant pressure conditions, although it does affect the bubble volume appreciably in the intermediate region.

F. EFFECT OF SURFACE TENSION OF THE LIQUID AND THE WETTING PROPERTIES OF THE ORIFICE

Tate's law makes use of bubble formation in the measurement of the surface tension of a liquid, and hence the fact that surface tension influences the bubble volume is obvious. At flow rates tending to zero, the bubble volume is such that the upward force due to buoyancy is balanced by the downward force of surface tension. So, an increase in surface tension should result in an increased bubble volume. This is borne out by the investigations of Datta, Napier, and Newitt (D4) and others (B3, C8, H7). All the above authors studied bubble formation at low gas-flow rates and under nearly constant flow conditions.

Davidson and Schuler (D8), however, find that under the conditions of constant flow, surface tension has no effect on the bubble volume. Similar results are reported by Siemes and Kaufmann (S11). Kumar and Kuloor (K16) predicted that the surface tension effects are large at low flow rates and continuously diminish as the flow rate is increased.

The conclusions have been verified by Ramakrishnan, Kumar, and Kuloor (R1). The results obtained from two liquids of surface tension values 72 and 41 dynes per centimeter are shown in Fig. 5. The values of bubble volume in the two liquids are seen to be different at low flow rates but merge with each other at higher flow rates, indicating that the contribution of surface tension to the bubble volume is negligible at higher flow rates.

At constant pressure conditions, Quigley, Johnson, and Harris (Q1) find that for higher flow rates, the effect of surface tension on bubble volume is negligible. These authors may not have adequately accounted for the large difference in the densities of the two liquids—water and carbon tetrachloride —used by them. Davidson and Schuler find that under constant pressure conditions, surface tension does appreciably affect the bubble volume.

The work currently being conducted by Satyanarayan, Kumar, and Kuloor (S3) indicates that the effect of surface tension is more involved than hitherto appreciated. Some of their data are presented in Figs. 6. and 7. They find that at very small orifice diameters or at very large flow rates, the surface tension variation has negligible influence on the bubble volume. For higher orifice diameters, the influence is more pronounced at small flow rates, as is evident from Fig. 7.

Thus, generally stating, for both constant-flow and constant-pressure conditions, surface tension affects the bubble volume at low flow rates and has negligible influence at high flow rates. This statement is again an oversimplification, because surface tension effects are less evident for highly viscous liquids than for less viscous liquids.

## G. Effect of Liquid Viscosity

A considerable amount of contradiction exists regarding the influence of this variable. Schurmann (S6), while studying bubble formation in various liquids from porous earthenware, concluded that viscosity is the principal factor which determines the bubble volume. Similar conclusions have been drawn by Davidson and Schuler (D8), who find that an increase in viscosity causes a marked increase in the bubble size.

In glaring contradiction to the above results are the findings of Datta *et al.* (D4). These investigators used a number of aqueous glycerine solutions having a wide range of viscosities and found that for orifice diameters of 0.036 to 0.63 cm, a hundredfold increase in viscosity caused a decrease in bubble volume of about 10%. All the above conclusions were arrived at under constant flow conditions, although they are equally applicable to constant pressure conditions also.

The results reported by Quigley *et al.* (Q1) and Coppock and Meiklejohn (C8) are in between the above two extremities. The former authors report that the bubble volume increases slowly with viscosity, whereas the latter find the viscosity to be ineffective.

The above results seem to be contradictory and irreconcilable. This is not so because the effect of viscosity is associated with those of flow rate, surface tension, and orifice diameter. Since the effect of viscosity is negligible when the flow rate tends to zero, even a large difference in the viscosities of the two fluids under consideration does not, at small flow rates, show the influence of viscosity. This is precisely the case in the investigations of Datta *et al.* (D4). What was mistakenly interpreted as the influence of viscosity, was probably, in reality, the influence of the reduced surface tension: though the viscosity had been increased a hundredfold, the surface tension was simultaneously reduced by about 5 dynes per centimeter. At the extremely small flow rates ($<0.1$ cm$^3$/sec) employed, the effect of viscosity was presumably negligible.

Schurmann (S6) and Davidson and Schuler (D8) used extremely small orifices, so that the downward surface-tension force was negligible, and viscosity was the controlling factor in bubble formation Hence these investigators concluded that viscosity has a great influence on the bubble formation.

All the above conclusions and the data reported will be discussed in detail later, when a general model is proposed for bubble formation under constant flow conditions. It will then be shown that each of the above observations is correct although the conclusions drawn from them are applicable only in limited range.

## H. Effect of Liquid Density

When the density of a liquid is increased, the buoyancy force corresponding to a specific size of the bubble increases, whereas the surface-tension force may remain constant. Thus, for a definite amount of surface-tension force, the bubble volume obtained is smaller.

Benzing and Myers (B3), Coppock and Meiklejohn (C8), and Davidson and Schuler (D8) agree that the bubble volume decreases with increasing density of the liquid. Quigley *et al.* (Q1) disagree with the above and observe that

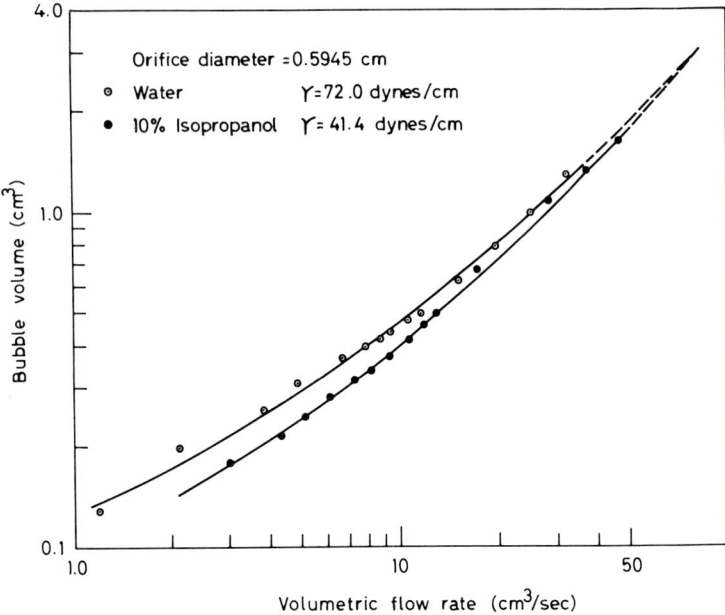

FIG. 5. Effect of surface tension on bubble volume under constant flow conditions.

the liquid density has no effect on the final bubble volume. This discrepancy can be explained as follows:

The increased density of a liquid does increase the buoyancy at all levels of bubble formation but its effect is most pronounced for liquids of high viscosity and at low flow rates. A downward force in addition to that of surface tension is created by the bubbles which expand or move, because a certain amount of the liquid is carried along with them. Although the volume of the liquid being carried along is independent of the liquid density, the downward force increases with increasing density. This net effect of an increase in the downward force is in direct opposition to what takes place at zero flow rates. The first set of investigators mentioned above used either very small flow rates or very high viscosities. Under these conditions, the increased downward force resulted in their observing a decrease in the bubble volume with increasing density. On the other hand, Quigley et al. (Q1), who employed high flow rates and low viscosities, were led to observe the noninfluence of density on the bubble volume.

I. Effect of Gas Properties

There has been no systematic study of the influence of gas properties. The gas density appears in the analysis either in the virtual mass term or in

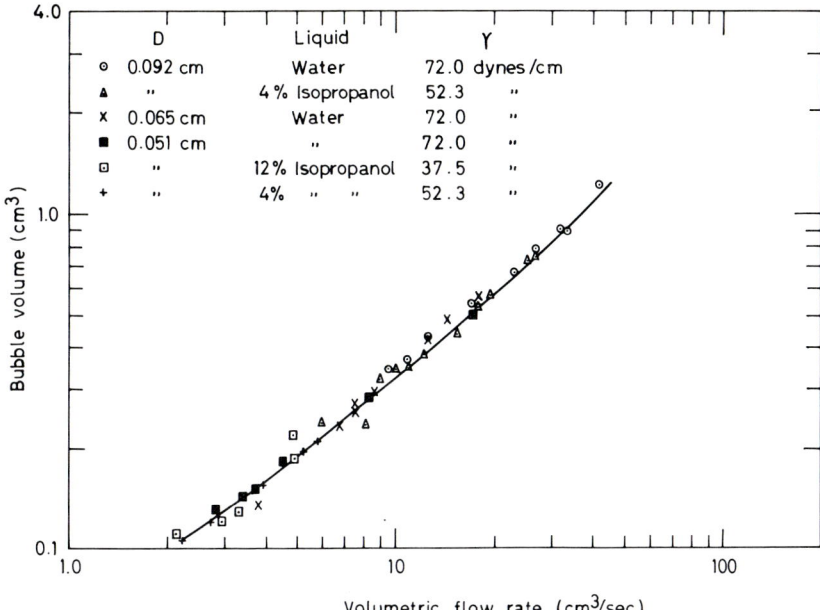

FIG. 6. Effect of surface tension on bubble volume for small orifice diameters under constant pressure conditions.

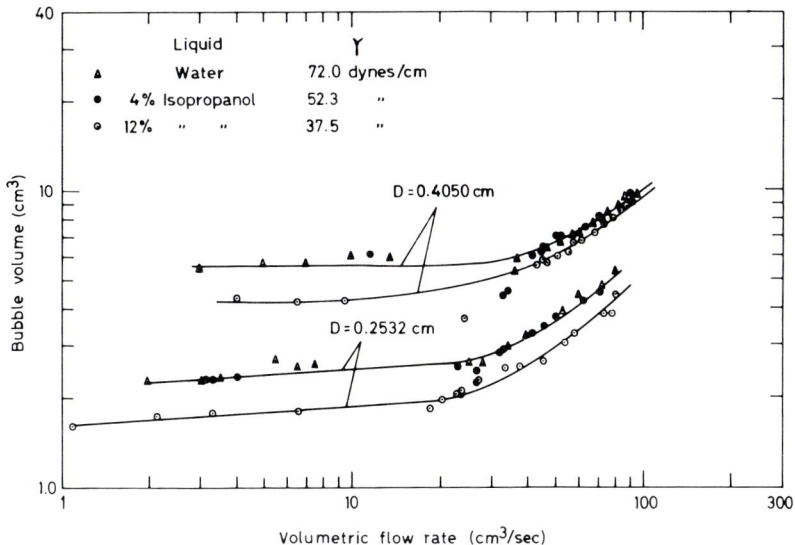

FIG. 7. Effect of surface tension on bubble volume for large orifice diameters under constant pressure conditions.

the buoyancy term. It is generally omitted from the analysis on the assumption that it is negligible when compared with the liquid density (i.e., $\rho_l - \rho_g \approx \rho_l$; $\rho_g + \frac{11}{16}\rho_l \approx \frac{11}{16}\rho_l$).

The viscosity of the gas has again a negligible effect on the bubble volume. Hence, the gas properties can reasonably be considered to be ineffective.

## J. Effect of Gas-Flow Rates

In the initial stages of bubble formation studies the frequency of bubble formation was considered as an important parameter. Later on, the frequency was observed to be decided by the flow rate and the bubble volume. Though the flow rate could be controlled independently, the frequency could not be controlled. More important is the fact that when flow rate is increased continuously for a particular system, a stage is reached when the frequency of bubble formation remains essentially constant, whereas the bubble volume changes. Thus the bubble volume can assume an infinite number of values for a constant frequency. On the other hand, the bubble volume is a single-valued function of the flow rate provided all the other conditions are maintained constant. Thus, for purposes of developing a correlation, it is more reasonable to employ the gas-flow rate, which is an independent variable, rather than the frequency, which is a dependent variable. Hence the wide range of gas flow rate as an operating variable.

There exists a general agreement on the pattern of bubble-volume variation when the flow rate is increased. As the flow rate is gradually increased from zero, the bubble volume remains fairly independent of the flow rate, whereas the frequency increases. The only exceptions to the above observation are provided by the results of Maier (M2), Datta et al. (D4), and Krishnamurthi, Kumar, and Kuloor (K8). They find that the bubble volume first decreases with increasing flow rate, passes through a minimum, and then starts increasing. This behavior is restricted to capillaries when inviscid liquids like water are used. They also find that with increasing viscosity the extent of decrease in bubble volume becomes increasingly less until, for very highly viscous liquids, there is no decrease at all.

Suspecting that constant-flow conditions are not realized even for capillaries and that the effect is due most probably to the chamber volume instead of kinetic effects as explained by the earlier investigators, data were collected using empty capillaries and capillaries filled with glass powder. The results show that for glass-powder-filled capillaries, the minimum is absent (K8).

Further, the constancy of bubble volume with flow rate at small flow rates which is observed by many investigators is true only for inviscid liquids having high surface-tension effects. If highly viscous liquids ($\mu > 500$ cp) are used, the bubble volume increases very fast with the flow rate. On increasing the flow rate, at first both the bubble volume and the frequency increase, but later on a stage is reached when the frequency remains essentially constant whereas the bubble volume continues to increase. Though these regions are observed for all the systems studied, the conditions under which one region ends and the other begins are not clear. There has been no systematic investigation of this aspect of bubble formation, though Davidson and Amick (D5) find that the flow rate at the transition to constant-frequency bubbling increases rapidly with orifice diameter.

K. Effect of Continuous-Phase Velocity

Maier (M2) has given quantitative data showing that the continuous-phase velocity results in a reduction in bubble size. During a study of bubble formation from vertical nozzles, Krishnamurthy *et al.* (K13) observed a decrease in the bubble volume resulting from an increase in buoyancy caused by the continuous-phase velocity. These authors developed equations based on drag considerations which can predict the bubble volume when the continuous phase has a velocity. But, in their study, the continuous-phase velocity is so directed as to decrease the bubble volume, and hence the results cannot be generalized.

If, however, the velocity tends to decrease the buoyancy (as in downward flows), there is every likelihood that the final bubble volume may be larger than that obtained without the continuous-phase velocity.

L. Conclusions

In general, it can be said that there exists a considerable amount of discrepancy in literature regarding the influence of various factors mentioned above, on the bubble volume. These discrepancies are listed in Table I for easy reference.

Most of the above discrepancies have arisen because of the lack of appreciation by various investigators of the effects of the chamber volume, the types of flow associated with it, and of the interaction of the several variables considered, such as the viscosity, surface tension, and flow rate. A proper understanding of such interactions is essential if the discrepancies listed in Table I are to be completely explained.

IV. Bubble Formation under Constant Flow Conditions

A general model applicable to the formation of bubbles under all kinds of operating conditions is not yet available. The complex nature of the problem may very well delay the emergence of such a theory. Even under constant

TABLE I

THE INFLUENCE OF LIQUID PROPERTIES ON BUBBLE VOLUME AS REPORTED BY VARIOUS INVESTIGATORS[a]

| Investigators | Viscosity | Reported effect of: Surface tension | Density |
|---|---|---|---|
| 1. Datta et al. (D4) | Negative (small) | Positive | — |
| 2. Quigley et al. (Q1) | Positive (small) | None | None |
| 3. Coppock and Meiklejohn (C8) | None | Positive | Negative |
| 4. Davidson and Schuler (D8) | Positive (large) | None— constant flow Positive— constant pressure | Negative |
| 5. Benzing and Myers (B3) | None | Positive | Negative |
| 6. Siemes and Kaufmann (S11) | Positive (large) | None | None |

[a] "Negative" means the bubble volume decreases with increase in the value of the property.

flow conditions, the theories available in literature are restricted to special situations only, which are discussed below.

### A. BUBBLE FORMATION IN INVISCID LIQUIDS

Some investigators have excluded the surface-tension force from the analysis of bubble formation since its effects have been found to be negligible at high flow rates.

*1. Bubble Formation in Inviscid Liquids Neglecting Surface Tension Effects*

The omission of the surface-tension force reduces the problem to a hydrodynamical one. Two models are available for this situation.

*a. Davidson and Schuler's Model (D9).* The bubble is assumed to be forming at a point source where the gas is supplied. As the bubble is formed, it moves upward with a velocity which is decided by the various forces acting on it. Detachment is considered to take place when the center of the bubble has covered a distance equal to the sum of the radius of the bubble and the radius of the orifice. For small orifice diameters, the extra distance of the orifice radius is neglected, resulting in a considerable simplification of the analysis. An idealized sequence of bubble formation is shown in Fig. 8.

At each stage of bubble formation, the bubble is assumed to be spherical.

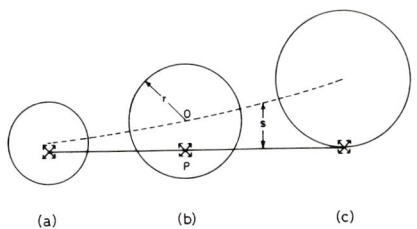

FIG. 8. Idealized sequence of bubble formation in a liquid—Davidson and Schuler's model (D8).

For evaluating the upward movement of the bubble during its formation, the upward force considered is that of buoyancy, which can be written as $V\rho_l g$ when $\rho_g \ll \rho_l$. Since the bubble is surrounded by the liquid, a part of the liquid is carried along with the bubble. For an irrotational flow, as in the present case when a sphere is moving away from a wall, the volume of the liquid carried along with a bubble is 11/16 of the volume of the sphere (M4). Thus, the rate of change of momentum of the bubble due to buoyancy is $(d/dt)[(11/16)V\rho_l v]$. On writing the velocity of the bubble $v$ as $ds/dt$, the final equation of motion for the bubble becomes

$$V\rho_l g = (d/dt)[(11/16)V\rho_l(ds/dt)] \tag{1}$$

As the flow remains constant throughout the bubble formation, the volume of the bubble at any time can be written as

$$V = Qt \tag{2}$$

Equation (2) is substituted in Eq. (1) and the latter integrated by using the following boundary conditions:
At $t = 0$, $(ds/dt) = 0$ and $s = 0$
Thus,

$$s = 4gt^2/11 \tag{3}$$

At detachment, i.e., at $t = t_c$

$$s = r_F = (3Qt_c/4\pi)^{1/3} \tag{4}$$

Substituting Eq. (4) in Eq. (3), we have

$$t_c = (11/4g)^{3/5}(3Q/4\pi)^{1/5} \tag{5}$$

The final bubble volume $V_F$ is given by

$$V_F = Qt_c = 1.378(Q^{6/5}/g^{3/5}) \tag{6}$$

This equation closely resembles the empirical expression of Van Krevelen and Hoftijzer (V1) for the bubble formation in inviscid liquids, provided that the gas density is negligible compared to the liquid density. Their relationship is

$$V_F = 1.722(Q^{6/5}/g^{3/5}) \tag{7}$$

Davidson and Schuler (D9) have given another equation which takes into consideration the residual bubble that forms the nucleus of the succeeding bubble. If this residual volume is $V_0$, then instead of writing $V = Qt$, we have to write

$$V = Qt + V_0 \tag{8}$$

Proceeding as before, we obtain

$$s = \frac{16g}{11}\left[\frac{t^2}{4} + \frac{V_0 t}{2Q} - \frac{V_0^2}{2Q^2}\ln\left(\frac{Qt + V_0}{V_0}\right)\right] \tag{9}$$

The bubble is now assumed to detach when its center has covered a distance equal to the sum of the radius of the final bubble and the radius of the orifice. If the radius of the orifice is $R$ and if it is assumed that $V_0 = (4\pi R_0^3/3)$, then the time of bubble formation can be obtained by plotting on the same axes Eq. (9) and $(r + R)$ as a function of time from Eq. (8).

The quoted authors (D9) collected data on bubble volumes in water, aqueous glycerol, and petroleum ether. They have used Eq. (6) for verifying bubble volumes obtained for flow rates up to 3 cm$^3$/sec. They find that theory and experiment agree excellently only in the flow range of 1.5 to 3.0 cm$^3$/sec and not below 1.5 cm$^3$/sec. This discrepancy has been qualitatively explained by them on the basis of surface tension effects, but there is no quantitative explanation. Although the equation has not been verified from 3 to 15 cm$^3$/sec, the authors feel that it would be applicable. Beyond 20 cm$^3$/sec, the experimental values have been compared with those obtained by using Eqs. (8)–(9) and considerable deviation has been observed.

The deviation of the experimental values from the theoretical ones has been explained as being due partly to the upward current induced by the bubbles in the liquid surrounding the orifice and partly to the distortion in the spherical shape of the forming bubbles.

Davidson and Harrison (D6) have brought about a change in Eq. (1) by using a value of 1/2 in the place of 11/16 for the virtual mass term. Proceeding in exactly the same manner as earlier, they obtained

$$V_F = 1.138(Q^{6/5}/g^{3/5}) \tag{10}$$

Equation (10) has been tested for the data of a large number of investigators. The data are plotted in Fig. 9 along with the Davidson and Harrison line (D6). The agreement is good enough to obtain approximate values, though in most of the cases the theoretically predicted volumes are smaller than those obtained experimentally.

b. *Kumar and Kuloor's Model* (K18). In this model, the bubble formation is assumed to take place in two stages, viz. the expansion stage and the detachment stage. The bubble is assumed to stay at the orifice in the first

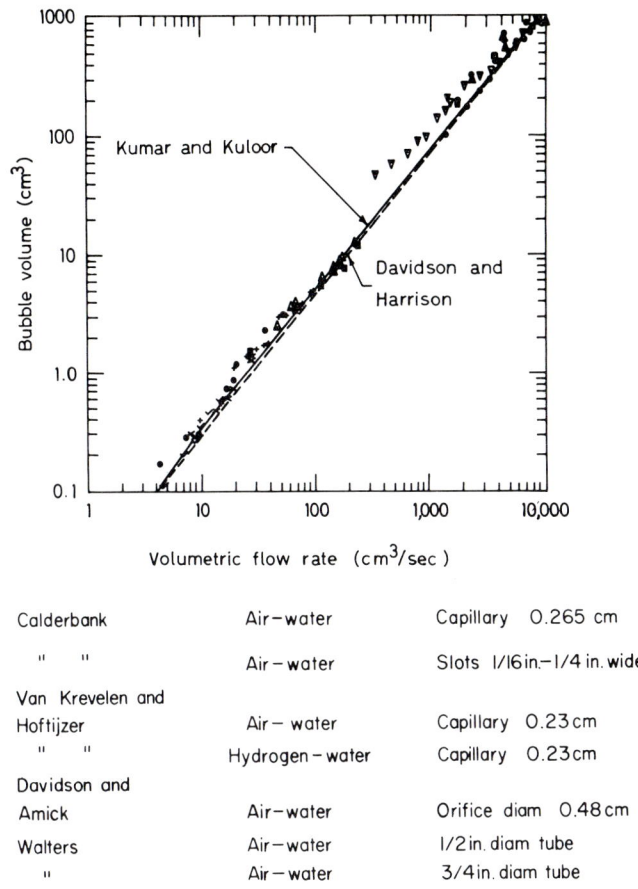

| | | | |
|---|---|---|---|
| + | Calderbank | Air–water | Capillary 0.265 cm |
| △ | " " | Air–water | Slots 1/16 in.–1/4 in. wide |
| ● | Van Krevelen and Hoftijzer | Air–water | Capillary 0.23 cm |
| × | " " | Hydrogen–water | Capillary 0.23 cm |
| ■ | Davidson and Amick | Air–water | Orifice diam 0.48 cm |
| ▽ ○ | Walters | Air–water | 1/2 in. diam tube |
| □ ⊖ | " | Air–water | 3/4 in. diam tube |
| ◐ ▲ | " | Air–water | 1 in. diam tube |
| ∨ | Present investigation | Air–water | Orifice diam 0.05 cm |

FIG. 9. Data on bubble formation in inviscid liquids.

stage, whereas in the second stage it is assumed to travel away from the tip until it detaches itself. An idealized picture of the bubble formation according to this model is shown in Fig. 10. Figure 10a shows the expansion stage where the bubble expands while its base remains fixed. This stage stops as soon as the bubble base starts moving away from the nozzle tip. This marks the beginning of the second stage, which is terminated completely when the bubble detaches from the neck. During the second stage, the bubble not only moves away from the gas source, but also goes on expanding due to the incoming gas.

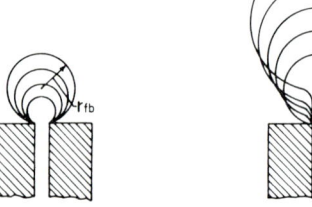
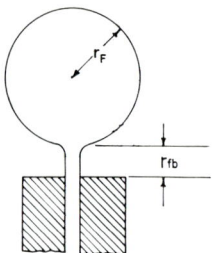

Expansion stage       Detachment stage       Condition of detachment

FIG. 10. Idealized sequence of bubble formation—Kumar and Kuloor's model (K18).

These two stages can also be anticipated from theoretical considerations. When the flow rates are vanishingly small, the surface tension is the main downward force and buoyancy the upward one. The bubble base must remain fixed to the nozzle tip until the buoyancy force exceeds the surface tension force. Apart from this, there is a drag force due to the expansion rate because an expanding bubble is equivalent to a moving bubble (H4). When the orifice diameter tends to zero and the surface tension force is negligible, the resisting force is due solely to the drag of the expanding bubble. This force is large even for small flow rates because the bubble grows from a point source where its rate of expansion is very large. As the bubble grows, the drag due to expansion goes on decreasing, whereas the buoyancy force increases continuously. The first stage comes to an end when the buoyancy force equals the downward forces.

The end of the expansion stage marks the beginning of the detachment stage. Now, the buoyancy force is higher than the downward resisting forces. The other upward force resulting from the kinetic energy of the gas is so small as to be negligible (D8). The main aspect in this stage consists of the motion of an expanding bubble and can be quantitatively expressed through Newton's law of motion. From this, the time of detachment $t_c$ is evaluated as the time required by the bubble-base to cover a distance equal to the radius of the bubble from the first stage. On multiplying $t_c$ with the volumetric flow rate, the extra volume entering the bubble during the second stage is obtained.

The final volume $V_F$ can be evaluated by finding individually the value of the force-balance bubble volume $V_{fb}$ and the volume $Qt_c$ entering the bubble during detachment, and then adding the two. Thus,

$$V_F = V_{fb} + Qt_c \tag{11}$$

(i) Evaluation of $V_{fb}$: the force-balance bubble volume is determined by making use of the equality of the upward buoyancy force with the downward drag force due to expansion.

$$\text{The upward buoyancy force} = V(\rho_l - \rho_g)g \tag{12}$$

The downward expansion force can arise from two sources, viz. the drag and the rate of change of momentum of the bubble and a part of the surrounding liquid. For an irrotational flow, the drag force on a bubble forming in an inviscid fluid is zero. The drag force is present if there is a fully developed wake behind the forming bubble. The forming bubble, however, is similar to a sphere being accelerated from rest, as pointed out by Davidson and Schuler (D9). For such a situation, the flow is essentially irrotational and the wake is not established until the body has covered an appreciable distance (G1, P4).

The other force resulting from expansion can be computed from the rate of change of momentum of the forming bubble, i.e., $(d/dt_e)(Mv_e)$. Here $M$ is the virtual mass of the bubble and is the sum of the mass of the gas in the bubble and that of 11/16'ths of its volume of the liquid surrounding it, and $v_e$ is the velocity of expansion of the bubble. Thus,

$$M = [\rho_g + (11/16)\rho_l]Qt_e \tag{13}$$

or if $\rho_g \ll \rho_l$,

$$M = (11/16)\rho_l Qt_e \tag{14}$$

The upper part of the bubble moves with a velocity equal to the rate of change of its diameter, whereas its base remains stationary. Thus, the average velocity of the bubble can be assumed to be that with which the center of the bubble is ascending, which is equal to the rate of change of the bubble radius:

$$v_e = \frac{dr}{dt_e} = \frac{Q}{4\pi r^2} = \frac{QV^{-2/3}}{4\pi(3/4\pi)^{2/3}} \tag{15}$$

Further,

$$\frac{d}{dt_e}(Mv_e) = M\frac{dv_e}{dt_e} + v_e\frac{dM}{dt_e} \tag{16}$$

The values of $dv_e/dt_e$ and $dM/dt_e$ found by the differentiation of Eqs. (13) and (15), respectively, are

$$dM/dt_e = [\rho_g + (11/16)\rho_l]Q \tag{17}$$

and

$$\frac{dv_e}{dt_e} = -\frac{Q^2 V^{-5/3}}{6\pi(3/4\pi)^{2/3}} \tag{18}$$

Substituting various terms in Eq. (16) and simplifying, we obtain

$$\frac{d}{dt_e}(Mv_e) = \frac{Q^2[\rho_g + (11/16)\rho_l]V^{-2/3}}{12\pi(3/4\pi)^{2/3}} \tag{19}$$

At the end of the expansion stage we have

$$V_{fb}(\rho_l - \rho_g)g = \frac{Q^2[\rho_g + (11/16)\rho_l]V_{fb}^{-2/3}}{12\pi(3/4\pi)^{2/3}} \quad (20)$$

Solving for $V_{fb}$, we obtain

$$V_{fb} = \left[\frac{Q^2[\rho_g + (11/16)\rho_l]}{12\pi(3/4\pi)^{2/3}(\rho_l - \rho_g)g}\right]^{3/5} \quad (21)$$

For $\rho_g \ll \rho_l$,

$$V_{fb} = \left[\frac{11Q^2}{192\pi(3/4\pi)^{2/3}g}\right]^{3/5} \quad (22)$$

(ii) Evaluation of the time of detachment, $t_c$: Equating the rate of change of momentum of the bubble to the net force, we obtain

$$(d/dt)(Mv') = (V_{fb} + Qt)(\rho_l - \rho_g)g \quad (23)$$

Here $v'$, the velocity of the center of the bubble, is made up of two parts, viz. the velocity of the bubble base due to movement and the velocity of the center due to expansion. Thus,

$$v' = v + dr/dt \quad (24)$$

Substituting Eq. (24) in Eq. (23) and simplifying, we obtain

$$M\frac{dv}{dt} + v\frac{dM}{dt} = (V_{fb} + Qt)(\rho_l - \rho_g)g - \frac{Q^2[\rho_g + (11/16)\rho_l](V_{fb} + Qt)^{-2/3}}{12\pi(3/4\pi)^{2/3}} \quad (25)$$

The equation of motion is more useful in the above form when expressed in terms of $v$, the velocity of the base, because the condition of detachment pertains to the bubble base and not to the bubble center. Equation (25) can be written as

$$dv/dT + v/T = P - NT^{-5/3} \quad (26)$$

where

$$P = \frac{(\rho_l - \rho_g)g}{Q[\rho_g + (11/16)\rho_l]}$$

$$N = \frac{Q}{12\pi(3/4\pi)^{2/3}}$$

and

$$T = (V_{fb} + Qt)$$

Equation (26), when solved using the boundary conditions $v = 0$ at $t = 0$ and $T = V_{fb}$ at $t = 0$, yields

$$v = \frac{PT}{2} - 3NT^{-2/3} + \left[3N(V_{fb}^{1/3}) - \frac{P}{2}(V_{fb}^2)\right]\frac{1}{T} \quad (27)$$

Expressing $v$ as $dx/dt$ and solving the resulting equation for a variation of bubble volume from $V_{fb}$ to $V_F$ when $x$ changes from zero to $r_{fb}$, we obtain

$$r_{fb} = \frac{P}{4Q}(V_F^2 - V_{fb}^2) - \frac{9N}{Q}(V_F^{1/3} - V_{fb}^{1/3})$$

$$+ \frac{1}{Q}\left(3N V_{fb}^{1/3} - \frac{P}{2}V_{fb}^2\right)(\ln V_F - \ln V_{fb}) \quad (28)$$

Equation (28) permits a direct evaluation of the final bubble volume, $V_F$, without first calculating the time of detachment. Calculations made on the basis of the above equation are presented in the form of a curve in Fig. 11 using Davidson and Schuler's (D9) data.

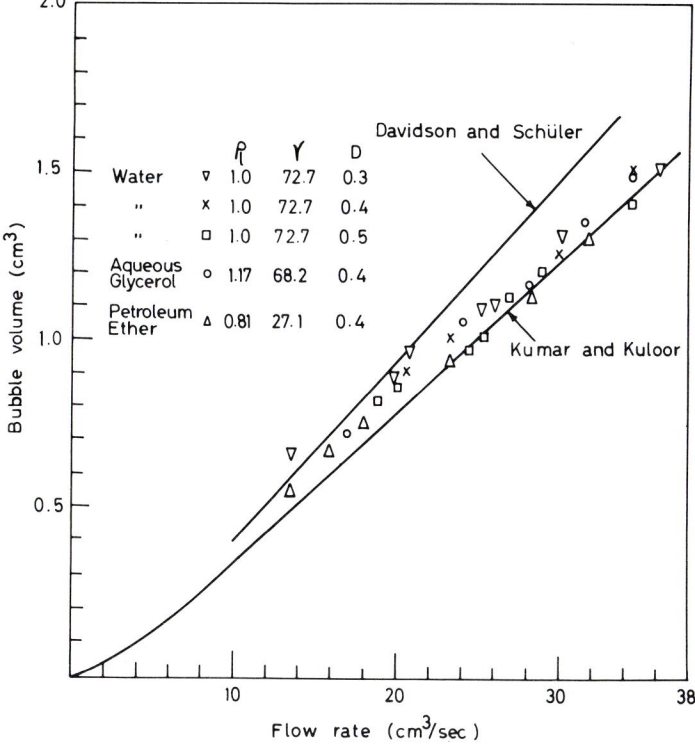

FIG. 11. Comparison of Kumar and Kuloor's model (K18) with the data and model of Davidson and Schuler (D9) for bubble formation in inviscid liquids.

The authors have collected data for flow rates up to 15 cm/³sec using water as the liquid, and have found the model to be applicable. Figure 9 shows a comparison made between the theoretical values of Kumar and Kuloor (K18), the data obtained at higher flow rates by various investigators, and the Davidson and Harrison (D6) equation.

Equation (28) can be simplified at the cost of a certain amount of accuracy. The second term on its right-hand side denotes the difference in the radii of the final bubble and the force-balance bubble, and its value is normally small. Similarly, the third term on the right-hand side is a multiplication of two small numbers and is hence very small. As these two terms have algebraically opposite signs, their difference can be neglected when compared with the first term on the right-hand side. With all these simplifications and by expressing $r_{fb}$ in terms of $V_{fb}$, Eq. (28) reduces to

$$V_F = 0.976(Q^{6/5}/g^{3/5}) \tag{29}$$

This simplified equation is of the same form as the equations of Van Krevelen and Hoftijzer (V1) for air-water system and the theoretical equation of Davidson and Schuler (D9) for inviscid liquids. It is interesting to observe that the various equations differ only in the value of the constant although they are based on different mechanisms.

To indicate the error introduced by the simplification of Eq. (28), the values of bubble volume at different flow rates calculated by both Eqs. (28) and (29) have been presented in Table II. The simplification is seen to give rise to errors of 10 to 30%, hence restricting the use of Eq. (29) to cases where only approximate values are needed.

## 2. Bubble Formation in Inviscid Liquids with Surface Tension

At vanishingly small flow rates, the bubble volume is decided mainly by surface tension. Thus,

$$V_F = \frac{2\pi R \gamma \cos \theta}{(\rho_l - \rho_g)g} \tag{30}$$

At very large flow rates, surface tension effects, though present, have a relatively smaller contribution. Between these two extremes is a region of flow rates where neither the flow rate nor the surface tension can be neglected and the final bubble volume is highly sensitive to both.

Kumar and Kuloor (K16) have extended their model to include surface tension. Here again, evaluation of the bubble volume comprises calculation of $V_{fb}$ and $t_c$. $V_{fb}$ is evaluated as before, but also taking the surface tension force $\pi D \gamma \cos \theta$ into account. Thus, the final equation is

$$V_{fb} - \frac{11Q^2}{192\pi(3/4\pi)^{2/3}g} V_{fb}^{-2/3} = \frac{\pi D \gamma \cos \theta}{\rho_l g} \tag{31}$$

## TABLE II

Values of Bubble Volume Calculated by Rigorous [Eq. (28)] and Simplified [Eq. (29)] Equations for Inviscid Liquids without Surface-Tension Effect

| No. | Volumetric flow rate (cm³/sec) | Bubble volume calculated from the | | Ratio = (i)/(ii) |
|---|---|---|---|---|
| | | (i) Simplified equation | (ii) Rigorous equation | |
| 1  | 1.0     | 0.0156   | 0.0181    | 0.8625 |
| 2  | 2.5     | 0.0470   | 0.0568    | 0.8270 |
| 3  | 10.0    | 0.2482   | 0.3650    | 0.6810 |
| 4  | 15.0    | 0.4037   | 0.5650    | 0.7140 |
| 5  | 20.0    | 0.5702   | 0.7625    | 0.7480 |
| 6  | 25.0    | 0.7452   | 0.9050    | 0.8232 |
| 7  | 35.0    | 1.1160   | 1.4700    | 0.7890 |
| 8  | 100.0   | 3.9300   | 4.9000    | 0.8020 |
| 9  | 200.0   | 9.0360   | 9.8500    | 0.9160 |
| 10 | 1000.0  | 62.3400  | 73.1000   | 0.8512 |
| 11 | 10000.0 | 988.1000 | 1150.0000 | 0.8510 |

When $Q$ tends to zero, Eq. (31) simplifies to Eq. (30). If the surface-tension effect is negligible, then the right-hand side of Eq. (30) vanishes and Eq. (31) reduces to Eq. (22), which is applicable to inviscid liquids without surface tension.

In order to calculate the final bubble volume, we again write the equation of motion of the bubble center which by, inclusion of the surface-tension effect becomes

$$M(dv'/dt) + v'(dM/dt) = (V_{fb} + Qt)\Delta\rho g - \pi D\gamma \cos\theta \qquad (32)$$

where $\Delta\rho = (\rho_l - \rho_g)$. The following substitutions are now made:

$$\frac{(\rho_l - \rho_g)g}{Q[\rho_g + (11/16)\rho_l]} = P$$

$$\frac{Q}{12\pi(3/4\pi)^{2/3}} = N$$

$$\frac{\pi D\gamma \cos\theta}{Q[\rho_g + (11/16)\rho_l]} = J$$

The final equation for the evaluation of $V_F$ becomes

$$r_{fb} = \frac{P}{4Q}(V_F^2 - V_{fb}^2) - \frac{9N}{Q}(V_F^{1/3} - V_{fb}^{1/3}) - \frac{J}{Q}(V_F - V_{fb})$$

$$+ \frac{1}{Q}\left[\left(JV_{fb} + 3NV_{fb}^{1/3} - \frac{P}{2}V_{fb}^2\right)\right](\ln V_F - \ln V_{fb}) \quad (33)$$

This equation has been verified (K16) by using two orifices of diameters 0.05 cm and 0.59 cm. The values of $D\gamma \cos\theta$ for the two are 3.6 and 31.9, respectively. The data for the 0.05 cm diameter orifice can be adequately expressed by Eq. (28). It is necessary to take into account the surface tension for the latter case through Eq. (33).

Though the effects of surface tension are present throughout the range of flow rates, their contribution to the bubble volume decreases as the flow rate increases. This is evident from Fig. 12 wherein for different orifice diameters data, with and without surface tension effects, are plotted for the air-water system. The dotted lines indicate that the surface tension has been taken into account, whereas the full lines indicate the omission of surface tension from the analysis. In each case, the dotted line is seen to tend towards the full line, thus showing that for large flow rates the surface tension need not be considered in the evaluation of bubble volume. Further, for inviscid liquids the region where the dotted line approaches the full line depends on both the surface tension and the orifice diameter. The higher the value of the product of the two, the higher is the flow rate at which the surface tension effects become negligible, as can be seen in Fig. 12.

Thus, we conclude that the surface-tension effects can be neglected only at higher flow rates and not at lower ones. The error caused by neglect at low flow rates can be quite large, its magnitude depending on the orifice diameter. Equation (33) can be used for inviscid liquids both in the presence and absence of surface-tension effects.

The use of Eq. (33) as it stands, involves trial and error method and is time-consuming. At the cost of some accuracy, the equation can be considerably simplified.

The second and the fourth terms on the right-hand side of Eq. (33) are not only very small but also have opposite signs. Their difference is still smaller when compared to the rest of the terms. Thus, for a wide range of experimental conditions these two terms can be neglected. The simplified equation can be written as

$$r_{fb} = (P/4Q)(V_F^2 - V_{fb}^2) - (J/Q)(V_F - V_{fb}) \quad (34)$$

Equation (34) is of second degree in $V_F$ and can be solved directly. Thus,

$$V_F = \frac{B \pm (B^2 - 4AC)^{1/2}}{2A} \quad (35)$$

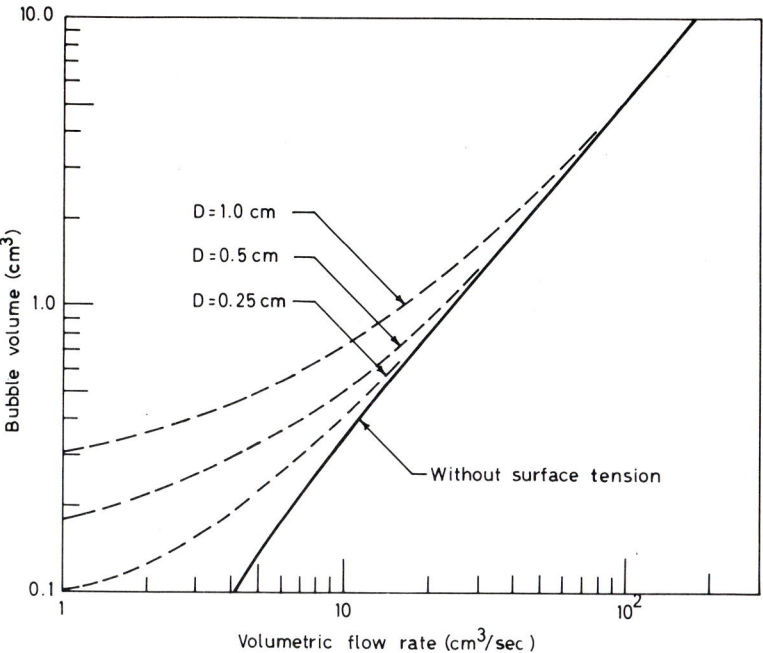

FIG. 12. Influence of surface tension on bubble volume for inviscid liquids.

where

$$A = P/4Q$$
$$B = J/Q$$

and

$$C = (J/Q)V_{fb} - (P/4Q)V^2_{fb} - r_{fb}$$

Equations (33) and (35) have been compared, and the values obtained by the two equations for arbitrary flow rates above 5 cm³/sec are presented in Table III. This table shows that the error introduced in the evaluation of bubble volumes by the use of the simplified Eq. (35) is of the order of 15%. Thus, Eq. (35) can be employed only when approximate values are required.

B. BUBBLE FORMATION IN HIGHLY VISCOUS LIQUIDS

1. *Davidson and Schuler's Model* (D8)

This model makes use of the same concepts as for inviscid liquids except that the viscous resistance is now taken into consideration. Further, the

## TABLE III

VALUES OF BUBBLE VOLUME CALCULATED BY RIGOROUS [EQ. (33)] AND SIMPLIFIED [EQ. (35)] EQUATIONS FOR INVISCID LIQUIDS WITH SURFACE-TENSION EFFECT[a]

| No. | Volumetric flow rate ($cm^3$/sec) | Bubble volume calculated from the | | Ratio = (i)/(ii) |
|---|---|---|---|---|
| | | (i) Simplified equation | (ii) Rigorous equation | |
| 1 | 5 | 0.23 | 0.26 | 0.88 |
| 2 | 10 | 0.39 | 0.39 | 1.00 |
| 3 | 20 | 0.78 | 0.71 | 1.09 |
| 4 | 100 | 4.38 | 3.65 | 1.20 |

[a] $\gamma = 71.7$ dyn/cm; $D = 0.25$ cm.

inertia of the liquid can be neglected at low flow rates, whereas it cannot be neglected for high flow rates. Thus, two separate equations have to be used for the above two cases in order to determine the bubble size.

*a. Small Flow Rates.* In this case, the bubble is assumed to be moving at any instant with the Stokes velocity appropriate to its size. The bubble is assumed to be spherical and unaffected by the presence of other bubbles. Further, the gas momentum and the liquid circulation are assumed to be negligible.

The velocity of the bubble at any instant can be written as

$$v = ds/dt = 2gr^2\rho_l/9\mu \qquad (36)$$

Based on Eq. (36) and the condition of detachment, the final equation is

$$V_F = (4\pi/3)^{1/4}(15\mu Q/2\rho_l g)^{3/4} \qquad (37)$$

This equation has been tested by Davidson and Schuler (D8) for an orifice radius of 0.0334 cm, liquid viscosity ranging from 515 to 1040 cp, and a flow range of about 0 to 2.5 $cm^3$/sec.

It may, however, be pointed out that the lowest flow rate obtained by them is about 0.25 $cm^3$/sec. Below this value, Krishnamurthi (K6) has found that the surface tension is very important and should be taken into account if bubble volumes are to be predicted accurately by theory.

*b. Large Flow Rates.* The momentum of the surrounding liquid is also considered here. The equation for bubble movement then becomes

$$Mg = (11/16)[d(Mv)/dt] + 6\pi r\mu v \qquad (38)$$

We know that $M = \rho_l Qt$
Hence,
$$Qgt = \frac{11}{16}Qt\frac{dv}{dt} + \frac{11}{16}vQ + 6\pi\frac{\mu}{\rho_1}\left(\frac{3Qt}{4\pi}\right)^{1/3}v \tag{39}$$

The above equation is applicable when the bubble detaches at $s = r_F$. But if the orifice diameter is large, the authors recommend the use of the relation $s = (r_F + R)$ to obtain a better value of the time of bubble formation. Then Eq. (39) is modified to

$$(Qt + V_0)g = \frac{11}{16}(Qt + V_0)\frac{dv}{dt} + \frac{11}{16}Qv + 6\pi\frac{\mu}{\rho_1}\left(\frac{3}{4\pi}\right)^{1/3}(Qt + V_0)^{1/3}v \tag{40}$$

Rearranging, we obtain

$$Q\frac{dv}{dV} + \left(\frac{Q}{V} + \frac{A}{V^{2/3}}\right)v = \frac{16}{11}g \tag{41}$$

where
$$V = (Qt + V_0)$$
and
$$A = \frac{96\pi}{11}\left(\frac{\mu}{\rho_l}\right)\left(\frac{3}{4\pi}\right)^{1/3}$$

On integrating Eq. (41) between the limits $V = V_0$ and $V = V$, and by assuming that at $V = V_0$, $v = 0$, we obtain

$$v = Q\frac{ds}{dV} = \frac{48g}{11Q}\left[\frac{V^{2/3}}{C} - \frac{5V^{1/3}}{C^2} - \frac{20}{C^3} - \frac{60V^{-1/3}}{C^4} + \frac{120V^{-2/3}}{C^5} - \frac{120V^{-1}}{C^6}\right] - \frac{L}{V\exp(CV^{1/3})} \tag{42}$$

where
$$L = \frac{48g}{11Q}\exp(CV_0^{1/3})\left[\frac{V_0^{5/3}}{C} - \frac{5V_0^{4/3}}{C^2} + \frac{20V_0}{C^3} - \frac{60V_0^{2/3}}{C^4} + \frac{120V_0^{1/3}}{C^5} - \frac{120}{C^6}\right]$$

and
$$C = 3A/Q$$

Equation (42) is again integrated from $V = V_0$ where $s = 0$ to $V = V$, resulting in

$$s = \frac{48g}{11Q^2}\left[\frac{3V^{5/3}}{5C} - \frac{15V^{4/3}}{5C} + \frac{20V}{C^3} - \frac{90V^{2/3}}{C^4} + \frac{360V^{1/3}}{C^5} - \frac{120\ln V}{C^6}\right]_{V_0}^{V} - \frac{L}{Q}\left[\ln V + 3\sum_{n=1}^{\infty}\frac{(-CV^{1/3})^n}{n.n!}\right]_{V_0}^{V} \tag{43}$$

A graph of $s$ versus $t$ can be plotted by making use of Eq. (43). The time required to reach a volume such that $s = (r_F + R)$ is thus determined. The final bubble volume is obtained by multiplying the calculated time with the volumetric flow rate of the gas.

The authors have verified Eq. (43) for an orifice diameter of 0.192 cm. The results, given in Table IV, show good agreement between the theoretical and the experimental values. The data for the range of 2.5 cm$^3$/sec to 14 cm$^3$/sec are not reported by the authors and it is not known which of the equations is applicable.

## 2. Kumar and Kuloor's Model (K19)

The model for inviscid liquids is equally well applicable to viscous liquids also, provided that the resistance due to viscous drag is included in the analysis. As a first approximation, the viscous drag may be evaluated by a Stokes resistance term, since the bubble is not followed by a wake. Thus we proceed as before, first evaluating the force-balance bubble volume $V_{fb}$ and then the total bubble volume by reference to the detachment stage.

The extra viscous resistance due to expansion to be included in the analysis is

$$6\pi r \mu v_e = \frac{3\mu Q}{2(3/4\pi)^{1/3}} V^{-1/3} \tag{44}$$

The equation for $V_{fb}$ then becomes

$$V_{fb} = \frac{11 Q^2 V_{fb}^{-2/3}}{192 \pi g (3/4\pi)^{2/3}} + \frac{3Q}{2g} (3/4\pi)^{-1/3} \left(\frac{\mu}{\rho_l}\right) V_{fb}^{-1/3} \tag{45}$$

It is interesting to note that Eq. (45) reduces to the inviscid case when the viscosity of the liquid $\mu$ is equal to zero. On rearranging and calculating various constants in Eq. (45), we obtain

$$V_{fb} = 4.85 \times 10^{-5} Q^2 V_{fb}^{-2/3} + 2.47 \times 10^{-3} (\mu/\rho_l) Q V_{fb}^{-1/3} \tag{46}$$

The equation of motion for the bubble center can now be written as

$$(d/dt)(Mv') = (V_{fb} + Qt)(\Delta\rho)g - 6\pi r \mu v' \tag{47}$$

where $v'$, the velocity of the bubble center, is the velocity of the center due to expansion, $dr/dt$, plus the velocity of the base, $v$. Substituting this value of $v'$ in Eq. (47) and simplifying, we obtain

$$M \frac{dv}{dt} + v \frac{dM}{dt} = (V_{fb} + Qt)(\Delta\rho)g - \frac{Q^2[\rho_g + (11/16)\rho_l]}{12\pi(3/4\pi)^{2/3}} (V_{fb} + Qt)^{-2/3}$$
$$- 6\pi r \mu v_e - 6\pi r \mu v \tag{48}$$

TABLE IV

COMPARISON OF THEORETICAL AND EXPERIMENTAL VALUES OF
DAVIDSON AND SCHULER (D8) FOR VISCOUS LIQUIDS AT
LARGE FLOW RATES[a]

| No. | Flow rate (cm³/sec) | Kinematic viscosity (stokes) | Bubble volume (cm³) | | Theory |
|---|---|---|---|---|---|
| | | | Experiment | Theory | Experiment |
| 1 | 18 | 7.9 | 2.1 | 1.9 | 0.90 |
| 2 | 24 | 7.9 | 2.6 | 2.5 | 0.96 |
| 3 | 33 | 7.9 | 3.2 | 3.3 | 1.03 |
| 4 | 15 | 6.5 | 1.6 | 1.6 | 1.00 |
| 5 | 26 | 6.5 | 2.4 | 2.4 | 1.00 |
| 6 | 35 | 6.5 | 3.0 | 3.1 | 1.03 |
| 7 | 14 | 5.4 | 1.3 | 1.3 | 1.00 |
| 8 | 20 | 5.4 | 1.8 | 1.7 | 0.95 |
| 9 | 26 | 5.4 | 2.2 | 2.2 | 1.00 |

[a] Orifice diameter = 0.192 cm.

It is to be noted that on expressing the motion of the center in terms of the motion of the base of the bubble, the various terms of Eq (47) have been split into two parts, viz. (i) that associated with the movement of the base due to the free motion of the bubble (this would be the only term if the bubble does not change its size during its movement), and (ii) that associated with the movement of the bubble center due to expansion. The inertial term has split into left-hand-side terms and the second term on the right-hand side, whereas the viscous drag term has split into the third and fourth terms on the right-hand side of Eq. (48).

After the evaluation of various terms of Eq. (48), it can be written as

$$(V_{fb} + Qt)\left(\rho_g + \frac{11}{16}\rho_l\right)\frac{dv}{dt} + \left(\rho_g + \frac{11}{16}\rho_l\right)Qv$$

$$= (V_{fb} + Qt)(\rho_l - \rho_g)g - \frac{3\mu Q}{2(3/4\pi)^{1/3}}(V_{fb} + Qt)^{-1/3}$$

$$- \frac{Q^2[\rho_g + (11/16)\rho_l]}{12\pi(3/4\pi)^{2/3}}(V_{fb} + Qt)^{-2/3}$$

$$- 6\pi(3/4\pi)^{1/3}\mu(V_{fb} + Qt)^{1/3}v \qquad (49)$$

Although the above equation can be solved as such, the solution is very complicated and needs the help of a computer to be practically applicable. Hence, it has been slightly simplified. Though the buoyancy force is directly proportional to the volume of the bubble, the resisting force is proportional only to its cube root. Thus, $r$ can be assumed to be a nonvariable. This is considered to be equal to 1.25 $r_{fb}$ rather than $r_{fb}$ since 1.25 $r_{fb}$ gives approximately an average value of $r$ during the expansion stage. Equation (49) now becomes

$$dv/dT + A(v/T) = B - CT^{-4/3} - DT^{-5/3} \tag{50}$$

where

$$A = 1 + \frac{(1.25)6\pi(3/4\pi)^{1/3}\mu V_{fb}^{1/3}}{Q[\rho_g + (11/16)\rho_l]}$$

$$\approx 1 + \frac{96}{11}\frac{(1.25)\pi r_{fb}\mu}{Q\rho_l}$$

$$B = \frac{(\rho_l - \rho_g)g}{Q[\rho_g + (11/16)\rho_l]}$$

$$\approx \frac{16g}{11Q}$$

$$C = \frac{3\mu}{2(3/4\pi)^{1/3}[\rho_g + (11/16)\rho_l]}$$

$$\approx \frac{24}{11(3/4\pi)^{1/3}}\frac{\mu}{\rho_l}$$

$$D = \frac{Q}{12\pi(3/4\pi)^{2/3}}$$

and $T = (V_{fb} + Qt)$

The solution of Eq. (50) with the boundary condition, $T = V_{fb}$ or $v = 0$ at $t = 0$, is

$$v = \frac{1}{T^A}\left[\frac{B}{A+1}(T^{A+1} - V_{fb}^{A+1}) - \frac{C}{A - 1/3}(T^{A-1/3} - V_{fb}^{A-1/3})\right.$$

$$\left. - \frac{D}{A - 2/3}(T^{A-2/3} - V_{fb}^{A-2/3})\right] \tag{51}$$

$v$ can be expressed as $dx/dt$ and the resulting equation is solved again for $x$ varying from 0 to $r_{fb}$ (condition of detachment) when $T$ changes from $V_{fb}$ to $V_F$. The final solution is

$$r_{fb} = \frac{B}{Q(A+1)} \left[ \frac{1}{2}(V_F^2 - V_{fb}^2) - \frac{1}{-A+1}(V_F^{-A+1}V_{fb}^{A+1} - V_{fb}^2) \right]$$

$$- \frac{C}{Q(A-1/3)} \left[ \frac{3}{2}(V_F^{2/3} - V_{fb}^{2/3}) \right.$$

$$\left. - \frac{1}{-A+1}(V_F^{-A+1}V_{fb}^{A-1/3} - V_{fb}^{2/3}) \right]$$

$$- \frac{D}{Q(A-2/3)} \left[ 3(V_F^{1/3} - V_{fb}^{1/3}) \right.$$

$$\left. - \frac{1}{-A+1}(V_F^{-A+1}V_{fb}^{A-2/3} - V_{fb}^{1/3}) \right] \quad (52)$$

As $A$ is quite large and $D$ is quite small in most of the situations of practical interest, Eq. (52) can be simplified as

$$r_{fb} = \frac{B}{2Q(A+1)}(V_F^2 - V_{fb}^2) - \frac{3C}{2Q(A-1/3)}(V_F^{2/3} - V_{fb}^{2/3}) \quad (53)$$

The authors have verified Eq. (53) by using their data and find the agreement to be good.

The data of Davidson and Schuler (D8) for higher flow rates have also been tested. Their experimental values are given in Table IV along with those calculated by this model. The authors collected data at flow rates higher than those of Davidson and Schuler (D8) and plotted them as shown in Fig. 13, where the solid line represents the values calculated through Eq. (53).

## C. Model of Ramakrishnan, Kumar, and Kuloor (R1)

This model is an extension of the concepts developed by Kumar and Kuloor (K16, K19). The first stage equation for this situation is

$$V_{fb}^{5/3} = \frac{11Q^2}{192\pi(3/4\pi)^{2/3}g} + \frac{3\mu Q V_{fb}^{1/3}}{2(3/4\pi)^{1/3}g\rho_l} + \frac{\pi D\gamma \cos\theta V_{fb}^{2/3}}{g\rho_l} \quad (54)$$

Evaluating various constants, we obtain

$$V_{fb}^{5/3} = (4.85)(10^{-5})Q^2 + (2.47)(10^{-3})\left(\frac{\mu}{\rho_l}\right)QV_{fb}^{1/3} + \frac{\pi D\gamma \cos\theta V_{fb}^{2/3}}{g\rho_l} \quad (55)$$

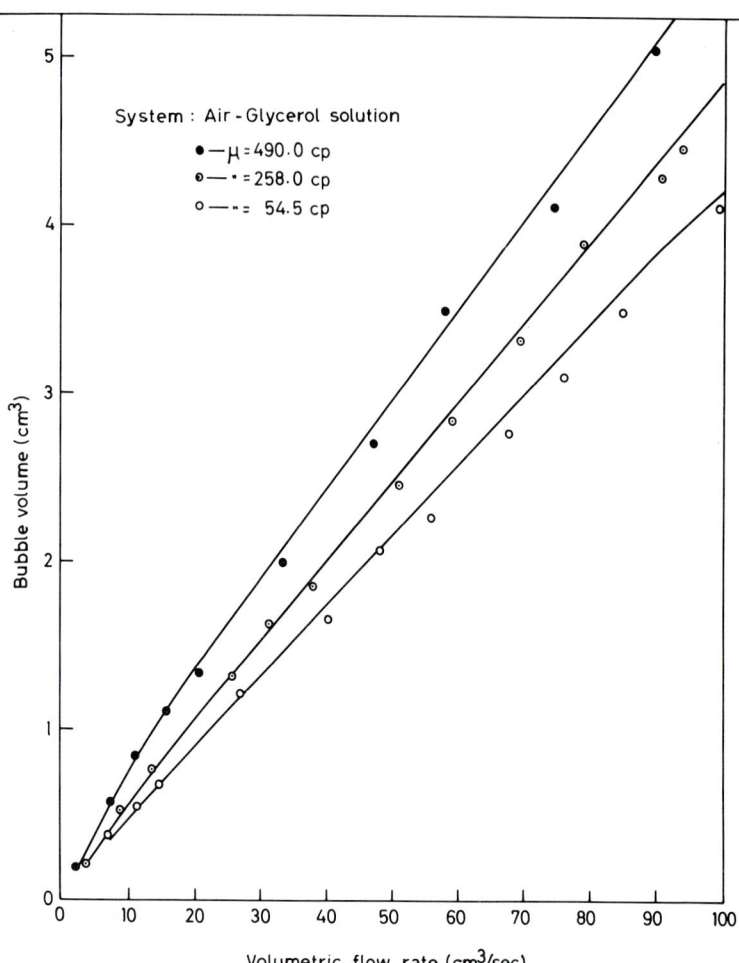

FIG. 13. Comparison of the model (K19) with the experimental data for bubble formation in viscous liquids without surface-tension effect.

Equation (54) is quite general in nature, and can be used for the evaluation of $V_{fb}$ in any situation when the bubble formation is taking place under constant flow conditions. Thus, if the last term on the right-hand side of Eq. (54) is omitted, the equation reduces to that of the viscous case without surface-tension effects. Similarly, if the second term on the right-hand side is dropped, the resulting equation is the one for the inviscid case with surface tension. Further, if both the second and the third terms on the right-hand side are deleted, the equation reduces to that of the inviscid case without surface tension.

The equation of motion for the bubble center now differs from that for the viscous case only so far as the inclusion of the surface tension force. Thus,

$$M\frac{dv}{dt} + v\frac{dM}{dt} = (V_{fb} + Qt)\Delta\rho g - \frac{Q^2[\rho_g + (11/16)\rho_l](V_{fb} + Qt)^{-2/3}}{12\pi(3/4\pi)^{2/3}}$$
$$- 6\pi r\mu v_e - 6\pi r\mu v - \pi D\gamma\cos\theta \quad (56)$$

As indicated before, $r$ for the purpose of viscous resistance can be considered to be constant and equal to $1.25 r_{fb}$.

$$dv/dT + A(v/T) = B - CT^{-4/3} - ET^{-1} - DT^{-5/3} \quad (57)$$

where

$$A = 1 + \frac{(1.25)(6\pi)(3/4\pi)^{1/3}V_{fb}^{1/3}\mu}{Q[\rho_g + (11/16)\rho_l]}$$

$$\approx 1 + \frac{(1.25)(96\pi r_{fb}\mu)}{11\rho_l Q}$$

$$B = \frac{(\rho_l - \rho_g)g}{Q[\rho_g + (11/16)\rho_l]} \approx \frac{16g}{11Q}$$

$$C = \frac{3\mu}{2(3/4\pi)^{1/3}[\rho_g + (11/16)\rho_l]}$$

$$D = \frac{Q}{12\pi(3/4\pi)^{2/3}}$$

$$E = \frac{\pi D\gamma\cos\theta}{Q[\rho_g + (11/16)\rho_l]} \approx \frac{16\pi D\gamma\cos\theta}{11Q\rho_l}$$

$$T = (V_{fb} + Qt)$$

The solution of Eq. (57) is formally

$$v = \left[\int R\exp\left(\int A\,dT/T\right) + K\right]\exp\left(-\int A\,dT/T\right)$$

where $R$ is the right-hand member of the equation and $K$ is a constant of integration; whence

$$v = \frac{1}{T^A}\left[\frac{B}{A+1}(T^{A+1} - V_{fb}^{A+1}) - \frac{E}{A}(T^A - V_{fb}^A)\right.$$
$$\left. - \frac{C}{A - 1/3}(T^{A-1/3} - V_{fb}^{A-1/3}) - \frac{D}{A - 2/3}(T^{A-2/3} - V_{fb}^{A-2/3})\right] \quad (58)$$

that is

$$Q\frac{dx}{dT} = \frac{B}{A+1}\left(T - \frac{V_{fb}^{A+1}}{T^A}\right) - \frac{E}{A}\left(1 - \frac{V_{fb}^A}{T^A}\right)$$
$$- \frac{C}{A - 1/3}\left(T^{-1/3} - \frac{V_{fb}^{A-1/3}}{T^A}\right) - \frac{D}{A - 2/3}\left(T^{-2/3} - \frac{V_{fb}^{A-2/3}}{T^A}\right) \quad (59)$$

Integrating for $x$ between the limits 0 and $r_{fb}$ and $T$ between the limits $V_{fb}$ and $T$, we get

$$Qx\Big|_0^{r_{fb}} = \frac{B}{2(A+1)}(V_F^2 - V_{fb}^2) - \frac{E}{A}(V_F - V_{fb})$$
$$- \frac{3C}{2(A-1/3)}(V_F^{2/3} - V_{fb}^{2/3}) - \frac{3D}{(A-2/3)}(V_F^{1/3} - V_{fb}^{1/3})$$
$$- \left(\frac{V_F^{-A+1}}{-A+1} - \frac{V_{fb}^{-A+1}}{-A+1}\right)\left[\frac{B}{(A+1)}V_{fb}^{A+1} - \frac{E}{A}V_{fb}^A\right.$$
$$\left. - \frac{C}{A-1/3}V_{fb}^{A-1/3} - \frac{D}{A-2/3}V_{fb}^{A-2/3}\right] \tag{60}$$

A large number of calculations show that the last two terms of Eq. (60) are extremely small and can be neglected. Then the equation simplifies to

$$r_{fb} = \frac{B}{2Q(A+1)}(V_F^2 - V_{fb}^2) - \frac{E}{AQ}(V_F - V_{fb})$$
$$- \frac{3C}{2Q(A-1/3)}(V_F^{2/3} - V_{fb}^{2/3}) \tag{61}$$

Knowing the force balance bubble volume $V_{fb}$ and the other system charctaeristics, the final bubble volume $V_F$ can be directly calculated from Eq. (61).

This general equation presents a number of noteworthy features. If the surface tension term $E$ is dropped, then the equation corresponds exactly to the equation for highly viscous liquids without surface-tension effects.

Similarly, all the cases discussed earlier become the special cases of this general equation, when appropriate terms are omitted. As this equation contains all the variables of importance, the verification of the data of the earlier investigators and explanations of the contradictory observations reported in literature should be possible by a judicious discussion of it.

### 1. *Effect of Surface Tension*

Considering the case of vanishingly small flows ($Q \approx 0$), both the first and the second terms on the right-hand side of Eq. (55) which contain $Q$, become zero, and the force balance bubble volume is obtained directly by equating the surface tension force with the buoyancy force. Even in the detachment stage, $Qt_c$ is zero because $Q \approx 0$. Thus, for zero flow rate, the equation reduces to the well-known Tate's law, which is used for finding the surface tension of liquids by the pendent drop or the forming bubble method.

The same situation exists when the viscosity is large but $Q$ is extremely small. The second term on the right-hand side of Eq. (55) is still small when compared to the third term and the bubble volumes are decided by the surface tension. As the flow is increased, the various terms which contain $Q$ go on increasing but the surface tension term remains constant. Thus, the surface tension effect goes on decreasing with increasing flow rate both for highly viscous and inviscid liquids. For highly viscous liquids, however, the relative magnitude of the surface tension force in the overall resisting force becomes less. Thus, from the general model, the following conclusions can be drawn regarding the influence of surface tension on the bubble volume:

*a.* At vanishingly small flow rates, the bubble volume is decided entirely by the surface tension and buoyancy forces.

*b.* For liquids of low viscosity, the surface tension is important at low flow rates, but assumes a continuously lesser importance as the flow rate is increased. Finally, a stage is reached when its contribution to the bubble volume becomes negligible.

*c.* For highly viscous liquids, as the flow rate is increased, the surface tension loses its importance at much smaller flow rates than those for inviscid liquids.

In order to verify the importance of surface tension, a large amount of data has been collected for bubble formation in liquids having different surface tension but nearly constant viscosity. Liquids of both low and high viscosity have been used. The equation is seen to agree excellently with the data obtained over a wide range of flow rates. It was also observed that for highly viscous liquids the surface-tension effects become negligible at much smaller flow rates.

The existence of a considerable discrepancy in the results reported by various investigators, has already been mentioned. Datta *et al.* (D4), Coppock and Meiklejohn (C8), and Benzing and Myers (B3) find increase in surface tension to increase bubble volume (Table I). All these investigators have used liquids of low viscosities and extremely small flow rates (about 0.008 cm$^3$/sec). Under these conditions the influence of drag terms during both the stages is very small ($<5\%$), and hence the influence of surface tension shows up strongly. Table V compares the experimental data of several investigators with the theoretical values calculated from the general model. The model is seen to predict the actual trends well.

Quigley, Johnson, and Harris (Q1) find that the surface tension has no effect on the bubble size. A comparison of their data with the present general model is not possible because they conducted their experiments under constant pressure, not constant flow, conditions. However, the flow rates employed by them are high; in such cases, even for constant flow conditions, the influence of surface tension is small.

## TABLE V

Comparison of Experimental Data of Various Investigators with the Values Calculated from General Model Showing the Influence of Surface Tension[a]

| Surface tension (dyn/cm) | Density $\rho_l$ (gm/cm$^3$) | $D$ (cm) | $Q$ (cm$^3$/sec) | Bubble volume (cm$^3$) Experimental | Bubble volume (cm$^3$) Calculated |
|---|---|---|---|---|---|
| \multicolumn{6}{c}{1. Datta et al. (D4)} | | | | | |
| 73.10 | 0.9997 | 0.036 | 0.0122 | 0.0292 | 0.0346 |
| 64.60 | 0.9954 | 0.036 | 0.0099 | 0.0237 | 0.0306 |
| 49.96 | 0.9844 | 0.036 | 0.0083 | 0.0199 | 0.0240 |
| 32.39 | 0.9518 | 0.036 | 0.0054 | 0.0129 | 0.0160 |
| 22.88 | 0.7936 | 0.036 | 0.0052 | 0.0126 | 0.0136 |
| 73.10 | 0.9997 | 0.630 | 0.0446 | 0.1071 | 0.1552 |
| 64.60 | 0.9954 | 0.630 | 0.0412 | 0.0990 | 0.1378 |
| 49.96 | 0.9844 | 0.630 | 0.0410 | 0.0985 | 0.1086 |
| 32.39 | 0.9518 | 0.630 | 0.0400 | 0.0960 | 0.0737 |
| 22.88 | 0.7936 | 0.630 | 0.0398 | 0.0955 | 0.0629 |
| \multicolumn{6}{c}{2. Coppock and Meiklejohn (C8)} | | | | | |
| 33.4 | 0.690 | 0.061 | 0.025 | 0.0065 | 0.0110 |
| 54.0 | 0.989 | 0.061 | 0.025 | 0.0100 | 0.0123 |
| 72.0 | 1.000 | 0.061 | 0.025 | 0.0133 | 0.0159 |
| 33.4 | 0.690 | 0.091 | 0.025 | 0.0095 | 0.0159 |
| 54.0 | 0.989 | 0.091 | 0.025 | 0.0140 | 0.0178 |
| 72.0 | 1.000 | 0.091 | 0.025 | 0.0185 | 0.0231 |
| 33.4 | 0.690 | 0.102 | 0.025 | 0.0107 | 0.0177 |
| 54.0 | 0.989 | 0.102 | 0.025 | 0.0148 | 0.0198 |
| 72.0 | 1.000 | 0.102 | 0.025 | 0.0212 | 0.0258 |
| \multicolumn{6}{c}{3. Davidson and Schuler (D9)} | | | | | |
| 72.7 | 1.000 | 0.0668 | 0.5 | 0.0260 | 0.0333 |
| 72.7 | 1.000 | 0.0668 | 1.0 | 0.0365 | 0.0481 |
| 72.7 | 1.000 | 0.0668 | 1.5 | 0.0365 | 0.0632 |
| 72.7 | 1.000 | 0.0668 | 2.0 | 0.0500 | 0.0789 |
| 72.7 | 1.000 | 0.0688 | 2.5 | 0.0680 | 0.0952 |
| 27.1 | 0.810 | 0.0688 | 0.5 | 0.0090 | 0.0215 |
| 27.1 | 0.810 | 0.0688 | 1.0 | 0.0200 | 0.0350 |
| 72.7 | 1.000 | 0.4000 | 5.0 | 0.2000 | 0.3106 |
| 72.7 | 1.000 | 0.4000 | 10.0 | 0.4200 | 0.5210 |
| 72.7 | 1.000 | 0.4000 | 20.0 | 0.9000 | 0.9897 |
| 72.7 | 1.000 | 0.4000 | 30.0 | 1.3000 | 1.5068 |
| 27.1 | 0.810 | 0.4000 | 5.0 | 0.2000 | 0.2306 |
| 27.1 | 0.810 | 0.4000 | 10.0 | 0.4200 | 0.4348 |
| 27.1 | 0.810 | 0.4000 | 20.0 | 0.8500 | 0.9007 |
| 27.1 | 0.810 | 0.4000 | 30.0 | 1.1500 | 0.4173 |

[a] Viscosity, $\mu = 0$.

Davidson and Schuler (D8) and Siemes and Kaufmann (S11) also report a negligible influence of surface tension. These authors have used low orifice diameters and highly viscous liquids, both of which reduce the surface tension effects. For very small flow rates, however, the surface tension is effective even for highly viscous liquids, as found by Krishnamurthi, Kumar, and Kuloor (K8).

Thus, the apparently inconsistent results of various investigators regarding the influence of surface tension on the bubble volume are explainable by the difference in the experimental conditions employed.

## 2. *Effect of Viscosity*

There exists the maximum amount of contradiction regarding the role of viscosity in bubble formation. The present model indicates that for both the stages of expansion and of detachment—

(i) the effect of viscosity is large at higher flow rates,

(ii) the effect of viscosity is large for liquids of low surface tension and where orifices of small diameter are used.

Thus, if the liquid is highly viscous, then the drag force is so predominant that the bubble volume is highly sensitive to viscosity. Again, if the viscosity and the flow rate are both small, then variations in the viscosity do not affect the bubble volume appreciably.

The actual data collected for liquids of various viscosities are presented in Fig. 14. The solid lines represent the theoretical values in the figure.

To recapitulate the discrepancies in literature, Datta *et al.* (D4) varied the viscosity of water from 0.012 to 1.108 poise and found that with an increase in the viscosity, the bubble volume decreased for all the nozzles used. This is in apparent contradiction to the observations of most of the other investigators. An effort can now be made to explain this discrepancy on the basis of the present model. Note is to be made of the extremely small volumetric flow rates employed by Datta *et al.* (D4). In fact, they are in the range where effects due to viscosity are negligible when compared to the effects of surface tension. Thus, though there is a hundredfold increase in the viscosity, it is accompanied by a large variation in the surface tension, which decreases from 72.8 to 65.7 dyn per centimeter. At the very small flow rates employed, the decrease in the bubble volume observed by Datta *et al.* (D4) seems more likely to be due to this decrease in the surface tension rather than to the hundredfold increase in the viscosity. Thus, the influence of surface tension has been mistakenly attributed to the effect of viscosity. The actual values of the bubble volumes obtained by these authors for a typical nozzle are given in Table VI along with those obtained by the application of the present model.

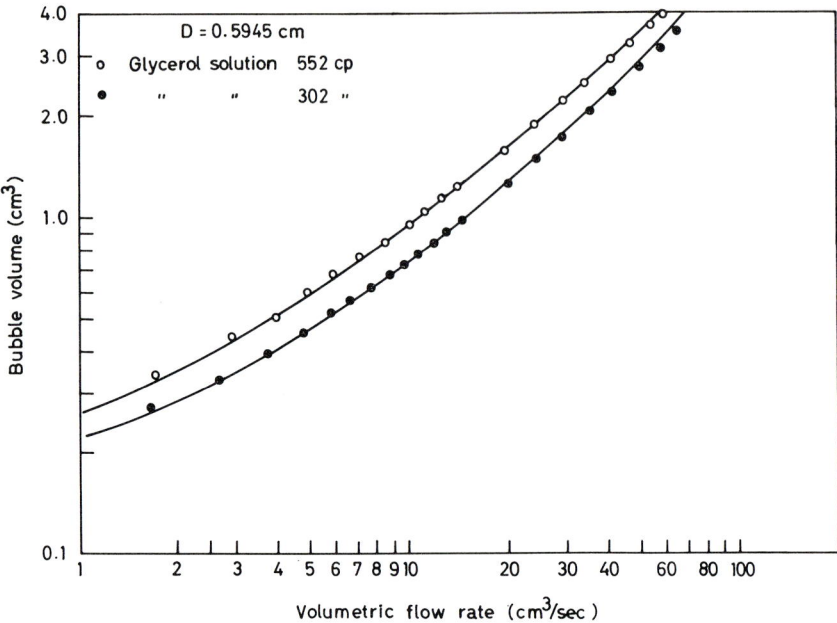

FIG. 14. Comparison of the general model (R1) with the data collected for bubble formation in viscous liquids.

It is interesting to note that the model predicts a decrease in the bubble volume in spite of the hundredfold increase in the liquid viscosity under the conditions of the experiments conducted by Datta *et al.* (D4).

On the other hand, Coppock and Meiklejohn (C8) find that the viscosity has negligible influence on the bubble volume. These authors have used liquids of very low viscosity and extremely small flow rates where the effect of viscosity is negligible. Calculations made for a set of data of Datta *et al.* (D4) are presented in Table VI. The predictions made by the general model are again seen to bear with the trends found experimentally.

The range of variables investigated by Davidson and Schuler (D8) is once again such that the influence of viscosity is predominant. This has already been discussed in the section dealing with the effect of surface tension.

## 3. *Effect of Density*

Only two main opinions merit our attention in this case. They are: (i) the bubble volume decreases with an increase in the liquid density, and (ii) the density has no effect.

Referring to Eq. (55), it is easily seen that if both $\mu$ and $Q$ are small, that of surface tension and that of buoyancy are the only important forces, and an

## TABLE VI

COMPARISON OF EXPERIMENTAL DATA OF DATTA et al. (D4) WITH THE VALUES CALCULATED FROM GENERAL MODEL SHOWING THE INFLUENCE OF VISCOSITY

| Viscosity $\mu$ (poise) | Surface tension $\gamma$ (dyn/cm) | Density $\rho_L$ (gm/cm³) | $D = 0.036$ cm | | | $D = 0.141$ cm | | | $D = 0.388$ cm | | |
|---|---|---|---|---|---|---|---|---|---|---|---|
| | | | $Q$ (cm³/sec) | Bubble volume (cm³) | | $Q$ (cm³/sec) | Bubble volume (cm³) | | $Q$ (cm³/sec) | Bubble volume (cm³) | |
| | | | | Experimental | Calculated | | Experimental | Calculated | | Experimental | Calculated |
| 0.012 | 72.8 | 0.9994 | 0.00810 | 0.0072 | 0.0107 | 0.06083 | 0.0292 | 0.06114 | 0.20500 | 0.0984 | 0.20723 |
| 0.154 | 68.3 | 1.1700 | 0.00787 | 0.0070 | 0.0077 | 0.05208 | 0.0250 | 0.03188 | 0.18120 | 0.0870 | 0.09653 |
| 0.235 | 67.6 | 1.1850 | 0.00787 | 0.0070 | 0.0077 | 0.05104 | 0.0245 | 0.03121 | 0.18120 | 0.0870 | 0.09137 |
| 0.497 | 66.4 | 1.2100 | 0.00765 | 0.0068 | 0.0075 | 0.04812 | 0.0231 | 0.03114 | 0.17170 | 0.0850 | 0.08762 |
| 1.108 | 65.7 | 1.2200 | 0.00765 | 0.0068 | 0.0074 | 0.04583 | 0.0220 | 0.03006 | 0.17500 | 0.0840 | 0.08645 |

increase in the density brings about a decrease in the bubble volume. Similar reasoning is applicable to the detachment stage also. If, on the other hand, $Q$ is very large and $\mu$ is negligible, then the second and the third terms on the right-hand side of Eq. (55) are small when compared with the first term, and so the bubble volume during expansion is independent of the liquid density. The same is true for the detachment stage also. Thus, the liquid density is noneffective for liquids of low viscosity and for high flow rates. This situation exists in the experiments carried out by Quigley et al. (Q1). A quantitative comparison is, however, not possible because their experiments have been conducted under constant pressure conditions.

On the other hand, if the first and third terms on the right-hand side of Eq. (55) are small (which is true for highly viscous liquids), the density still appears in the equation and its increase results in a decrease in the bubble volume. This situation is applicable to the experiments of Davidson and Schuler (D8). Since no data are reported which pertain to the effect of density alone, a comparison is not possible.

## V. Bubble Formation under Constant Pressure Conditions

The constant pressure condition arises when the chamber volume tends to infinity (in practice, more than about a liter), and the pressure in the gas chamber remains constant. As the pressure in the bubble varies with the extent of its formation, the pressure difference across the forming device also varies, thereby bringing about a condition of changing flow rates.

### A. Correlation of Hayes, Hardy, and Holland (H4)

These authors have assumed the bubble to be expanding at the orifice, and have used the force balance equation at the time of detachment. The various forces considered by these authors are buoyancy, force due to the addition of mass (P2), excess pressure force, surface tension force, drag force, and force due to the inertia of the liquid.

Considering all the above forces, and applying Newton's second law of motion, we obtain

$$v_0 \frac{dm}{dt} + V(\Delta\rho)g + \left(\frac{\pi D^2}{4}\right)(p_i - p) - F + \pi D\gamma - m'\frac{dv}{dt} = \frac{d(mv)}{dt} \quad (62)$$

To obtain bubble volume by Eq. (62), the various terms are evaluated at a time just prior to the release of the bubble. Thus,

$$(p_i - p) = 4\gamma/d \quad (63)$$

Further,

$$\frac{d(mv)}{dt} - v_0 \frac{dm}{dt} = \frac{-4Q^2 \rho_g}{\pi D^2}\left[1 - \frac{1}{12}(D/d)^2\right] \quad (64)$$

The drag of the bubble is expressed in terms of the drag coefficient as

$$F = C_D \rho_l v^2 A_f / 2 \quad (65)$$

Assuming the bubbles to be spherical, the frontal area $A_f$ of the bubble becomes $\pi d^2/4$. Further, by making use of the expression $v = Q/\pi d^2$, the drag force can be written in terms of Froude number, the drag coefficient, and the volume of the bubble. Thus

$$F = (C_D \rho_l/2)(Q^2/\pi^2 d^4)(\pi d^2/4) \quad (66)$$
$$= 3g\rho_l C_D N_{Fr} V/4$$

where

$$N_{Fr} = v^2/dg = Q^2/\pi^2 d^5 g$$

In Eq. (66), in addition to $V$, $C_D$ is also unknown; it has been empirically found by the authors from the experimental data on bubble formation.

The force due to the movement of the liquid surrounding the bubble is $m'(dv/dt)$. For a sphere moving in an infinite medium of an inviscid fluid, the mass of the liquid $m'$ is equal to half the mass of the displaced liquid. The authors, however, assumed merely a direct proportionality between $m'$ and the mass of the displaced fluid, instead of the above relationship, because they considered their flow not to be irrotational.

The force can be expressed as,

$$m'(dv/dt) = \tau V \rho_l (dv/dd)(dd/dt) \quad (67)$$

where $\tau$ is the proportionality factor, found empirically.

On evaluating various terms of Eq. (67) and substituting, we obtain

$$m'(dv/dt) = \tau \rho_l V(-2Q/\pi d^3)(2Q/\pi d^2) \quad (68)$$
$$= -4\rho_l g\tau N_{Fr} V$$

Now all the terms of the basic Eq. (62) have been evaluated; hence it can be written as follows:

At the time of detachment, when $V = V_F$,

$$V_F(\Delta\rho)g(1 - \psi) = \pi D\gamma(1 - D/d) - \phi \quad (69)$$

where

$$\psi = (3\rho_l/4 \, \Delta\rho)N_{Fr}(C_D - 16\tau/3)$$

and

$$\phi = (4Q^2 \rho_g/\pi D^2)[1 - (1/12)(D/d)^2]$$

Equation (69) contains unknown quantities like $\psi$, which have to be determined either theoretically or empirically. The term $(1 - \psi)$ is a function of both the Reynolds number and Froude group, and presumably can be correlated as a power relationship of the form

$$|(1 - \psi)| = K(N_{Re})^a (N_{Fr})^b (D/d)^c \tag{70}$$

Further, Eq. (69) gives rise to two types of situations depending on whether $\pi D\gamma(1 - D/d)$ is greater or smaller than $\phi$. Since $d$ is larger than $D$, the term $(1/12)(D/d)^2$ can be neglected in $\phi$, and so $\phi$ is simply equal to $(4Q^2 \rho_g/\pi D^2)$, which is the time rate of change of momentum of the gas added to the bubble just prior to its release.

The authors find that for $\pi D\gamma(1 - D/d) > \phi$, the bubble volume remains essentially constant, whereas for $\pi D\gamma(1 - D/d) < \phi$, the frequency remains essentially constant, and the extra flow goes to increase the bubble volume. Thus, two relationships are given, corresponding to each of the above two conditions. The following correlations are obtained by the solution of Eqs. (69) and (70):

For $\pi D\gamma(1 - D/d) > \phi$:

Since for this case the bubble volume remains essentially constant but the frequency changes with the flow rate, the authors have considered frequency as the dependent variable in the following correlation:

$$f = \frac{7.56 \times 10^{-3}}{\overline{N}_{Re}^3} (\overline{N}_{Fr} \overline{N}_{Re}^{-0.5})^{-0.172} (NQ\overline{N}_{Re}^3)^{0.525} \left[ \frac{D\overline{N}_{Re}^{-1}}{(6Q/\pi)^{1/3}} \right]^{-0.697} \tag{71}$$

where

$$\overline{N}_{Fr} = N_{Fr} f^{-5/3} = \frac{Q^{1/3}}{g\pi^{1/3} 6^{5/3}} \quad \text{Modified Froude group}$$

$$\overline{N}_{Re} = N_{Re} f^{-1/3} = \frac{Q^{2/3} \rho_l}{\pi^{2/3} 6^{1/3} \mu_l} \quad \text{Modified Reynolds number}$$

and

$$N = \frac{\pi}{6} \left| \left[ \frac{(\Delta\rho)g}{\pi D\gamma(1 - D/d) - \phi} \right] \right|$$

For $\pi D\gamma(1 - D/d) < \phi$:

Here, the bubble volume has been assumed to be the dependent variable and the correlation is:

$$V_F = (23.4/N)(N'_{Fr} N^{5/3})^{0.463} (N'_{Re} N^{1/3})^{-0.0764} (DN^{1/3})^{0.712} \tag{72}$$

where

$$N'_{Fr} = N_{Fr} d^5$$

and
$$N'_{Re} = N_{Re}\, d$$

The relationships (71) and (72) were tested by their proposers, who reported that the average percentage deviations of the calculated values of $f$ and $V_F$ from the experimental values are 16 and 10, respectively.

## B. Davidson and Schuler's Model (D8, D9)

The analysis used by these authors for constant flow conditions has been extended to constant pressure conditions also. The major change arises from the fact that now the flow is a function of the extent to which the bubble has already been formed. This has been introduced into the constant flow equation by means of an orifice equation.

### 1. Bubble Formation in Inviscid Fluids

The basic equation here is the same as that for constant flow conditions, involving the instantaneous movement of the bubble. Thus:

$$Vg = \frac{d}{dt}\left(\frac{11}{16} V \frac{ds}{dt}\right) \tag{73}$$

The flow rate $Q$, at which the bubble volume is changing, is evaluated by the equation

$$Q = dV/dt = K(P + \rho_l g s - 2\gamma/r)^{1/2} \tag{74}$$

where

$$P = (P_1 - \rho_l g h)$$

In Eq. (74), $K$, an experimental constant depending on the orifice, is determined by experiments with a steady state flow of gas through the orifice in the absence of the liquid.

Equations (73) and (74) taken together were solved (D8, D9) with the aid of a computer through numerical methods. The initial condition used was that at $t = 0$:

$$s = 0,\ ds/dt = 0,\ \text{and}\ V_0 = 4\pi R^3/3$$

The final bubble volume $V_F$ is evaluated by assuming the detachment to take place when $s = r_F + R$.

The influence of surface tension of the system and of the orifice radius were found to be small. Surface tension, however, has an effect on the minimum value of $P$ for bubbling to occur, since bubbling ceases if $P$ is less than $2\gamma/R$.

The equations were verified mainly with water as the experimental liquid. A part of the results, as given by the authors in tabular form, is reproduced in Table VII. From Table VII it is evident that the agreement between the theory and experiment is not very good.

## 2. Bubble Formation in Highly Viscous Liquids

Two separate sets of equations have been derived by Davidson and Schuler, one for small and one for large flow rates. Both cases are discussed below.

*a. Small Flow Rates.* The equation of the motion of the bubble remains the same as for constant flow conditions and corresponds to Stokes' equation, i.e.,

$$v = 2gr^2\rho_l/9\mu_l \quad (75)$$

The final equation, by taking the variation in flow rate into account, becomes:

$$KP^{1/2}t = (4\pi/3)(2\gamma/P)^3 \left[ \tan\theta \left( \sec^5\theta + \frac{5}{4}\sec^3\theta + \frac{15}{8}\sec\theta \right) \right.$$

$$\left. + \frac{15}{8} \ln \tan\left( \frac{\pi}{4} + \frac{\theta}{2} \right) \right] \quad (76)$$

where

$$\theta = \cos^{-1}\left[ \left( \frac{4\pi}{3V} \right) \frac{2v}{\rho} \right]^{1/2}$$

$V$ and $r$ can be evaluated for any value of $t$, from Eq. (76). But the equation cannot directly give the final bubble volume $V_F$, for evaluating which it is necessary to plot $v$ versus $t$, by making use of Eqs. (75) and (76). The distance $s$ travelled at any time $t$ can be found from the plot by graphical integration. From a plot of $s$ versus $r + r_0$, the value of $r + r_0$ corresponding to any $s$ is obtained, from which, in turn, $r_F$ and $V_F$ are calculated.

The above equations were verified by using aqueous glycerol solutions (D8). The model gives results which agree reasonably well with the experimental data.

*b. High Flow Rates.* Here the inertia of the surrounding liquid is taken into consideration in the movement of the bubble.

The equation of motion now assumes the form

$$Vg = 6\pi \frac{\mu_l}{\rho_l}(3V/4\pi)^{1/3}\frac{ds}{dt} + \frac{11}{16}V\frac{d^2s}{dt^2} + \frac{11}{16}\frac{ds}{dt}\frac{dV}{dt} \quad (77)$$

The equation of flow through the orifice is

$$dV/dt = K(P - 2\gamma/r - \rho_l gs)^{1/2} \quad (78)$$

## TABLE VII
### Data on Formation of Air Bubbles in Water under Constant Pressure Conditions

| S.No. | $D$ (cm) | $K$ (ml·cm$^{\frac{1}{2}}$/gm$^{\frac{1}{2}}$) | $P$ (gm/cm·sec$^2$) | Mean flow rate $Q$ (cm$^3$/sec) | | | Bubble volume $V$ (cm$^3$) | | |
|---|---|---|---|---|---|---|---|---|---|
| | | | | Expt. (D9) | Theory | | Expt. (D9) | Theory | |
| | | | | | Eq (73), (74) (D9) | Eq (89) (S3) | | Eq (73)(74) (D9) | Eq (89) (S3) |
| 1 | 0.298 | 1.90 | 951 | 32 | 67 | 65.5 | 2.3 | 3.5 | 3.29 |
| 2 | 0.298 | 1.90 | 1118 | 45 | 70 | 68.0 | 2.9 | 3.8 | 3.52 |
| 3 | 0.298 | 1.90 | 1323 | 61 | 76 | 73.2 | 3.4 | 4.2 | 3.78 |
| 4 | 0.374 | 3.06 | 779 | 33 | 102 | 86.4 | 3.4 | 6.1 | 5.89 |
| 5 | 0.374 | 3.06 | 877 | 47 | 105 | 89.7 | 4.1 | 6.4 | 6.13 |
| 6 | 0.374 | 3.06 | 1024 | 60 | 112 | 93.9 | 4.5 | 6.9 | 6.47 |
| 7 | 0.412 | 3.82 | 734 | 30 | 124 | 109.0 | 4.3 | 7.8 | 7.88 |
| 8 | 0.412 | 3.82 | 832 | 57 | 129 | 113.0 | 4.9 | 8.3 | 8.18 |
| 9 | 0.412 | 3.82 | 1006 | 68 | 141 | 118.2 | 5.7 | 9.1 | 7.98 |
| 10 | 0.460 | 4.90 | 632 | 25 | 156 | 135.0 | 5.6 | 10.7 | 10.73 |
| 11 | 0.460 | 4.90 | 739 | 60 | 163 | 140.0 | 6.9 | 11.4 | 11.18 |
| 12 | 0.460 | 4.90 | 790 | 68 | 169 | 142.8 | 7.1 | 11.7 | 11.39 |

Equations (77) and (78) cannot be solved analytically. A computer was used for obtaining the values of $s$, $V$, and $ds/dt$ for regularly increasing values of $t$: The initial values taken were

$$\text{at } t = 0, \quad s = 0, \quad V = (4\pi R^3/3), \quad \text{and} \quad (ds/dt) = 0$$

The volume $V_F$ of the bubble moving away from the orifice was taken to be the difference of the total volume at $s = r_F + R$, and the initial volume $V_0 = 4\pi r^3/3$. The volume $V_0$ formed the nucleus of the succeeding bubble. The model gives good agreement up to a flow range of 60 cm$^3$/sec.

## C. Model of Satyanarayan, Kumar, and Kuloor (S3)

These authors have extended the concepts developed by Kumar and Kuloor (K16, K18, K19) for bubble formation under constant flow conditions to the situations of constant pressure conditions.

### 1. Bubble Formation in Inviscid Fluids

As before, the force balance bubble volume is first found, and its volume is then used to determine the final bubble volume.

The procedure is the same as for constant flow conditions. The only forces to be considered are the inertia force, the surface-tension force, and the buoyancy force. The viscous effects are neglected. Making a force balance, we obtain:

$$V_{fb}(\rho_l - \rho_g)g = \frac{d(Mv_e)}{dt_e} + 2\pi R\gamma \cos\theta \tag{79}$$

The term $d(Mv_e)/dt_e$ must be determined before the above equation can be used for evaluating $V_{fb}$. Thus

$$\frac{d(Mv_e)}{dt_e} = M\frac{dv_e}{dt_e} + v_e\frac{dM}{dt_e} \tag{80}$$

Further,

$$v_e = dr/dt_e = Q/4\pi r^2 \tag{81}$$

$Q$ is not a constant in this case; it depends on the stage of bubble formation. It can be determined from orifice equation, as

$$Q = K(P + \rho_l g r - 2\gamma/r)^{1/2} \tag{82}$$

where

$$P = (P_1 - \rho_l g h)$$

Substituting Eq. (82) in Eq. (81), we obtain

$$v_e = \frac{K(P - \rho_l gr - 2\gamma/r)^{1/2}}{4\pi r^2} \tag{83}$$

Thus,

$$\frac{dv_e}{dt_e} = \frac{K}{8\pi}(Pr^{-4} + \rho_l gr^{-3} - 2\gamma r^{-5})^{-1/2}(-4Pr^{-5} - 3\rho_l gr^{-4}$$

$$+ 10\gamma r^{-6})\frac{dr}{dt_e} \tag{84}$$

But $(dr/dt_e) = v_e$, and $v_e$ is known from Eq. (83). Thus,

$$\frac{dv_e}{dt_e} = \frac{-K^2}{32\pi^2 r^6}(4Pr + 3\rho_l gr^2 - 10\gamma) \tag{85}$$

Further,

$$\frac{dM}{dt_e} = K\left(\rho_g + \frac{11}{16}\rho_l\right)(P + \rho_l gr - 2\gamma/r)^{1/2} \tag{86}$$

Therefore,

$$v_e \frac{dM}{dt_e} = \frac{K^2[\rho_g + (11/16)\rho_l]}{4\pi r^2}(P + \rho_l gr - 2\gamma/r) \tag{87}$$

The various terms of Eq. (79) have now been evaluated so it can be written as

$$2\pi R\gamma \cos\theta + \frac{K^2[\rho_g + (11/16)\rho_l]}{4\pi r_{fb}^2}(P + \rho_l gr_{fb} - 2\gamma/r_{fb})$$

$$- \left(\frac{V_{fb}[\rho_g + (11/16)\rho_l]K^2}{32\pi^2 r_{fb}^6}\right)(4Pr_{fb} + 3\rho_l gr_{fb}^2 - 10\gamma)$$

$$= V_{fb}(\rho_l - \rho_g)g \tag{88}$$

Expressing $r_{fb}$ as $(3/4\pi)^{1/3}V_{fb}^{1/3}$ in Eq. (88), we have

$$C_1 V_{fb}^{-2/3}(P + 0.6203\rho_g g V_{fb}^{1/3} - 3.222 V_{fb}^{-1/3}\gamma)$$

$$- C_2 V_{fb}^{-1}(2.482 V_{fb}^{1/3}P - 1.149\rho_l g V_{fb}^{2/3} - 10\gamma)$$

$$+ 2\pi R\gamma \cos\theta = V_{fb}(\rho_l - \rho_g)g \tag{89}$$

where

$$C_1 = \frac{K^2[\rho_g + (11/16)\rho_l]}{4\pi(3/4\pi)^{2/3}} \approx 0.1422 K^2 \rho_l$$

and
$$C_2 = \frac{K^2[\rho_g + (11/16)\rho_l]}{32\pi^2(3/4\pi)^2} \approx 0.03822 K^2 \rho_l$$

Equation (89) represents an implicit relationship between the force balance bubble volume and the other variables.

To evaluate $V_F$, it is necessary to analyze the second stage of the bubble formation where the expanding bubble is moving away from the orifice base. The equation of motion for this case differs from that for constant flow conditions in that $Q$ here varies with time. The orifice equation, however, shows that $Q$ in this stage varies very little. It is evident that at higher values of $r$, the variation of the bubble volume at a definite rate causes increasingly small variation in $r$. As the change in the flow rate is due mainly to the variation in $r$ for a particular set of conditions, it can be considered to be negligible when the change in $r$ is small. Since this condition holds good in the second stage, $Q$ can be assumed to be constant, its value being $Q_{fb}$, which is the flow rate at the end of the force balance bubble stage.

The equations for the constant flow rate have already been derived in Section IV for various situations. For inviscid fluids with surface tension, the equation is reproduced below.

$$r_{fb} = \frac{P'}{4Q}(V_F^2 - V_{fb}^2) - \frac{9N}{Q}(V_F^{1/3} - V_{fb}^{1/3}) - \frac{J}{Q}(V_F - V_{fb})$$
$$\frac{1}{Q}\left(JV_{fb} + 3NV_{fb}^{1/3} - \frac{P'}{2}V_{fb}^2\right)(\ln V_F - \ln V_{fb}) \quad (90)$$

where
$$P' = \frac{(\rho_l - \rho_g)g}{Q[\rho_g + (11/16)\rho_l]}$$

$$N = \frac{Q}{12\pi(3/4\pi)^{2/3}}$$

and
$$J = \frac{\pi D \gamma \cos\theta}{Q[\rho_g + (11/16)\rho_l]}$$

$r_{fb}$ is evaluated by making use of Eq. (89). Using this value of $r_{fb}$, and the corresponding $Q_{fb}$ in Eq. (90), the value of $V_F$ can be determined.

From Eq. (89), the surface tension is seen to influence the bubble volume in two ways, viz. by varying the flow rate in the expanding bubble, and by causing a downward force at the periphery of the orifice. Even if the effect of the surface tension becomes negligible at the orifice tip due to either the small size of the orifice or the nonwettable character of the system, it still influences the bubble volume because of the variations in flow during the bubble formation.

The Eqs. (89) and (90) get simplified, when the surface tension effect at the orifice tip is small, to

$$C_1 V_{fb}^{-2/3}(P + 0.6203\rho_g g V_{fb}^{1/3} - 3.22 V_{fb}^{-1/3}\gamma)$$
$$- C_2 V_{fb}^{-1}(2.483 V_{fb}^{1/3} P - 1.149\rho_l g V_{fb}^{2/3} - 10\gamma)$$
$$= V_{fb}(\rho_l - \rho_g)g \tag{91}$$

and

$$r_{fb} = \frac{P'}{4Q}(V_F^2 - V_{fb}^2) - \frac{9N}{Q}(V_F^{1/3} - V_{fb}^{1/3})$$
$$+ \frac{1}{Q}\left(3NV_{fb}^{1/3} - \frac{P'}{2}V_{fb}^2\right)(\ln V_F - \ln V_{fb}) \tag{92}$$

Equations (89)–(92) have been verified by the authors, both for the final overall values and for the influence of individual variables. Their typical results for inviscid fluids are presented in Fig. 15, which shows a good agreement between the theoretical and the experimental values. The authors have obtained higher bubble volumes than those of Davidson and Schuler (D9) under otherwise similar conditions. This is probably due to absence of liquid circulation in the present condition.

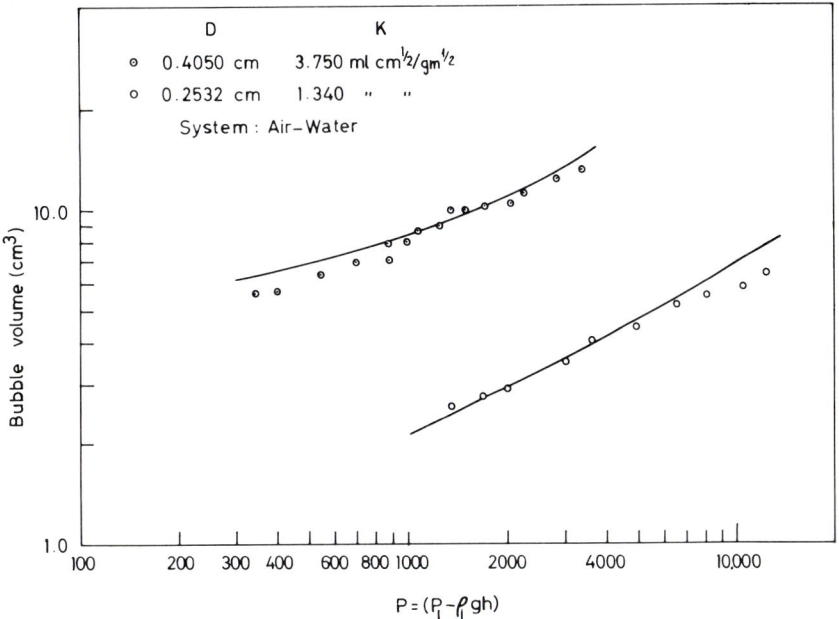

FIG. 15. Comparison of the model (S3) with the data collected for bubble formation in inviscid liquids under constant pressure conditions.

## 2. Satyanarayan, Kumar, and Kuloor Model for General Situation (S3)

The above discussion dealt with only that particular situation where the continuous phase approximated to an inviscid fluid. However, the equations thus derived can be easily modified to include the effects of viscosity of the of the continuous phase. Under constant pressure conditions also, viscosity of the continuous phase tends to increase the bubble volume by increasing the drag during both the expansion and detachment stages.

All the forces have already been considered except the viscous drag which is given by

$$\text{Viscous drag} = \frac{6\pi r \mu}{4\pi r^2} K(P + \rho_l g r - 2\gamma/r)^{1/2} \tag{93}$$

At the end of the expansion stage, when the bubble volume is $V_{fb}$ and its radius $r_{fb}$, the force balance becomes:

$$V_{fb}(\rho_l - \rho_g)g = \frac{K^2[\rho_g + (11/16)\rho_l]V_{fb}^{-2/3}}{4\pi(3/4\pi)^{2/3}} \left[P + \rho_l g(3/4\pi)^{1/3}V_{fb}^{1/3}\right.$$

$$\left. - \frac{2\gamma V_{fb}^{-1/3}}{(3/4\pi)^{1/3}}\right] - \frac{V_{fb} K^2[\rho_g + (11/16)\rho_l]V_{fb}^{-2}}{32\pi^2(3/4\pi)^2}$$

$$[4P(3/4\pi)^{1/3}V_{fb}^{1/3} - 3\rho_l g(3/4\pi)^{2/3}V_{fb}^{2/3} - 10\gamma]$$

$$+ \frac{3K\mu V_{fb}^{-1/3}}{2(3/4\pi)^{1/3}} \left[P + \rho_l g(3/4\pi)^{1/3}V_{fb}^{1/3}\right.$$

$$\left. - \frac{2\gamma}{(3/4\pi)^{1/3}V_{fb}^{1/3}}\right]^{1/2} + 2\pi R \gamma \cos\theta \tag{94}$$

The volumetric flow rate $Q$ remains essentially constant, since the changes in the bubble radius are quite small in this stage. The value of this $Q$ can be taken as $Q_{fb}$, which is the flow rate at the end of the expansion stage. The analysis for this stage, therefore, is the same as that for constant flow conditions. Thus:

$$r_{fb} = \frac{B}{2Q(A+1)}(V_F^2 - V_{fb}^2) - \frac{E}{AQ}(V_F - V_{fb})$$

$$- \frac{3C}{2Q(A-1/3)}(V_F^{2/3} - V_{fb}^{2/3}) - \frac{3D}{Q(A-2/3)}(V_F^{1/3} - V_{fb}^{1/3}) \tag{95}$$

where

$$A \approx 1 + 1.25 \frac{96\pi r_{fb} \mu}{11\rho_l Q}$$

$$B \approx 16g/11Q$$

$$C = \frac{3\mu}{2(3/4\pi)^{1/3}[\rho_g + (11/16)\rho_l]}$$

$$D = \frac{Q}{12\pi(3/4\pi)^{2/3}}$$

and

$$E = \frac{16\pi D\gamma \cos\theta}{11Q\rho_l}$$

The set of Eqs. (94) and (95) are quite general in nature, and are applicable to bubble formation under constant pressure conditions, in the range of single bubbling.

If the viscosity term is removed, then these equations reduce to the ones given earlier for inviscid fluids. Thus, these general equations are applicable to inviscid fluids, both with and without the surface tension at the orifice tip.

Verification of the equation for viscous liquids is shown in Fig. 16, where the solid lines correspond to theoretically calculated values.

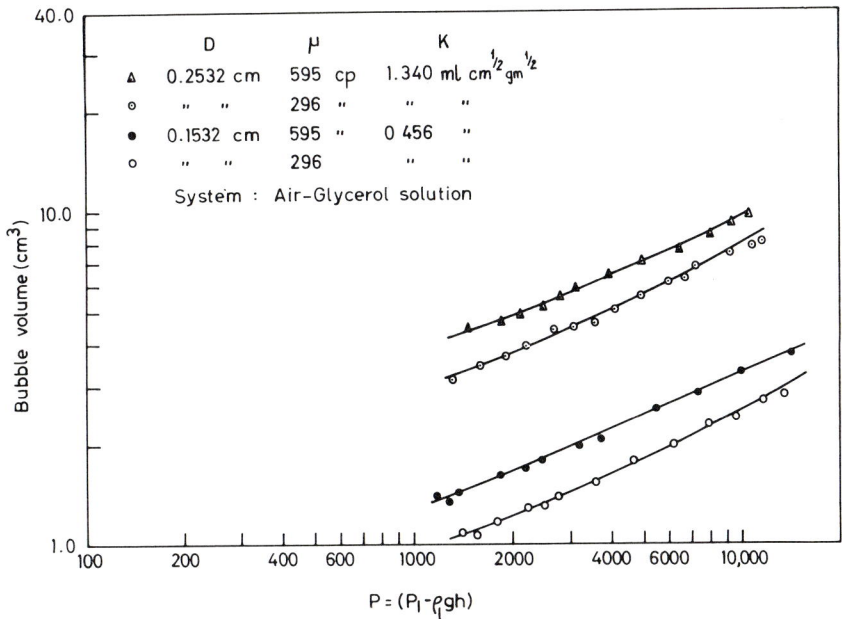

FIG. 16. Comparison of the model (S3) with the data collected for bubble formation in viscous liquids under constant pressure conditions.

The range of variables in which the equations have been tested is given in Table VIII. The agreement in overall range is found to be good.

TABLE VIII

Range of Variables Covered for Testing the Model of Satyanarayan, Kumar, and Kuloor (S3)

| Orifice diameter $D$ (cm) | Surface tension $\gamma$ (dyn/cm) | Viscosity $\mu$ (cp) | Density $\rho_l$ (gm/cm³) | Flow rate $Q$ (cm³/sec) | $P$ (gm/cm · sec²) |
|---|---|---|---|---|---|
| 0.051–0.405 | 35–72 | 90–600 | 0.990–1.24 | 0–260 | 500–25,000 |

## VI. Bubble Formation in Non-Newtonian Fluids

The study of bubble formation in non-Newtonian fluids has not been reported in literature in spite of the great industrial uses of these fluids. Recently, Subramaniyan and Kumar (S16) have studied bubble formation under constant flow conditions in fluids following the Ostwald-de-Waele rheological model. The model of Kumar and Kuloor (K16, K18, K19) has been extended to take into consideration the drag variation caused by the complexity of the rheological equation.

For Ostwalde-de-Waele fluids,

$$\tau = -\{k|[(1/2)\Delta : \Delta]^{1/2}|^{n-1}\}\Delta \tag{96}$$

The creeping flow of a sphere in a fluid following Eq. (96) has been extensively studied (S13, T2, W1, W2). The drag coefficient is given by

$$C_D = 24\, X_n/Re' \tag{97}$$

where $Re'$ is the modified Reynolds number, defined as

$$Re' = (2r)^n \rho_l v^{2-n}/k \tag{98}$$

$X_n$ is a function of the flow index $n$. For Newtonian fluids, the value of $X_n$ is unity.

The total drag on the sphere worked out through Eq. (97) becomes

$$F_D = \frac{24\pi X_n k}{2^{n+1} r^{n-2} v^{-n}} \tag{99}$$

The force balance bubble volume equation, on the introduction of the drag term given by Eq. (99) becomes

$$V_{fb} \Delta \rho g = \pi D \gamma \cos \theta + \frac{Q^2[\rho_g + (11/16)\rho_l]V_{fb}^{-2/3}}{12\pi(3/4\pi)^{2/3}}$$

$$+ \frac{24X_n k Q^n V_{fb}^{-(3n-2)/3}}{2^{3n+1}\pi^{n-1}(3/4\pi)^{(3n-2)/3}} \quad (100)$$

The values of $n$ and $k$ are evaluated from Capillary Viscometer data. The upper and lower bounds of $X_n$ are known (W2). The mean of the two bounds is used as the actual $X_n$ value for computational purposes.

The final equation for the second stage now becomes

$$\frac{dv}{dT} + \frac{v}{T} + AT^{-(n+1)/3}v^n = B - CT^{-5/3} - DT^{-(3n+1)/3} - ET^{-1} \quad (101)$$

where

$$A = \frac{24\pi X_n k}{2^{n+1}(3/4\pi)^{(n-2)/3}[\rho_g + (11/16)\rho_l]Q}$$

$$B = \frac{(\rho_l - \rho_g)g}{[\rho_g + (11/16)\rho_l]Q}$$

$$C = \frac{Q}{12\pi(3/4\pi)^{2/3}}$$

$$D = \frac{24X_n k Q^{n-1}}{2^{3n+1}\pi^{n-1}(3/4\pi)^{(3n-2)/3}[\rho_g + (11/16)\rho_l]}$$

$$E = \frac{\pi D \gamma \cos \theta}{[\rho_g + (11/16)\rho_l]Q}$$

Equation (101) is numercially solved by the Runge-Kutta method. The agreement between the experimental values and theory is seen from Fig. 17, where the solid lines indicate the values obtained by theory. The liquid used for the data presented in Fig. 17 had $n = 0.617$ and $k = 0.0059$.

It is possible to frame equations for various models of non-Newtonian fluids by incorporating the adequate drag terms into (100) and (101), but the experimental data for verifying these equations are still not available.

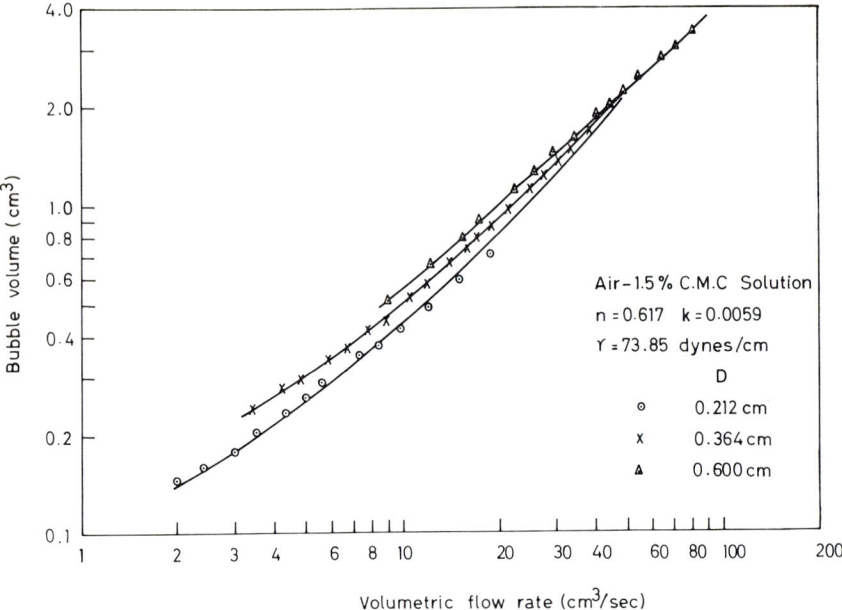

FIG. 17. Comparison of the model (S16) with the collected data for bubble formation in non-Newtonian liquids under constant flow conditions.

## VII. Bubble Formation in Gas Fluidized Beds

### A. Introduction

The two-phase theory of aggregative fluidization considers the bed to be made up of two parts, viz. (i) a particulate phase wherein the gas flow rate corresponds to that required for incipient fluidization, and (ii) a bubble phase which conveys the extra gas through the bed in the form of bubbles. This theory, which is applicable up to flow rates corresponding to approximately $10U_0$, is supported by many investigators (B2, M6, R8, T3, Y1). The bubbles which are formed at the distributor rise through the bed, exchanging the fluidizing gas and undergoing coalescence, rupture, etc. It is necessary to determine the sizes of the bubbles produced at the distributor, as their size distribution is important in a study of bubble dynamics, and also the extent of transfer operations and chemical reaction occurring in the bed. In order to understand the mechanism of bubble formation, single nozzles are normally employed.

The investigations of Davidson *et al.* (D7) enable a theoretical analysis of the phenomenon of bubble formation in fluidized beds. They found that, as far as the movement of the bubbles is concerned, the fluidized bed behaves like an inviscid fluid with zero surface tension. This has been confirmed by Bloore and Botterill (B5) who compared their results with those obtained by Calderbank (C1) for an air-water system and concluded that there is a marked similarity in the behavior of the gas bubbles in liquid systems and in fluidized systems, although the actual mechanism of bubble formation in these two systems must be quite dissimilar. In these experiments the bubbles were formed by injecting a measured stream of gas through a nozzle placed in a gently fluidized bed of ballotini (0.8 mm diam.).

The authors (B5) report that for flow rates between 50 and 250 cm³/sec, the bubble frequency remains essentially constant at 22 bubbles per second, whereas at lower flow rates, the frequency shows a marked tendency to rise.

A more thorough experimental investigation was carried out by Harrison and Leung (H3).

The conclusions drawn are as follows:

1. The bed depth has no influence on the size of the bubble produced. This indicates that the bubbles are formed under either constant flow or constant pressure conditions. In the intermediate region, Padmavathy, Kumar, and Kuloor (P1) have shown that the bubble volume in an air-water system is highly sensitive to the variation in the depth of the liquid column above the bubble forming nozzle. As the bed has no surface tension, no variation of flow is expected during bubble formation, and the conditions of constant flow are approximated. This explanation is due to present authors.

2. The particle size of the solids does not influence the bubble size.

3. The bubble frequency is independent of the main fluidizing velocity.

4. The bubble formation in fluidized beds follows a trend very similar to that for bubble formation in inviscid liquids.

From the above discussion, it is seen that the bed acts as an inviscid fluid without surface tension, and that the bubble formation takes place under constant flow conditions. Thus, the theories which are applicable to the above conditions should also be applicable to bubble formation in fluidized beds.

## B. Model of Davidson and Harrison (D6)

These authors have modified the model of Davidson and Schuler (D9). The modification is in the value of liquid being carried with the bubble, which is assumed to be 1/2 the bubble volume, instead of the earlier 11/16th the bubble volume. The resulting equation for the bubble volume is now written as

$$V_F = 1.138(Q^{6/5}/g^{3/5}) \tag{102}$$

The authors have tested this equation with the data of Harrison and Leung (H3) and those of Bloore and Botterill (B5) and found a reasonably satisfactory agreement between the experimental and the calculated values.

## C. Model of Kumar and Kuloor (K17)

This model is a modification of the model developed by Kumar and Kuloor (K18) for bubble formation in inviscid fluids in the absence of surface-tension effects. The need for modification arises because the bubble forming nozzles actually used to collect data on bubble formation in fluidized beds differ from the orifice plates in that they do not have a flat base. Under such conditions the bubble must be assumed to be moving in an infinite medium and the value of 1/2 is more justified than the value 11/16.

The force balance bubble volume at the end of the expansion stage is given by

$$V_{fb} = \left[\frac{Q^2}{24\pi(3/4\pi)^{2/3}g}\right]^{3/5} \tag{103}$$

In Eq. (103) the value of 1/2 has been used for evaluating the virtual mass of the bubble.

The final equation for the bubble volume, which results from the analysis of the second viz. the detachment stage, is

$$(3/4\pi)^{1/3}(V_{fb})^{1/3} = (P/4Q)(V_F^2 - V_{fb}^2) - (9N/Q)(V_F^{1/3} - V_{fb}^{1/3})$$
$$+ (1/Q)[3NV_{fb}^{1/3} - (P/2)V_{fb}^2](\ln V_F - \ln V_{fb}) \tag{104}$$

where

$$P = \frac{(\rho_l - \rho_g)g}{Q(\rho_g + \tfrac{1}{2}\rho_l)} \approx \frac{2g}{Q}$$

and

$$N = \frac{Q}{12\pi(3/4\pi)^{2/3}}$$

The authors (K17) find that Eq. (104) explains the data of Bloore and Botterill (B5) and Harrison and Leung (H3) quite well.

On introducing various simplifications, Eq. (104) becomes

$$V_F = 0.806\, Q^{6/5}/g^{3/5} \tag{105}$$

Though both the above equations represent the data with reasonable accuracy the conditions present at the distributor agree better with the assumption that 11/16th of the volume of the liquid is being carried by the bubble. It is therefore recommended that, for calculation of bubble formation at the distributor, Eqs. (22) and (28) be used.

## VIII. The Effect of Orifice Geometry on Bubble Size

Circular orifices are normally employed in the study of bubble formation at submerged orifices because they admit an accurate assessment of the surface-tension effect. In industrial operations, however, the use of noncircular slots for bubble formation is quite frequent. A few investigators (C5, G2, J2, K4, S10, V4), who used porous solids for bubble formation have employed the average pore size for purposes of correlation. Spells and Bakowski (S15) and Cross and Rider (C9) have used rectangular slots for bubble formation in water, but only with a view to finding the influence of slot submergence on the air flow rate. Johnson et al. (J3) have indicated the influence of orifice geometry on bubble formation by observing that the bubbles produced through drilled orifices vary in size from those obtained through punched orifices.

Krishnamurthi, Kumar, and Datta (K7) employed a circular orifice of arbitrarily chosen dimensions as the standard, and constructed two sets of noncircular orifices having either the perimeter or the area equal to that of the standard orifice. The configurations chosen were an equilateral triangle, a square, and a rectangle. The system used by these authors was air-water, and their studies were confined to extremely small flow rates ($<0.5$ cm$^3$/sec). Their results indicate that noncircular orifices do not utilize their entire perimeter for bubble formation, and, for equi-sided orifices at low frequencies of formation, the bubble is formed as if from a circle inscribed in the noncircular orifice. In this range, the perimeter and the area are important in determining the final bubble size.

Further investigations by the above authors (K5), using an air-glycerine solution system (density $= 1.22$ g/ml, viscosity $= 76.2$ cp and surface tension $= 61.8$ dyn/cm) confirmed the above conclusions for highly viscous liquids also. The authors have further correlated their results by the empirical relation (K8):

$$(V_F)_{n.c.} = (V_F)_i (R)^{0.52} \tag{106}$$

where

$(V_F)_{n.c}$ = the bubble volume from the noncircular orifice,
$(V_F)_i$ = the bubble volume from the inscribed circle, and
$R$ = the ratio of the area of the noncircular orifice to that of the inscribed circle.

Ramakrishnan, Kumar, and Kuloor (R1) have extended the investigations of Krishnamurthi et al. (K5, K7, K8) to more geometries, much higher flow rates (up to 100 cm$^3$/sec), liquids of still higher viscosity, and to completely constant flow conditions.

For both the standard orifices studied here, the equal-area orifices are found to give the same bubble volume versus $Q$ curve as that of the standard orifice for all the regular equi-sided geometries and all the systems investigated. This fact permits the evaluation of the bubble volume from a noncircular orifice by using the equation for the circular orifice with an equivalent diameter. Thus it is not necessary to develop separate equations for each geometry. Figure 18 shows the results obtained for various geometries and

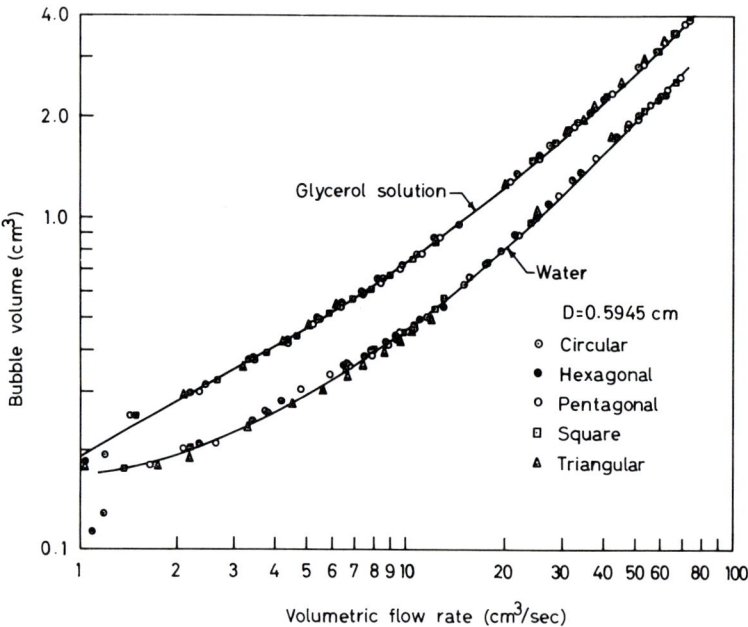

FIG. 18. Effect of orifice geometry on bubble formation for orifices based on area equivalence.

for the circular orifice based on area equivalence. For both the air-water and the air-glycerine systems ($\mu = 288$ cp) all the geometries yield single curves.

Figure 19, which contains correspondingly results based on perimeter equivalence, shows that the various geometries yield different curves. The triangular orifice yields the lowest bubble volumes under a given set of conditions. The bubble volume increases as the number of sides of the orifice increases, the maximum being obtained from the circular orifice. The difference in values is most evident at low flow rates.

The difference arises from the fact that under constant flow conditions, the major function of the orifice is to determine the effect of the surface tension force. The change in velocity, which can be caused by changing the area of the

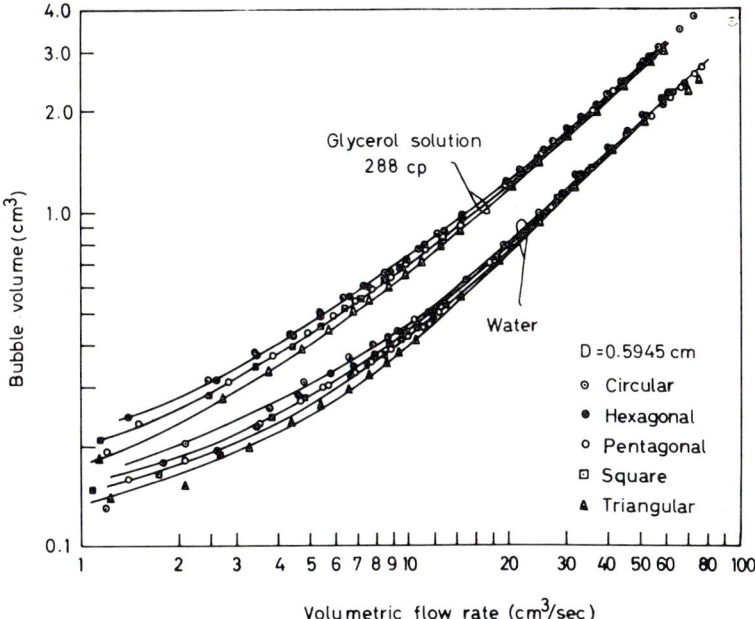

FIG. 19. Effect of orifice geometry on bubble formation for orifices based on perimeter equivalence.

orifice, is of little importance in bubble formation, because of the negligible kinetic energy of the gas stream. For noncircular orifices of equal area, the total perimeter varies with the number of sides. The fewer the number of sides, the greater is the perimeter; thus, the equilateral triangle has the largest perimeter among all the geometries studied. The bubble, however, does not form at the edges, and the more acute the angle the larger is the section which does not participate in bubble formation. This again is maximum for the triangle. The remaining perimeter may be called "effective perimeter." Thus, the unused perimeter decreases with increasing number of sides. The overall effect seems to be such that the effective perimeter remains the same.

Noncircular orifices made on the basis of perimeter equivalence always have a smaller effective perimeter than the circular orifice. Since the deficiency is greater for geometries with fewer numbers of sides, the bubble volumes obtained from triangular orifices are the least.

The influence of orifice geometry is closely associated with that of overall surface-tension force. From Sections IV and V, however, it is evident that the surface-tension effects are limited to small flow rates, the actual value of the flow rate where the effect becomes absent being dependent on viscosity of liquid. Similar effects are evident from Fig. 19, where the various geometries yield distinct curves at low flow rates, which merge into a single curve at high flow rates.

A comparison of Fig. 18 with Fig. 19 throws further light on the phenomenon. The results obtained from area and perimeter equivalence are seen to be wide apart at low flow rates, but are the same at high flow rates. In general, the orifice geometry has no influence at high flow rates ($>80 \text{ cm}^3/\text{sec}$), and the bubble volume can be evaluated by using the equations given in Section IV for situations where the surface tension is not important. For smaller flow rates, the general model that includes the surface-tension effect has to be employed, with the modification that for a noncircular orifice, the diameter must be replaced by the expression

$$D_e = (4A/\pi)^{1/2} \tag{107}$$

where $D_e$ is the effective diameter of a hypothetical equivalent circular orifice. and $A$ is the area of the noncircular orifice.

The modified equations apply quite well to the regular geometries, as is evident from Figs. 18 and 19, wherein the solid lines represent the values obtained from these equations.

The situation may be different for irregular geometries like rectangle, etc., as is found by Krishnamurthi *et al.* (K7), who observe that the bubble volumes obtained for a rectangle made on the basis of area equivalence are higher than those obtained from a circular orifice.

Some experiments conducted under constant pressure conditions indicate that the above equivalence does not hold for this situation.

## IX. The Influence of Orifice Orientation on Bubble Formation

### A. INTRODUCTION

Almost all the results reported in literature pertain to studies with horizontally oriented orifices, and the discussions in the preceding sections are applicable only to such orifices. In industrial applications, however, the orifices are usually oriented at various angles, the most common (as in the case of the bubble plates) being $\pi/2$ from the horizontal.

The first quantitative attempt (K12) in this direction was made with vertical orifices, under constant flow conditions. Here, the bubble formation is considered to be occurring in two distinct steps. In the first stage, the bubble is assumed to expand at the tip, moving vertically at the same time. As the bubble is formed at an angle to the vertical, a vertical component of the surface tension force will be operative during this stage. The first stage is assumed to end when the buoyancy of the growing bubble just counterbalances the vertical component of the surface tension force. The bubble volume corresponding to this condition is denoted by $V_S$.

In the second stage, at any particular instant, the bubble is assumed to be moving with a Stokes terminal velocity corresponding to its size at that instant. Thus,

$$\frac{dx}{dt} = \frac{CV^{2/3}(\rho_l - \rho_g)}{\mu} \tag{108}$$

The second stage is assumed to end when the base of the bubble has covered a distance equal to the orifice diameter. Thus,

$$V_F = \frac{CD\gamma}{(\rho_l - \rho_g)g} + V_m \tag{109}$$

Integrating Eq. (108) within the limits of the orifice diameter, viz. from 0 to $D$, and substituting $Q_t$ for $V$, we obtain

$$V_F = \frac{5}{3}\frac{\mu QD}{C(\rho_l - \rho_g)} + V_S^{5/3} \tag{110}$$

This equation, however, does not adequately account for the influence of viscosity, and was hence modified empirically to

$$V_F^{5/3} = 0.069\frac{\mu}{(\rho_l - \rho_g)}QD(\mu_w/\mu)^{0.87} + \left[1.22(\mu/\mu_w)^{0.09}\frac{D\gamma}{(\rho_l - \rho_g)g}\right]^{5/3} \tag{111}$$

Equation (111) is reported to be applicable up to flow rates of about 0.05 cm³/sec, for liquids having viscosity up to 185 cp. But apart from the approach being semiempirical, the flow rates studied are in too low a range to be of any practical importance. The problem is further complicated by the treatment of each angle as a separate case.

## B. Models of Kumar, Kuloor, and Co-workers

Experimental data for much higher flow rates and for various orifice orientations have been collected by the above investigators, under both constant flow and constant pressure conditions. The data of the above workers, for liquids of different physical properties under constant flow conditions, show that for any definite set of conditions, the bubble size does not decrease continuously with increasing angle of orientation. The data for a viscous liquid are presented in Fig. 20. The orifice oriented at 15° yields higher bubble volumes than the one oriented horizontally. Similarly, the vertically oriented orifice yields higher bubble volumes than that oriented at 60°, under otherwise

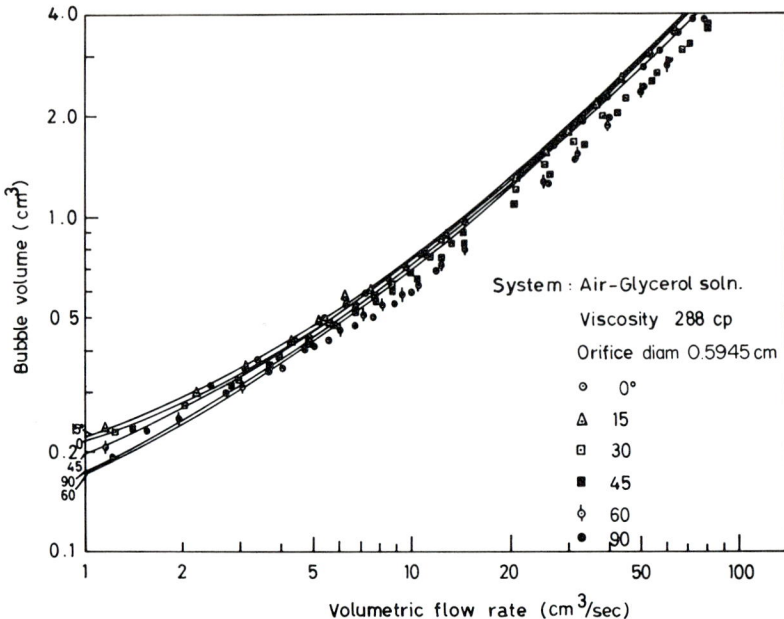

FIG. 20. Effect of orientation of orifice on bubble formation under constant flow conditions.

identical conditions. Further, the behavior varies with the flow rate and the viscosity of the liquid.

These results are explained by extending the models of Kumar and Kuloor discussed earlier in Sections IV and V. At any orientation, the bubble is again considered to be formed in two distinct stages, but the conditions for the termination of the two stages are different in this case.

During the first stage, the bubble initially expands perpendicular to the orifice, but later, both the expansion and the movement are essentially in the vertical direction for all the orifice orientations. The sequence of formation for three different orientations is evident from Fig. 21a,b. This sequence has been idealized in Fig. 21c. Obviously, it is difficult to obtain completely vertical expansion and ascent in the case of the vertical orifice if the expansion takes place exactly at the orifice tip. It is observed through high-speed photography that in this case the bubble neck extends slightly in the first stage itself and the surface-tension force is therefore not zero. The neck makes an angle of about 60° with the horizontal, and hence the surface tension influence for this case is the same as that for an orifice of 60° orientation. Thus the first stage remains essentially the same as that for horizontal orifices except that only a component of surface tension is now operative. This component is $\pi D \gamma \cos \phi$ up to 60° and remains constant afterwards.

The first stage is assumed to end when the vertical components of the expansion drag and the surface-tension force together become equal to the buoyancy.

The second stage commences, and its characteristics in the present model are the same as before, except for the condition for detachment which occurs when the bubble has covered a distance equal to $r_{fb} \cos \phi + D/2 \sin \phi$. This is the most general case. For horizontal orientation, $\phi = 0$, and hence this distance reduces to $r_{fb}$.

Below are given the modified equations for both constant flow and constant pressure conditions.

1. *Constant Flow Conditions*

The modified equation for $V_{fb}$ has to include the vertical components of the expansion drag and the surface-tension forces. Hence, for viscous fluids we have

$$V_{fb}^{5/3} = \frac{11Q^2}{192\pi(3/4\pi)^{2/3}g} + \frac{3\mu Q V_{fb}^{1/3}}{2(3/4\pi)^{1/3}g\rho_l}$$

$$+ \frac{\pi D \gamma V_{fb}^{2/3}}{g\rho_l} \cos \phi \qquad (112)$$

In the second stage, the bubble now covers a distance equal to $(r_{fb} \cos \phi + D/2 \sin \phi)$. Thus,

$$(r_{fb} \cos \phi + D/2 \sin \phi) = \frac{B}{2Q(A+1)}(V_F^2 - V_{fb}^2) - \frac{C}{AQ}(V_F - V_{fb})$$

$$- \frac{3D}{2Q(A-1/3)}(V_F^{2/3} - V_{fb}^{2/3}) \qquad (113)$$

where

$$A = 1 + \frac{6(1.25)\pi(3/4\pi)^{1/3}V_{fb}^{1/3}\mu}{Q[\rho_g + (11/16)\rho_l]}$$

$$B = (\rho_l - \rho_g)g/Q\left(\rho_g + \frac{11}{16}\rho_l\right)$$

$$C = \pi D \gamma \cos \phi/Q\left(\rho_g + \frac{11}{16}\rho_l\right)$$

$$D = 3\mu/2(3/4\pi)^{1/3}\left(\rho_g + \frac{11}{16}\rho_l\right)$$

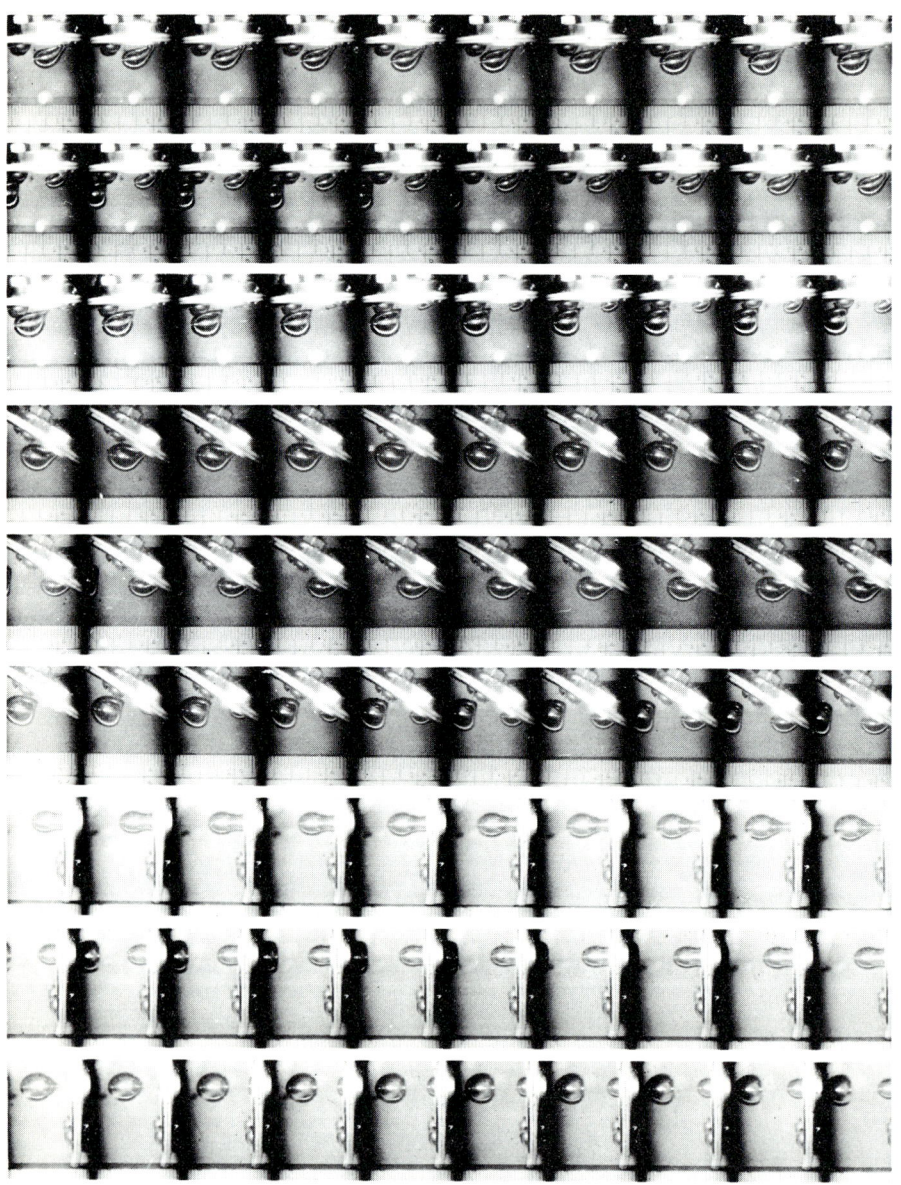

FIG. 21(a). Bubble formation at oriented orifices in viscous systems.

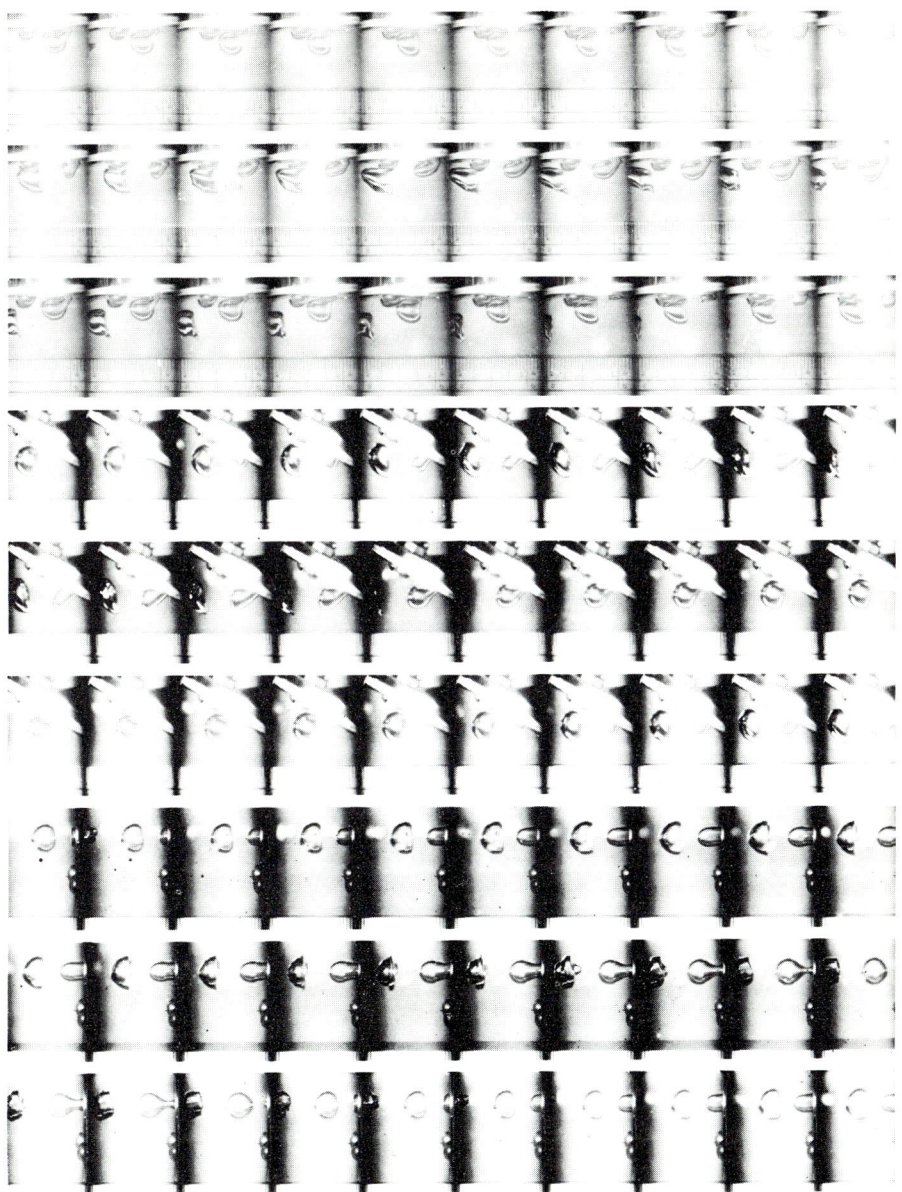

Fig. 21(b). Bubble formation at oriented orifices in inviscid systems.

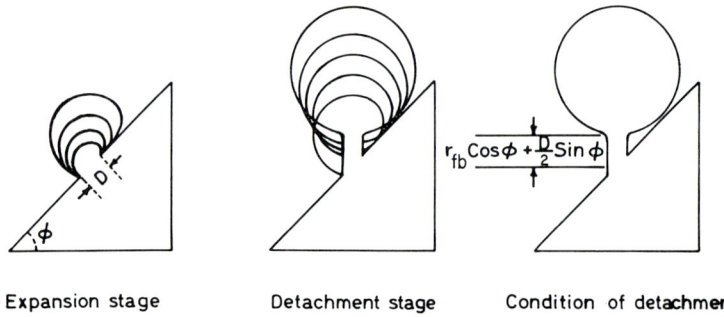

FIG. 21(c). Idealized sequence of bubble formation at inclined orifices.

The increase in $\phi$ tends to reduce the first stage volume because of the smaller net downard force, whereas it increases the second stage volume because of the distance covered by the bubble base. The final volume is the result of both these conflicting factors.

This is very much marked for an orifice of diameter 0.60 cm, in an air-glycerol ($\mu = 288$ cp) system, as shown in Fig. 20. At small flow rates, the bubble volume from vertical orifices are relatively small (though greater than those from 60°), but as the flow rate is increased the curve for this orifice crosses all the other curves, and virtually joins that for the horizontal orifice.

The solid curves in Fig. 20 represent the values obtained by applying Eqs. (112) and (113).

## 2. Constant Pressure Conditions

The results obtained for these conditions exhibit the same kind of irregular behavior as observed for constant flow conditions.

The model proposed for constant flow conditions is again used here to explain the data. The modified equations for inviscid fluids are:

$$V_{fb}(\rho_l - \rho_g)g = \frac{K^2[\rho_g + (11/16)\rho_l]}{4\pi(3/4\pi)^{2/3}} V_{fb}^{-2/3}\left[P + \rho_l g(3/4\pi)^{1/3}V_{fb}^{1/3} - \frac{2\gamma V_{fb}^{-1/3}}{(3/4\pi)^{1/3}}\right]$$

$$- \frac{K^2[\rho_g + (11/16)\rho_l]}{32\pi^2(3/4\pi)^2}$$

$$[4P(3/4\pi)^{1/3}V_{fb}^{1/3} - 3\rho_l g(3/4\pi)^{2/3}V_{fb}^{2/3} - 10\gamma] + 2\pi R\gamma \cos\phi \quad (114)$$

and
$$(r_{fb} \cos \phi + D/2 \sin \phi)$$
$$= (P'/4Q)(V_F^2 - V_{fb}^2) - (9N/Q)(V_F^{1/3} - V_{fb}^{1/3})$$
$$- (J/Q)(V_F - V_{fb}) + (1/Q)$$
$$[JV_{fb} + 3NV_{fb}^{1/3} - (P'/2)V_{fb}^2](\ln V_F - \ln V_{fb}) \qquad (115)$$

where
$$P' = (\rho_l - \rho_g)/Q\left(\rho_g + \frac{11}{16}\rho_l\right)$$
$$J = \pi D\gamma \cos \phi/Q\left(\rho_g + \frac{11}{16}\rho_l\right)$$
$$N = Q/12\pi(3/4\pi)^{2/3}$$

Similarly, for viscous liquids:
$$V_{fb}(\rho_l - \rho_g)g = \frac{K^2[\rho_g + (11/16)\rho_l]}{4\pi(3/4\pi)^{2/3}} V_{fb}^{-2/3}\left[P + \rho_l g(3/4\pi)^{1/3}V_{fb}^{1/3}\right.$$
$$\left. - \frac{2\gamma V_{fb}^{-1/3}}{(3/4\pi)^{1/3}}\right] - \frac{K^2[\rho_g + (11/16)\rho_l]V_{fb}^{-1}}{32\pi^2(3/4\pi)^2}[4P(3/4\pi)^{1/3}V_{fb}^{1/3}$$
$$- 3\rho_l g(3/4\pi)^{2/3}V_{fb}^{2/3} - 10\gamma] + \frac{3K\mu V_{fb}^{-1/3}}{2(3/4\pi)^{1/3}}$$
$$\left[P + \rho_l g(3/4\pi)^{1/3}V_{fb}^{1/3} - \frac{2\gamma V_{fb}^{-1/3}}{(3/4\pi)^{1/3}}\right]^{1/2} + 2\pi R\gamma \cos \phi \qquad (116)$$

and
$$[r_{fb} \cos \phi + (D/2)\sin \phi] = \frac{B}{2Q(A+1)}(V_F^2 - V_{fb}^2) - (C/AQ)(V_F - V_{fb})$$
$$- \frac{3D}{2Q(A-1/3)}(V_F^{2/3} - V_{fb}^{2/3}) \qquad (117)$$

where
$$A = 1 + \frac{6(1.25)\pi(3/4\pi)^{1/3}V_{fb}^{1/3}\mu}{Q[\rho_g + (11/16)\rho_l]}$$
$$B = (\rho_l - \rho_g)g/Q\left(\rho_g + \frac{11}{16}\rho_l\right)$$
$$C = \pi D\gamma \cos \phi/Q\left(\rho_g + \frac{11}{16}\rho_l\right)$$
$$D = 3\mu/2(3/4\pi)^{1/3}\left(\rho_g + \frac{11}{16}\rho_l\right)$$

The above equations have been tested by the authors for data collected over a wide range of variables ($Q = 200$ cm³/sec, $\mu$ up to 500 cp, $D$ up to 0.40 cm) and found to be applicable.

## X. The Influence of Continuous Phase Velocity on Bubble Size

Depending on its subtraction from or addition to the buoyancy force, the continuous phase velocity can either increase or decrease the bubble volume. Normally, this velocity is such that the bubble detaches prematurely from the nozzle tip. Maier (M2) has shown that the shear force experienced by the bubble, which causes its premature detachment, is a maximum when the continuous phase flows at right angles to the nozzle axis.

Krishnamurthy et al. (K13) have confirmed the above conclusion and have developed an expression for evaluating the bubble volume under conditions of flow when the corresponding volume for static conditions is given. These authors used capillaries of different diameters ground at the tip as nozzles. The capillaries were arranged horizontally, and the fluid travelled vertically so as to add to the buoyancy. The liquid viscosity was varied from 1 to 30 cp and the surface tension from 62 to 70 dyn/cm.

The results showed that for all nozzle diameters the decrease in bubble volume with an increase in the continuous phase velocity is nonlinear. Further, for any particular velocity, the reduction in the bubble volume is greater as the liquid viscosity is increased.

The reduction in bubble volume due to the continuous phase velocity is attributed to an extra upward drag force, which adds to the buoyancy. Thus,

$$F_B = F_B' + F_D \tag{118}$$

where

$$F_B = V_F(\rho_l - \rho_g)g$$

and

$$F_B' = V_F'(\rho_l - \rho_g)g.$$

$F_D$ was evaluated from the drag coefficient at bubble Reynolds number and the projected area of the bubble. As the Reynolds number varied from 2 to 700, the drag coefficient $C_D$ was evaluated by the Schiller and Naumann (S4) equation:

$$C_D = (24/Re')(1 + 0.15Re'^{0.687}) \tag{119}$$

Expressing the various terms of Eq. (118) in terms of volume and simplifying, we obtain

$$V_F' + \frac{11.7}{g\rho_l}\mu u V_F'^{1/3} + 2.035(\mu/\rho_l)^{0.313}\frac{u^{1.687}}{g}V_F'^{0.562} = V_F \tag{120}$$

Equation (120) permits the evaluation of $V_F'$ from the values of $V_F$ and the other pertinent parameters. As an analytical solution of Eq. (120) is not possible, numerical procedures have to be used. The equation can, however, be rendered cubic and solved analytically if the power of $V_F'$ in the third term on the left-hand side is changed from 0.562 to 0.667. The error thus caused is found to be negligible (about 1%). The equation then becomes

$$V_F' + \frac{11.7\mu u}{g\rho_l} V_F'^{1/3} + 2.035(\mu/\rho_l)^{0.313} + \frac{u^{1.687}}{g} V_F'^{2/3} = V_F \qquad (121)$$

Equation (121) is solved by Cardon's method and found to predict the data with an average deviation of $\pm 6\%$.

The above analysis assumes a constant downward force which is possible only when the surface tension is the main resisting force. This restricts the utility of Eq. (121) to cases where the gas flow rates employed are extremely small.

## XI. Formation of Bubbles at Multiple Orificed Plates

When the flow rates used are small or the interslot distances large, the bubble interaction does not take place and the bubbles are formed as if from single orifices. In this region the fractional hold-up increases with the increasing gas velocity because the bubble volume increases with the gas flow rate without corresponding change in the velocity of the rise of the bubble. The limit is reached when fractional gas hold-up reaches about 0.60. This corresponds to the condition of close packing of spheres. Under these conditions, the neighboring orifices restrict the bubble diameter in the horizontal direction, and the vertical deformation in bubbles occurs, if the diameter of the bubble (from a single orifice) is greater than the pitch of the perforation.

As the gas flow rate is further increased, Calderbank and Rennie (C4) consider two possibilities. The bubbles may increase in size but deform into polyhedra to yield cellular foams containing very little liquid. Alternatively, the foam may break due to kinetic energy of the gas and bubbles may burst into much smaller units, resulting in froth. Higher vapor velocities (>15.2 ft/sec) generate froth (T1). Cellular foams are formed for low interslot distances, low gas flows, low submergence, and small diameter holes, whereas froths are formed for high flow rates and submergences (C4, T1).

Calderbank and Rennie (C4) and Rennie and Evans (R5) have found Sauter mean bubble diameters both photographically and from foam densities by $\gamma$-ray absorption technique. Their bubble size could be predicted by the equation of Leibson et al. (L2) with a frothing system, for orifice Reynolds numbers between 2000 and 10,000. Thus,

$$d = 0.713 Re_o^{-0.05} \text{cm} \qquad (122)$$

The hold-up can be evaluated by the relationship developed by Crozier (C10):

$$\ln \frac{1}{1 - H_g} = 0.715 v_s (\rho_g)^{1/2} + 0.45 \qquad (123)$$

where $v_s$ is the superficial gas velocity in feet per second, and $\rho_g$ is the gas density in pounds per cubic foot.

By making use of Eqs. (122) and (123), the interfacial area $a$ can be calculated by

$$a = 6 H_g / d \qquad (124)$$

The work of Calderbank and Rennie (C4) has been criticized by Sargent and Macmillan (S2) on the basis that the liquid flow conditions used by Calderbank and Rennie (C4) are not found in distillation columns. They (S2) consider that cellular foams are formed for dilute aqueous solutions only when low gas flow rates are employed. By using an $n$-pentane-isopentane system, Macmillan (M1) found that for all gas flow rates, froths with densities less than 0.15 were formed and the froth densities were independent of the factor $v_s(\rho_g)^{1/2}$ but dependent on tray geometry. The associated problem of foam stability has also attracted considerable attention (A1, D3, Z1).

## XII. Formation of Drops

Drops, like bubbles, can be formed by nozzles, and also by microburettes, spinning disks, or swirl chambers. The various methods have been discussed by Batchelor and Davis (B1) in their monograph. The simplest method of forming drops is from nozzles and is most commonly employed; the present review is restricted to this mode only.

### A. INFLUENCE OF VARIABLES ON DROP SIZE

When a dispersed phase is passed through a nozzle immersed in an immiscible continuous phase, the most important variables influencing the resultant drop size are: the velocity of the dispersed phase, viscosity and density of continuous phase, and the density of the dispersed phase (G2, H1, H5, M3, N1, P5, R3, S5). In general, an increase in continuous-phase viscosity, nozzle diameter, and interfacial tension increases the drop volume, whereas the increase in density difference results in its decrease. However, Null and Johnson (N4) do not find the influence of continuous-phase viscosity significant and exclude this variable from their analysis. Contradictory findings are also reported regarding the influence of volumetric flow rate of the dispersed phase. Hayworth and Treybal (H5) find that when drop volume is plotted versus volumetric flow rate, the curve has one maximum. Null and Johnson (N4), on the other hand, report two peaks in the $V_F$ versus $Q$ curve. Kumar, Kuloor, and co-workers (K2, R3) find one maximum in the flow range studied by them. The maximum is shifted (R3) when the viscosity of the continuous phase is increased and no maximum is observed when the continuous-phase viscosity is raised to quite high values ($>5$ poise). The maximum has also been found to be absent when the dispersed-phase viscosity is increased (K2), though it is possible that if higher flow rates are investigated, the maximum would appear.

The influence of dispersed-phase viscosity was found to be negligible by Hayworth and Treybal (H5), but found to be significant (K2) when a greater range of dispered-phase viscosity was investigated. From the graph of Hayworth and Treybal, the influence of interfacial tension appears, as in the case of bubbles, to be more at low flow rates than at high flow rates.

At vanishingly small flow rates, the drop volume calculated by the equalization of the buoyancy with interfacial tension gives drop volumes higher than those observed experimentally, because of the residual drop. Harkins' (H2) correction has therefore to be applied to calculated drop volumes under these conditions. Thus,

$$V_S = \frac{\pi D \gamma}{\Delta \rho g} \psi(D/d) \qquad (125)$$

This correction has been incorporated in most of the models developed for drop formation.

### B. Hayworth and Treybal Model

The model is based on the assumption that the drop detachment does not take place until the velocity $v$ of the drop is larger than the velocity of the dispersed phase in the nozzle, and also until the buoyancy force equals the force due to interfacial tension.

Another force acting on the drop is due to the kinetic energy of the dispersed-phase stream, which adds to the buoyancy force. Each of these forces has been considered to be contributing to the final drop volume. Thus, the final volume becomes

$$V_F = V_s + V_v - V_K \qquad (126)$$

where $V_v$ is the volume representing the force experienced by the moving drop, and $V_K$ is the negative volume resulting from buoyancy. The term $V_S$ is given by Eq. (125). The expressions for $V_v$ and $V_K$ are

$$V_v = 0.5236 \left[ \frac{18 K v'_c \mu_c}{(\Delta \rho) g} \right]^{3/2} \tag{127}$$

where $K$ is a correction to Stokes law, and

$$V_K = \frac{V_F \rho_d v'^2_c}{2 \Delta \rho g d} \tag{128}$$

Substituting Eqs. (125), (127), and (128) in Eq. (126), we obtain

$$V_F + V_F^{2/3} \left[ \frac{0.403 \rho_d v'^2_c}{g \Delta \rho} \right] = \frac{\pi D \gamma}{\Delta \rho g} \psi(D/d)$$

$$+ 0.5236 \left[ \frac{18 K v'_c \mu_c}{\Delta \rho g} \right]^{3/2} \tag{129}$$

Equation (129) has been simplified to avoid trial and error procedure by taking $\psi(D/d) = 0.655$ as a first approximation and expressing $K$ empirically by

$$K = \frac{4.05 D^{0.747}}{v'^{0.635}_c \mu_c^{0.814}} \tag{130}$$

Introducing the above simplification in Eq. (129), we obtain

$$V_F + 4.11(10^{-4}) V_F^{2/3} \left[ \frac{\rho_d v'^2_c}{\Delta \rho} \right] = 21(10^{-4})(\gamma D / \Delta \rho)$$

$$+ 1.069(10^{-2}) \left[ \frac{D^{0.747} v'^{0.365}_c \mu_c^{0.186}}{\Delta \rho} \right]^{3/2} \tag{131}$$

Equation (131) can be used for the evaluation of $V_F$. The use of this equation has been simplified by having a graphical solution, which represents the drop diameter graphically as a function of two variables,

$$J = 0.00411 \left[ \frac{\rho_d v'^2_c}{\Delta \rho} \right]$$

and

$$H = 0.0021 \left( \frac{\gamma D}{\Delta \rho} \right) + 0.01069 \left[ \frac{D^{0.747} v'^{0.365}_c \mu_c^{0.186}}{\Delta \rho} \right]^{3/2}$$

The $\mu_c$ here is in centipoise.

The average deviation reported by the authors is 7.5%.

## C. Null and Johnson Model

These investigators have considered the drop formation from the viewpoint of the geometry of the forming drop, as actually observed with the aid of stroboscopic light. Two steps of drop formation, viz. the growth at the tip and the detachment stage, have been recognized. Before breakage the drop is considered as a sphere atop, and tangent to, a right truncated cone passing through the circumference of the nozzle.

The final graphical correlation given by the authors is in terms of the following two groups:

$$\left(\frac{D^2 \Delta\rho}{8\gamma}\right)^{1/2} \quad \text{and} \quad \left(\frac{8v_c'^2 \rho_d}{Dg \Delta\rho}\right)^{1/2}$$

A prediction accuracy of 10 to 30% is claimed.

## D. Models of Kumar, Kuloor, and Co-workers

### 1. Model of Rao, Kumar, and Kuloor (R3)

This model considers the drop formation to take place in two stages, the expansion stage when the drop inflates at the nozzle tip and the detachment stage when the drop rises, forms a neck, and finally gets detached from the nozzle. The first stage is assumed to end when the buoyancy becomes equal to the interfacial tension force. For the termination of the second stage two conditions have been used, which result in two values of time of detachment. The lower of the two values is employed for calculation.

The static drop volume is given by the equation

$$V_S = \frac{2\pi R \gamma \phi(R/V^{1/3})}{(\rho_c - \rho_d)g} \tag{132}$$

For evaluation of $t_c$, two models have been proposed.
where $\phi(R/V^{1/3})$ is a correction factor suggested by Harkins and Brown (H2).

When the static drop stage is passed, the drop starts rising with varying velocity, but still maintaining its connection with the nozzle through a neck. As further liquid is pumped into the drop during the time of detachment $t_c$ also, the final drop volume becomes

$$V_F = V_S + Qt_c \tag{133}$$

For evaluation of $t_c$, two models have been proposed.

*a. Model I.* At the time of release, the detaching drop leaves behind a hemispherical residual drop. This model considers that the drop detachment occurs when the rate of growth of the hemisphere is equal to the velocity of the moving drop.

The rate of growth of the hemisphere is found to be

$$dr/dt = \tfrac{1}{2}v'_c \tag{134}$$

where $v'_c$ is the velocity of the dispersed phase in the nozzle.

The movement of the drop in the second stage is governed by the equation of motion. Thus,

$$d(Mv)/dt = Q(\rho_c - \rho_d)gt + Qv_c\rho_d - 6\pi r\mu v \tag{135}$$

Here, the virtual mass $M$ of the drop can be assumed to be constant and equal to that of the static drop, because there is not much growth before detachment. The mass term is written as

$$M = V_S\rho_d + (11/16)V_S\rho_c \tag{136}$$

where $(11/16)V_S\rho_c$ is the correction which accounts for the inertia of the continuous phase and which assumes the flow around the drop to be irrotational and unseparated, as suggested by Davidson and Schuler (D9). The equation of motion now becomes

$$M(dv/dt) = Q(\rho_c - \rho_d)gt + Qv_c\rho_d - 6\pi r\mu v \tag{137}$$

The above equation is a first-order linear differential equation which on solution gives

$$v = \frac{C}{A} + B\left(\frac{t}{A} - \frac{1}{A^2}\right) + \left(\frac{B}{A^2} - \frac{C}{A}\right)C\exp(-At) \tag{138}$$

where

$$A = 6\pi r\mu/M$$
$$B = Q(\rho_c - \rho_d)g/M$$
$$C = Qv'_c\rho_d/M$$

The integration constant for Eq. (138) is calculated by imposing the condition $v = 0$, when $t = 0$. At detachment, the condition $v = (1/2)v'_c$ is applied and the corresponding time of detachment $(t = t_c)$ is evaluated. This model is useful only when the distance between the detachment drop and the residual drop is negligible.

*b. Model II.* This model is applicable to highly viscous continuous phases. This model assumes necking to take place before final detachment. The length of the neck is equal to the diameter of the static drop. Thus the time of detachment is that required by the ascending drop to cover a distance equal to the diameter of the static drop.

For viscous liquids, all the terms except $B/A$ of Eq. (138) are small and can be neglected. Thus,

$$v = dx/dt = Bt/A \tag{139}$$

Equation (139) is solved with the boundary condition:

$$x = 0 \quad \text{at} \quad t = 0$$

The time of detachment is evaluated from the condition $x = d_s$ when $t = t_c$. Thus,

$$t_c = (2Ad_s/B)^{1/2} \tag{140}$$

The final volume is evaluated by making use of Eqs. (132), (133), and (140).

## 2. Inclusion of Dispersed Phase Viscosity

Kalyanasundaram, Kumar, and Kuloor (K2) found the influence of dispersed phase viscosity on drop formation to be quite appreciable at high rates of flow. The increase in $\mu_d$ results in an increase in drop volume. To account for this, the earlier model was modified by adding an extra resisting force due to the tensile viscosity of the dispersed phase. The tensile viscosity is taken as thrice the shear viscosity of the dispersed phase, in analogy with the extension of an elastic strip where the tensile elastic modulus is represented by thrice the shear elastic modulus for an incompressible material. The actual force resulting from the above is given by $3\pi R\mu_d v$.

The first stage is evaluated by

$$V_S = (2\pi R\gamma)/(\rho_c - \rho_d)g \tag{141}$$

The equation governing the second stage becomes

$$M(dv/dt) = Q(\rho_c - \rho_d)gt + Qv_c\rho_d - (6\pi r_s \mu + 3\pi R\mu_d)v \tag{142}$$

Expressing $v$ as $dx/dt$ and solving the equation with the boundary conditions at $t = 0$, $v = 0$, and $x = 0$, and putting $t = t_c$ for $x = d_s$,

$$\left[\left(\frac{B}{A^3} - \frac{C}{A^2}\right)\Big/e^{At_c}\right] - \left(\frac{B}{2A}\right)t_c^2 + \left(\frac{B}{A^2} - \frac{C}{A}\right)t_c$$
$$= \left(\frac{B}{A^3} - \frac{C}{A^2}\right) - d_s \tag{143}$$

where

$$A = (6\pi r_s \mu)/M$$
$$B = (Q\Delta\rho g)/M$$
$$C = (Qv_c\rho_d)/M$$

Equation (143) permits the evaluation of the time of detachment $t_c$ by trial and error procedure.

The final drop volume can then be obtained by using Eq. (133).

## E. Comparison of Various Models

The equation of Hayworth and Treybal (H5) is semi-empirical and is based on a force balance made by expressing the various contributing forces acting on the drop as fractions of the total drop volume. This procedure is probably not wholly justified, since the exact instant at which the forces act is not known, nor is their quantitative contribution to the total volume. The model also neglects the influence of dispersed-phase viscosity.

The model of Null and Johnson (N4) neglects the viscosity of the continuous phase which has been found to be effective (K2).

Null and Johnson (N4) compared their experimental data with the available models and found a maximum deviation of 377% with the model developed by Hayworth and Treybal (H5) and 94% with their own model.

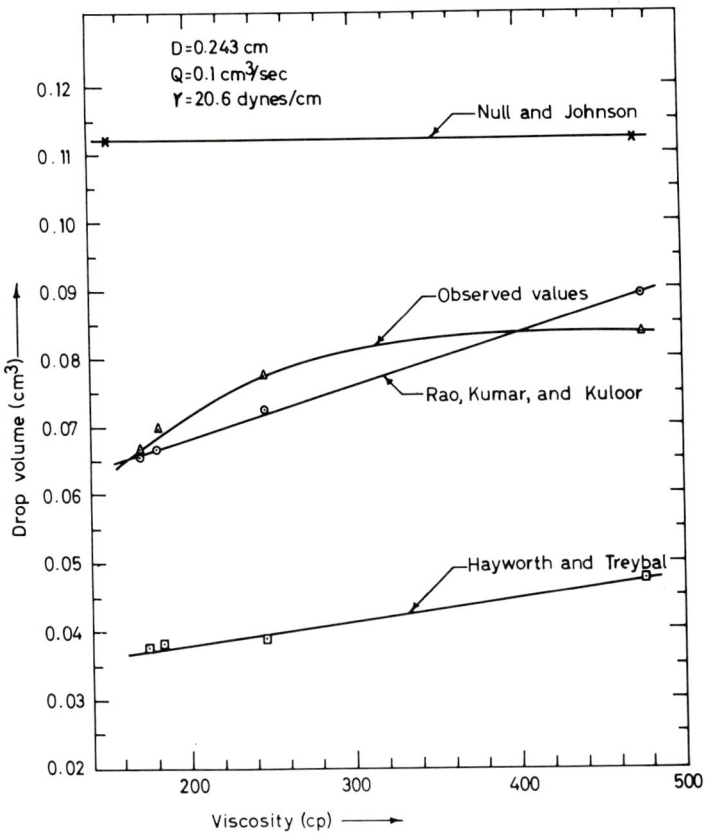

FIG. 22. Effect of viscosity of continuous phase on drop volume.

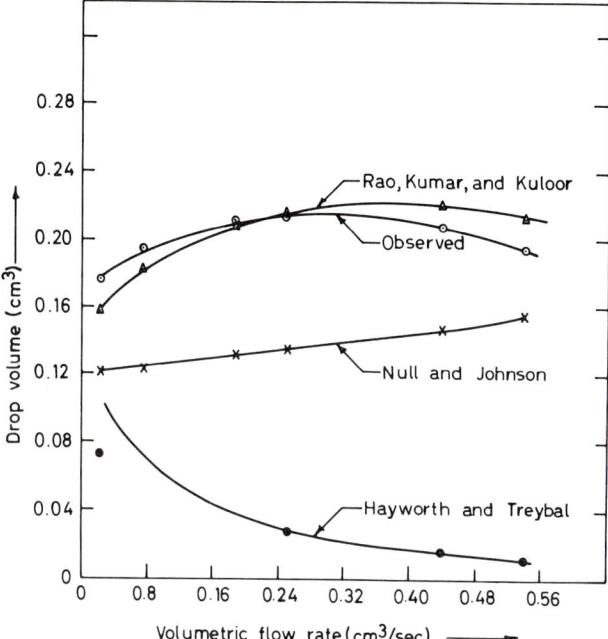

FIG. 23. Effect of volumetric flow rate on drop volume.

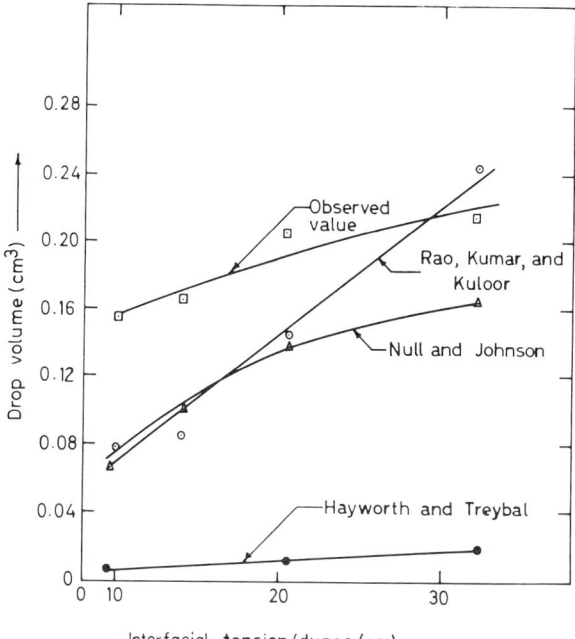

FIG. 24. Effect of interfacial tension on drop volume.

Rao et al. (R3) have compared their experimental results with various models regarding the prediction of the influence of various parameters on drop volume. Their findings are presented in Figs. 22–24. Their model seems to predict the influence of viscosity of the continuous phase and volumetric flow rate on the drop volume better than the other models, but there is considerable deviation for interfacial-tension effect. The interfacial-tension change was brought about by these investigators by the addition of the surface active agent Lissapol; the surface active agents have a tendency to concentrate at the surface rather than in the bulk (C8). Thus the interfacial tension value measured and used by the authors in their calculations may have been much lower than the one operating during drop formation in the liquid bulk.

However, the model of Rao et al. (R3) does not consider the influence of dispersed-phase viscosity. Further, the maximum size of the drop is limited to static drop size, which is true only for low flow rates.

The improvement in this model (K2) takes the dispersed-phase viscosity into consideration and predicts better than the earlier models for situations when the dispersed phase is viscous. A typical set of values is shown in Fig. 25, from which it is seen that the model predicts better results in high flow range only. At lower flow rates, the predicted values are higher because the drop detaches at the nozzle tip itself and the application of Harkins and Brown's (H2) correction becomes important, which has been neglected in the model.

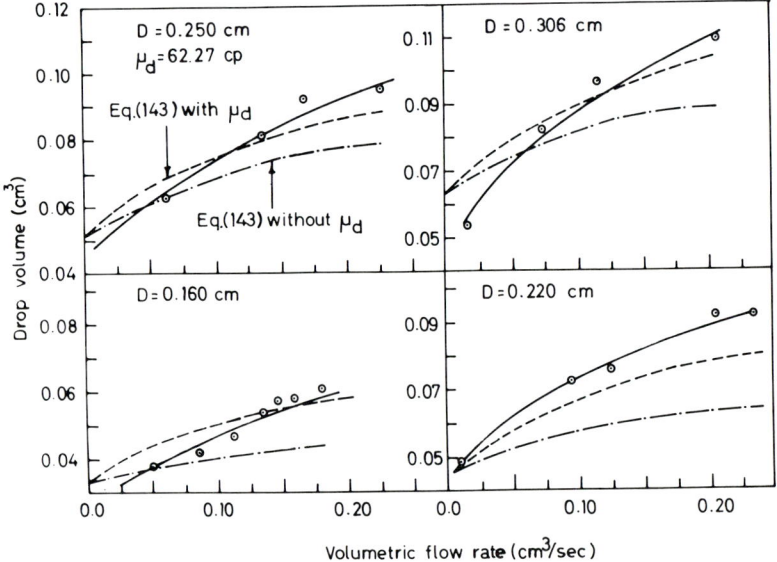

FIG. 25. Effect of viscosity of dispersed phase on drop volume.

The verification of these models at higher flow rates has not been made. Hence they can be applied only in the ranges of conditions at which they have been tested. However, the model of Hayworth and Treybal (H5) has been found reasonably applicable for drop formation during spraying in the absence of mass transfer.

## XIII. Drop Formation in Non-Newtonian Fluids

Though most of the industrial fluids show non-Newtonian characteristics, the drop formation studies in them have not been reported. The results will very strongly depend on whether the non-Newtonian fluid forms the dispersed or continuous phase.

### A. Drop Formation in Power-Law Fluids

Experiments have recently been conducted by Saradhy and Kumar (S1) for drop formation of benzene in a C.M.C.[1] solution which followed the

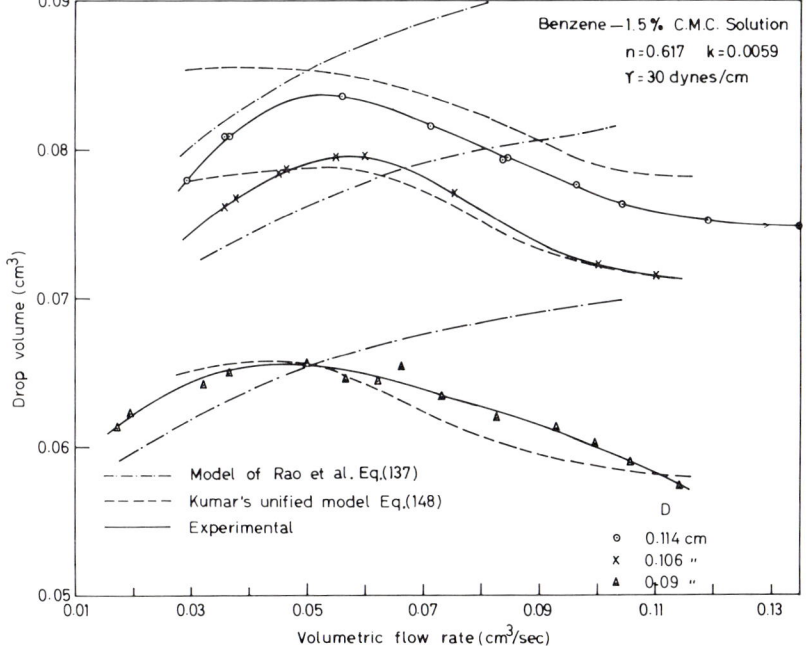

FIG. 26. Comparison of the model (S1) with the collected data for drop formation in non-Newtonian liquids.

[1] Carboxy Methyl Cellulose.

power-law model. The results follow drop-volume versus flow-rate curves which pass through maximum, and are similar to the ones obtained for Newtonian fluids (R3). Figure 26 presents a set of data for a solution having $n = 0.617$ and $k = 0.0059$, which shows a maximum for each of the three nozzles studied. The unified model (K14) has been extended to explain the data collected. The first stage of drop formation remains the same except for the modification of the drag term. The second stage must consider the equation of motion of a spherical bubble in a power-law fluid. This has been studied extensively (S13, T2, W1, W2). The drag coefficient is given by

$$C_D = 24 X_n / N'_{Re} \qquad (144)$$

where $X_n$ is a function of flow index $n$, and $N'_{Re}$ is the modified Reynolds number, defined as

$$N'_{Re} = (2r)^n \rho v^{2-n} / k \qquad (145)$$

$X_n$ is not known rigorously, but its upper and lower bounds have been calculated (W2).

From Eqs. (144) and (145) and the definition of $C_D$, the downward resisting force becomes

$$F_D = (24\pi X_n k v^n) / 2^{n+1} r^{n-2}) \qquad (146)$$

Introducing the above drag expression Eq. (146) in the force balance particle equation of unified model (K14), we obtain

$$V_{fb}(\rho_c - \rho_d)g = \frac{Q^2[\rho_d + (11/16)\rho_c] V_{fb}^{-2/3}}{12\pi(3/4\pi)^{2/3}} + \pi D \gamma \cos \theta$$

$$+ \frac{24 X_n k Q^n (V_{fb})^{-(3n-2)/3}}{2^{3n+1} \pi^{n-1} (3/4\pi)^{(3n-2)/3}}$$

$$- Q \rho_d v'_c \qquad (147)$$

Equation (147) neglects the influence of dispersed-phase viscosity by considering the dispersed phase to be inviscid, which is true for benzene.

The equation for detachment stage is

$$dv/dT + A(v^n / T^{(n+1)/3}) + v/T = B - DT^{-(3n+1)/3} - CT^{-1} - ET^{-5/3} \qquad (148)$$

where

$$v = Q(dx/dT)$$
$$T = V_{fb} + Qt$$
$$A = \frac{24\pi X_n k}{2^{n+1}(3/4\pi)^{(n-2)/3}[\rho_d + (11/16)\rho_c]Q}$$
$$B = \frac{(\rho_c - \rho_d)g}{Q[\rho_d + (11/16)\rho_c]}$$
$$C = \frac{Z}{Q[\rho_d + (11/16)\rho_c]}$$
$$D = \frac{24 X_n k Q^{n-1}}{2^{3n+1}\pi^{n-1}(3/4\pi)^{(3n-2)/3}[\rho_d + (11/16)\rho_c]}$$
$$E = \frac{Q}{12\pi(3/4\pi)^{2/3}} \; ; Z = -Q\rho_d v'_c + \pi D\gamma \cos\theta$$

Equation (148) is solved with boundary conditions at

$$t = 0, \; dx/dT = 0, \; T = V_{fb} \quad \text{and} \quad x = 0$$

The volume $V_F$ is the value of $T$ for $x = r_{fb}$. This is obtained numerically by using the Runge-Kutta method. The results are shown as dotted lines in Fig. 26, where they show excellent agreement in the range investigated. For calculation purposes, the value of $X_n$ has been used as the mean of the upper and the lower bounds found by Wasserman and Slattery (W2).

Use has also been made of the other model (R3) to explain the results. The calculated values through this model are also shown in Fig. 26. The model gives very poor agreement.

## B. Power-Law Drops in Newtonian Fluid

For this case, the first stage remains the same as before, except for the inclusion of the dispersed-phase viscosity, but the equation of motion for the second stage is changed to

$$dv/dT + A(v/T) = B - DT^{-4/3} - CT^{-1} - ET^{-5/3} - Fv^n \qquad (149)$$

where $B$, $C$, and $E$ remain the same as in Eq. (148) and $A$, $D$, and $F$ are

$$A = 1 + \frac{6\pi r_{fb}\mu}{[\rho_d + (11/16)\rho_c]Q}$$
$$D = \frac{6\pi r \mu v_e}{[\rho_d + (11/16)\rho_c]Q}$$
$$F = 3k\pi R^{2-n}$$

This equation is solved with the same boundary conditions as Eq. (148) and the time of detachment is calculated.

## C. Power-Law Drops in Power-Law Fluids

The equation governing the second stage for this situation becomes

$$dv/dT + A \frac{v^n}{T^{(n+1)/3}} + \frac{v}{T} = B - DT^{-(3n+1)/3} - CT^{-1} - ET^{-5/3} - Fv^n \qquad (150)$$

where

$$F = 3k\pi R^{2-n}$$

This equation too is solved with the same boundary conditions as Eq. (148).

A series of equations results when different combinations of fluids are used. There is no change for the first stage. All the terms of equation of motion remain the same except the force terms arising out of dispersed-phase and continuous-phase viscosities. The main information required for formulating the equations is the drag during the non-Newtonian flow around a sphere, which is available for a number of non-Newtonian models (A3, C6, F1, S13, S14, T2, W2). Drop formation in fluids of most of the non-Newtonian models still remains to be studied, so that whether the types of equations mentioned above can be applied to all the situations cannot now be determined.

## XIV. Drop Formation from Vertically Oriented Orifices

In the absence of surface tension influences, the drop formation at vertical orifices is expressed by the equation given for bubble formation. The force due to kinetic energy of the liquid is neglected as its component is zero in the vertical direction. The drop ascends right from the beginning according to the equation of motion and detaches when it has covered a distance equal to the diameter of the nozzle.

For a viscous liquid the equation of motion is

$$d(Mv)/dt = (\Delta\rho)gQt - 6\pi r\mu v \qquad (151)$$

Equation (151) reduces to the inviscid case when $\mu$ is put equal to zero. Expressing $v$ in terms of the distance covered,

$$(d/dt)[M(dx/dt)] = (\Delta\rho)gQt - 6(3/4\pi)^{1/3}\mu Q^{1/3}t^{1/3}(dx/dt) \qquad (152)$$

This equation, when solved between $x = 0$ to $D$ for $t = 0$ to $t_c$, yields

$$D = A\left[\frac{3t^{5/3}}{5B} - \frac{5t^{4/3}}{4B^2} + \frac{20t}{9B^3} - \frac{60t^{2/3}}{18B^4} + \frac{120t^{1/3}}{27B^5} - \frac{120\ln(3Bt^{1/3})}{81B^6}\right] \qquad (153)$$

where

$$A = \frac{(\rho_c - \rho_d)g}{[\rho_d + (11/16)\rho_c]}$$

and

$$B = \frac{6\pi(3/4\pi)^{1/3}\mu Q^{-2/3}}{[\rho_d + (11/16)\rho_c]}$$

Final drop volume $V_F = Qt$.
For the inviscid case, the final equation is

$$t = (2D/A)^{1/2} \qquad (154)$$

where

$$A = \frac{(\rho_c - \rho_d)g}{[\rho_d + (11/16)\rho_c]}$$

Final drop volume $V_F = Qt$.

No data are available for other orientations, but equations similar to those for bubble formation, taking the appropriate component of the force due to kinetic energy, will be applicable.

When surface-tension effect is significant, a component of this force becomes operative which has not been theoretically calculated (K12). The drop now has a static volume-value which must be added to the volume computed through equation of motion with appropriate boundary conditions. These equations have been found to be applicable at small flow rates by Shankar Srinivas and Kumar (S7).

## XV. Spraying and Atomization

### A. Liquid-Liquid Systems

Considerable attention has been paid in literature to the formation of liquid sprays in liquids and gases, through the disintegration of jets. The liquid-liquid systems have been studied mainly with a view to understanding the operation of spray extraction towers. Attempts to measure the drop sizes in liquid-liquid sprays have been made both at lower and higher flow rates (A2, J4, K3, L3, S8, W3). At low flow velocities through the nozzles, the data on drop size in the absence of mass transfer (L3, S8, W3) can be predicted to a rough approximation by the model of Hayworth and Treybal (H5) discussed in Section XII,B. Other models have not as yet been tested for spraying. The main factors influencing the drop sizes produced are: the nozzle diameter, volumetric flow rate of the dispersed phase, and the system properties.

When the flow rate is made exceedingly large, flooding is observed (B4, M5) when the whole column gets filled by closely packed spheres, and further increase in flow rate results in the rejection of the dispersed phase at the entrance region itself. In the vicinity of these conditions, the drop formation mechanism is bound to be influenced by the presence of dispersed phase. The conditions and the theory of flooding in the spray column have been worked out (M5), but apply only in the absence of mass transfer. From studying a wide range of flow rates, Keith and Hixson (K3) point out the various stages of jet formation.

At vanishingly small flow rates, the drops get detached at the nozzle tip. As the flow rate is increased, the jetting point is reached where a very short continuous neck of liquid exists between the nozzle tip and the point of detachment. Further increase in the flow rate lengthens the jet, whereas still further increase to the critical velocity gives the jet a ruffled appearance. Beyond this region, the jet recedes to the nozzle tip.

Below the jetting and the varicose point the drops obtained are of uniform size. Near the critical region, the nonuniformity is a maximum, whereas at still higher velocities the sizes become slightly more uniform, producing smaller drops. Near the critical region, a number of large drops appear which terminate the constant increase of surface area found below this region. Due to this phenomenon, an optimum flow rate is observed which gives the maximum area for unit volume of the dispersed phase.

The uniformity of the particles as measured by the standard deviation for a normal probability curve was found to be a function of flow rate, nozzle size (better for smaller nozzles), nozzle length (decrease in nozzle length decreases uniformity), etc. A decrease in interfacial tension is insufficient to cause change in uniformity.

The range of dispersed-phase velocity studied by Keith and Hixson (K3) is from 10 to 30 cm/sec which, according to those authors, is of industrial interest. The results obtained by them in the absence of mass transfer can be predicted roughly by extrapolation of the Hayworth and Treybal correlation. In the presence of mass transfer, the results obtained (F2), the drop size distribution, flooding, etc. are different from those observed in the absence of mass transfer. There is no reliable theory at present which can predict the drop size distribution in sprays, though rough approximations are possible when mass transfer is completely avoided.

## B. Liquid-Gas Systems

A number of techniques have been evolved to disperse liquids in gases in the form of fine droplets. The various atomizing techniques are jet injections, fan sprays, centrifugal nozzles, twin fluid atomizers, impinging jets, and rotary atomizers. Only the jet injections system is discussed here on the ground that the mechanism of its working is allied to the rest of this study. In the case of centrifugal nozzles and rotary atomizers, the process of drop formation is more due to the disintegration of liquid sheets than to the detachment of liquid drops from a liquid column. Spraying results in particle diameters predominantly between 100 and 1000 microns, whereas atomization results in droplets of 100 microns diameter and smaller (O1). The various factors influencing liquid disintegration are: the liquid properties and flow rate, pressure, nozzle design, and the condition of the gas into which the liquid is injected. The problem of jet breakage leading to droplet formation has been considerably studied since the work of Rayleigh (R4). Though the main force participating in jet disintegration is the surface tension, other system properties like the liquid density, its viscosity, and its velocity in the gas also contribute to the disintegration process.

Small disturbance causes the formation of crests and troughs on the jet, which intensify as long as there is a velocity difference between the gas and the wave surfaces. This can result in finally disintegrating the jet. The jet itself has turbulent eddies which contribute to jet disintegration. This factor has greater importance at high flow rates, so much so that a series of liquid ligaments form, which break into small droplets (D1, D2). Satellite droplets of 1 to 2 microns diameter are formed along with the main droplets. A further complicating factor is the secondary breakup of droplets already formed. The criterion for secondary breakup is found experimentally as $\rho_g v_j^2/\gamma = $ constant (L4), the maximum stable droplet size being $r_{st} = 8\gamma/(\rho_g v_j^2)$; when the air is also in motion, the mechanism of disintegration is different (W4).

Because of such factors as wave formation, jet turbulence, and secondary breakup, the drops formed are not of uniform size. Various ways of describing the distribution, including the methods of Rosin and Rammler (R9) and of Nukiyama and Tanasawa (N3), are discussed by Mugele and Evans (M7). A completely theoretical prediction of the drop-size distribution resulting from the complex phenomena discussed has not yet been obtained. However, for simple jets issuing in still air, the following approximate relation has been suggested (P3):

$$d_L = f\left(\frac{D^{0.5}\mu^{0.2}}{\rho_l^{0.2} p^{0.4}}\right) \tag{155}$$

where $d_L$ is the linear mean droplet diameter and $p$ is the pressure of the liquid.

The other kinds of atomizers have been extensively studied and the available literature has recently been compiled (O1).

## XVI. Unified Model for Drop and Bubble Formation

Though the topics of bubble and drop formation have been treated separately by almost all the investigators, the two phenomena are quite similar as far as the influence of most of the variables is concerned. Even in the formation observed visually, the expansion and the detachment stages are similar. The main differences between the two arise from such properties as density, viscosity, and velocity of the dispersed phase, which cannot be neglected during the drop formation but are unimportant during bubble formation. Further, for vanishingly small flow rates, the drop formation must take the residual drop into consideration. Kumar (K14) has extended the bubble formation concepts of Kumar and Kuloor (K16, K18, K19, R1) for constant flow conditions to take the above factors into consideration. The velocity of the dispersed phase through the nozzle has been considered to yield a kinetic force which adds to the buoyancy as in drop formation (R3). The dispersed-phase viscosity is considered to add a tensile force which resists detachment (K2).

The drop formation is considered to proceed exactly in the same fashion as the bubble formation under constant flow conditions, viz. the two step (the expansion and detachment) mechanism. The tensile force does not arise in the expansion stage because there is no neck formation.

The first-stage equation can now be written as

$$\underbrace{V_{fb}(\rho_c - \rho_d)g}_{(I)} = \underbrace{\frac{Q^2[\rho_d + (11/16)\rho_c]V_{fb}^{-2/3}}{12\pi(3/4\pi)^{2/3}}}_{(II)} + \underbrace{\frac{3\mu Q V_{fb}^{-1/3}}{2(3/4\pi)^{1/3}}}_{(III)} + \underbrace{\pi D \gamma \cos\theta}_{(IV)} - \underbrace{\frac{Q^2 \rho_d}{\pi R^2}}_{(V)} \quad (156)$$

Equation (156) is applicable for both bubble and drop formation. Thus, when term (V) is negligible, it describes bubble formation for all fluids when surface tension is important. On further removal of term (IV), it applies for viscous liquids in the absence of surface tension. If only terms (I) and (II) are retained, the equation applies to the inviscid case without surface tension.

The equation of motion for the second stage is modified by the inclusion of both the kinetic force and the tensile force due to the dispersed-phase viscosity.

The basic equation now becomes

$$dv/dT + A(v/T) = B - DT^{-4/3} - CT^{-1} - ET^{-5/3} \quad (157)$$

where

$$A = 1 + \frac{6\pi(1.25)(3/4\pi)^{1/3}V_{fb}^{1/3}\mu}{Q[\rho_d + (11/16)\rho_c]} + \frac{3\pi D\mu_d}{2Q[\rho_d + 11/16)\rho_c]}$$

$$B = \frac{(\rho_c - \rho_d)}{Q[\rho_d + (11/16)\rho_c]}$$

$$C = \frac{\pi D\gamma \cos\theta}{Q[\rho_d + (11/16)\rho_c]} - \frac{4Q\rho_d}{\pi D^2[\rho_d + (11/16)\rho_c]}$$

$$D = \frac{3\mu}{2(3/4\pi)^{1/3}[\rho_d + (11/16)\rho_c]}$$

and

$$E = \frac{Q}{12\pi(3/4\pi)^{2/3}}$$

Equation (157) also reduces to various special cases of bubble formation when particular terms become insignificant.

When the drop detachment takes place farther from the nozzle, the surface-tension term in Eq. (157) must be dropped.

The final solution with the same condition of detachment (distance covered by the base of the bubble or drop equals $r_{fb}$) becomes

$$r_{fb} = \frac{B}{2Q(A+1)}(V_F^2 - V_{fb}^2) - \frac{C}{AQ}(V_F - V_{fb})$$
$$- \frac{3D}{2Q(A-1/3)}(V_F^{2/3} - V_{fb}^{2/3}) - \frac{3E}{Q(A-2/3)}(V_F^{1/3} - V_{fb}^{1/3})$$
$$- \frac{1}{Q}\left(\frac{V_F^{-A+1}}{-A+1} - \frac{V_{fb}^{-A+1}}{-A+1}\right)$$
$$\left[\left(\frac{B}{A+1}V_{fb}^{A+1} - \frac{C}{A}V_{fb}^A - \frac{D}{A-1/3}V_{fb}^{A-1/3} - \frac{EV_{fb}^{A-2/3}}{A-2/3}\right)\right]$$

(158)

Equations (156) and (158) are applicable to bubble formation under constant flow conditions because all the pertinent equations can be obtained from Eqs. (156) and (158).

The equations have also been tested for drop formation under arbitrarily chosen conditions by making use of available data (K1, R2). The results are given in Table IX for comparison. The corresponding values calculated by

## TABLE IX
### Comparison of Drop Volume Data Available in Literature with Those Calculated by the Unified Model

| System | $D$ (cm) | $\mu$ (poise) | $\mu_d$ (poise) | $\gamma$ (dyn/cm) | $\rho_c$ (gm/cm³) | $\rho_d$ (gm/cm³) | $Q$ (cm³/sec) | Drop volume (cm³) Experimental | Drop volume (cm³) Calculated |
|---|---|---|---|---|---|---|---|---|---|
| *1. West et al. (W3)* | | | | | | | | | |
| Benzene-water | 0.162 | 0.009 | 0.006 | 32.2 | 0.9968 | 0.87 | 0.064 | 0.0952 | 0.1566 |
| | | | | | | | 0.087 | 0.0867 | 0.1599 |
| | | | | | | | 0.247 | 0.0692 | 0.1724 |
| *2. Null and Johnson (N4)* | | | | | | | | | |
| Benzene-water | 0.252 | 0.0101 | 0.0065 | 28.9 | 0.9972 | 0.8639 | 0.1472 | 0.1738 | 0.2211 |
| | | | | | | | 0.4370 | 0.1683 | 0.2552 |
| | | | | | | | 0.6460 | 0.1343 | 0.2499 |
| | | | | | | | 0.5310 | 0.1518 | 0.2559 |
| *3. Hayworth and Treybal (H5)* | | | | | | | | | |
| Benzene, alka-terge-water | 0.472 | 0.0101 | 0.0065 | 10.9 | 0.9982 | 0.8788 | 0.0910 | 0.1954 | 0.1711 |
| | | | | | | | 0.4900 | 0.2680 | 0.2391 |
| | | | | | | | 1.7500 | 0.3330 | 0.2279 |
| Benzene, alka-terge-water | 0.155 | 0.0101 | 0.0065 | 10.9 | 0.9982 | 0.8788 | 0.0155 | 0.0610 | 0.0514 |
| | | | | | | | 0.0624 | 0.0919 | 0.0599 |
| | | | | | | | 0.2457 | 0.1021 | 0.0571 |
| Benzene-water | 0.472 | 0.0101 | 0.0065 | 25.7 | 0.9982 | 0.8788 | 0.1435 | 0.3330 | 0.3899 |
| | | | | | | | 0.4550 | 0.4076 | 0.4632 |
| | | | | | | | 1.1375 | 0.4632 | 0.5456 |
| Benzene, alka-terge-water | 0.472 | 0.0101 | 0.0065 | 3.3 | 0.9982 | 0.8788 | 0.0770 | 0.0871 | 0.0599 |
| | | | | | | | 0.5600 | 0.2209 | 0.1081 |

| System | | | | | | | | | |
|---|---|---|---|---|---|---|---|---|---|
| **4. Rao et al. (R3)** | | | | | | | | | |
| Benzene-water | 0.243 | 0.009 | 0.0059 | 32.2 | 0.9968 | 0.87 | 0.0224 | 0.1750 | 0.2121 |
| | | | | | | | 0.0950 | 0.2050 | 0.2348 |
| | | | | | | | 0.3650 | 0.2190 | 0.2739 |
| | | | | | | | 0.5700 | 0.1890 | 0.2749 |
| Benzene-water | 0.110 | 0.009 | 0.0059 | 32.2 | 0.9968 | 0.87 | 0.0119 | 0.0633 | 0.0961 |
| | | | | | | | 0.0548 | 0.0842 | 0.1060 |
| | | | | | | | 0.0975 | 0.0864 | 0.1099 |
| | | | | | | | 0.1820 | 0.0691 | 0.1044 |
| Benzene-glycerine | 0.243 | 4.77 | 0.0059 | 20.7 | 1.2510 | 0.87 | 0.0061 | 0.0371 | 0.0568 |
| | | | | | | | 0.0499 | 0.0637 | 0.0858 |
| | | | | | | | 0.1120 | 0.0882 | 0.1106 |
| | | | | | | | 0.3380 | 0.1525 | 0.1734 |
| **5. Kalyanasundaram (K1)** | | | | | | | | | |
| Glycerol solution carbon tetrachloride | 0.160 | 0.0090 | 0.1536 | 27.05 | 1.590 | 1.184 | 0.005 | 0.0280 | 0.0362 |
| | | | | | | | 0.010 | 0.0290 | 0.0376 |
| | | | | | | | 0.030 | 0.0328 | 0.0405 |
| | | | | | | | 0.050 | 0.0362 | 0.0435 |
| | | | | | | | 0.100 | 0.0443 | 0.0482 |
| | | | | | | | 0.200 | 0.0575 | 0.0523 |
| Glycerol solution carbon tetrachloride | 0.220 | 0.0090 | 0.1536 | 27.05 | 1.590 | 1.184 | 0.005 | 0.0412 | 0.0493 |
| | | | | | | | 0.010 | 0.0427 | 0.0506 |
| | | | | | | | 0.030 | 0.0478 | 0.0547 |
| | | | | | | | 0.050 | 0.0520 | 0.0579 |
| | | | | | | | 0.100 | 0.0597 | 0.0640 |
| | | | | | | | 0.200 | 0.0698 | 0.0719 |

the individual models developed by the earlier investigators are also given (K2, R3). It is seen that this model gives values which are as good as the earlier models but has the advantage of being general. The model cannot, however, be applied to drop formation at very low flow rates because the Harkins correction, which has been neglected in Kumar's analysis, then becomes important.

## XVII. Bubble and Drop Size in Stirred Vessels

In stirred vessels the formation and maintenance of dispersions involves a balance of disintegration and coalescence of the dispersed-phase particles in a continuous phase that is typically in turbulent flow. The disintegration of a globule can occur either by viscous-shear forces or by turbulent pressure fluctuations. Clay (C7) has proposed turbulence as the primary cause of breakage. The various kinds of globule deformation resulting in final breakage have been discussed by Hinze (H6) who expresses the forces controlling deformation and breakup in terms of a Weber group and a viscosity group. Breakup occurs when the Weber group exceeds a critical value which, in turn, depends upon the mechanism of globule breakage. For a dispersion in turbulent flow the critical Weber group is nearly unity and the corresponding value of viscosity group is small. Calderbank (C2) has further elaborated the work of Hinze (H6) and shown that viscous forces may be ignored in the dispersions of gases and liquids. Shinnar and Church (S9), while discussing the particle size of agitated dispersions on the basis of Kolmogoroff's postulate of local isotropy, conclude that the gas bubbles do not become unstable due to oscillation of the globule as a whole, and the breakup of the bubbles is caused by the breakage of the surface layer by viscous shear.

Under noncoalescing conditions, the size of gas bubbles in gas-liquid dispersions can be obtained by balancing the surface-tension forces with those due to turbulent fluctuations, which results in (C3, H3):

$$d_m = k \left[ \frac{\gamma^{0.6}}{(W/G)^{0.4} \rho^{0.2}} \right] \quad (159)$$

where $d_m$ is the diameter of maximum surviving bubble in centimeters, $\gamma$ is the interfacial tension, $W$ is the horse power dissipated, and $G$ is the volume in cubic feet.

For a homogeneous and isotropic turbulent field, Hinze (H6) has shown that $k$ in Eq. (159) equals $[(1/2)(N_{We_{cr}})]^{3/5}$, where $N_{We_{cr}}$ is the critical Weber number of maximum bubble size capable of survival.

Detailed investigations of dispersions in agitated and baffled vessels have been made by Vermeulen, Williams, and Langlois (V3), Rodger, Trice, and

Rushton (R6), Rodriguez, Grotz, and Engle (R7), and by Calderbank (C3). Vermeulen *et al.* (V3) have recognized the physical properties of the two fluids, volume fraction of the dispersed phase, stirring speed, and factors associated with the agitator geometry as the pertinent variables influencing the particle diameter in the agitator. They have made a detailed study of the first three sets of variables. The influence of the agitator geometry and other mechanical factors has been studied by Rodriguez *et al.* (R7). Vermeulen *et al.* (V3) report that an increase in agitator speed reduces the particle diameter both for liquid-liquid and gas-liquid dispersions; the decrease in interfacial tension decreases particle size; the viscosity of the continuous phase has no influence for liquid-liquid systems but significant effect for gas-liquid systems where the increase in viscosity increases the bubble size; the dispersed-phase viscosity increases the drop size but its effect is inconclusive in gas-liquid systems because of the nearly equal viscosities of the gases used; the density differences are not important in liquid-liquid systems and the gas density is not important in gas-liquid systems; the mean diameter of the particle decreases with decrease in volume fractions of the dispersed phase both for liquid-liquid and gas-liquid systems. Further, the authors find that for some liquid-liquid and all the gas-liquid systems there is an increase in the particle diameter at points away from the impeller tip. Calderbank (C3) observes that definite distributions exist for interfacial area both in vertical and radial directions. The influence of various pertinent groups has been discussed by Rodger *et al.* (R6).

As completely theoretical analysis is not available for such systems, empirically found correlations involving pertinent dimensional groups have been proposed. Vermeulen *et al.* (V3) have listed ten sets of dimensionless variables which can possibly influence the nature of the dispersions obtained.

Out of these, six, viz. Weber number, Reynolds number, volume fraction $\lambda$, viscosity ratios, density ratios, and length ratios were finally used for correlation. The other four were used but found not to improve the correlation.

The final correlations of Vermeulen *et al.* (V3) are:
For liquid-liquid systems:

$$\frac{N^2 \rho' L^{4/3} d^{5/3}}{\gamma f_\lambda^{5/3}} = 0.016 \tag{160}$$

where $N$ equals angular velocity of the stirrer (revolutions per second or revolutions per minute)

$\rho' = 0.4\, \rho_c + 0.6\, \rho_d$

$L$ = paddle diameter (centimeters)

$f_\lambda$ = ratio of actual mean drop or bubble diameter to diameter at $\lambda = 0.1$

For gas-liquid systems:

$$\frac{N^{1.5} \, dL \rho^{0.5} \mu_D^{0.75}}{\gamma f_\lambda \mu^{0.25}} = 4.3 \times 10^{-3} \qquad (161)$$

The authors have also given a unified correlation applicable to both bubbles and drops.

Calderbank (C3) has given different correlations for drops and bubbles as well as for coalescing and noncoalescing systems. However, Jackson (J1) has pointed out that it is unlikely that an absolute classification of this type is possible in practice, where speed of coalescence is continuously varying.

Rodriguez *et al.* (R7) have presented correlations on the basis of power dissipation which can be useful for scale-up purposes.

Both Vermeulen *et al.* (V3) and Calderbank (C3) conclude that the mean particle size is a function of impeller tip speed, whereas Rodriguez *et al.* (R7) find it to be a function of power input per unit volume. Jackson (J1) explains this apparent discrepancy on the basis that the System of Rodriguez *et al.* (R7) was more coalescing in nature than the systems studied by the others (C3, V3). In a coalescing dispersion there is frequent circulation and redispersion which requires impeller power. It is further pointed out (J1) that, although the tip speed determines the mean particle size leaving the impeller, the particle size will also depend on the frequency of circulation which is a function of power input.

Thus, here again a final correlation taking into account all the factors is still not available.

## XVIII. Bubble Formation under Intermediate Conditions and at Sintered Disks

### A. Bubble Formation under Intermediate Conditions

When both the chamber pressure and the gas flow rate into the forming bubble are time dependent, the bubbles are said to be formed under intermediate conditions. The experiments conducted in this region yield results which in some respects are quite different from those obtained under constant flow or constant pressure conditions. A major difference is observed with respect to the influence of the depth of submergence on the bubble volume. Whereas the submergence has no influence under constant flow or constant pressure conditions, it has marked influence (P1) under intermediate conditions.

The importance of chamber volume as a parameter was recognized by Hughes *et al.* (H7) and by Davidson and Amick (D5). Hughes *et al.* (H7) gave conditions under which the chamber volume influence cannot be neglected. As the pressure inside the chamber fluctuates, there is a time lag between the instant of detachment and the start of the next bubble. Any attempt towards quantitative prediction must take this as well as the variation of flow into consideration.

Kurana and Kumar (K20) have written the general equations taking the variation of flow into consideration for the two step mechanism of bubble formation. The fluctuations of pressure and the corresponding variations in flow are treated by an electrical analog.

Their first and second stage equations, respectively, are

$$V_{fb}(\rho_l - \rho_g)g = \frac{[\rho_g + (11/16)\rho_l]}{12\pi(3/4\pi)^{2/3}V_{fb}^{2/3}}\left[3V_{fb}\left(\frac{dQ}{dt_e}\right)_{fb} + Q_{fb}^2\right]$$
$$+ \pi D\gamma + \frac{3\mu Q_{fb}}{2(3/4\pi)^{1/3}V_{fb}^{1/3}} \quad (162)$$

and

$$V_t\left(\rho_g + \frac{11}{16}\rho_l\right)\left(\frac{dv}{dt}\right) + Q_t\left(\rho_g + \frac{11}{16}\rho_l\right)v = V_t\Delta\rho g - 6\pi r\mu(v + v_e) + \pi D\gamma$$
$$- \frac{[\rho_g + (11/16)\rho_l]}{12\pi(3/4\pi)^{2/3}V_t^{2/3}}\left[3V_t\left(\frac{dQ_t}{dt}\right) + Q_t^2\right] \quad (163)$$

In the viscous resistance term in Eq. (163), the radius $r$ can be taken as $1.25\, r_{fb}$ without causing a great loss in accuracy. Equations (162) and (163) reduce to constant flow or constant pressure conditions, respectively, when $Q$ is constant or when it is expressed in terms of orifice equation.

To express the flow rate as a function of time, an electrical analog shown in Fig. 27 has been used.

$E_1$ is a voltage source which represents the pressure source and has an internal resistance $R_i$ which is analogous to the pipeline resistances, etc. Voltmeter $V$ represents a manometer. $R_T$ and $L_T$ are the tip resistance and inductance, respectively. $E_2$ is a voltage source connected in opposition to $E_1$ and represents the orifice submergence $h$. Ammeter $A$ is analogous to the flow meter, and the capacitor $C$ represents the chamber. The capacitor has been connected in parallel and not in series because in the latter case it would act as an open circuit. The chamber in the actual set-up permits the flow of gas through it, in addition to storing a part of the inflow of gas in it during a part of the cycle and releasing the stored amount during the remaining part of the cycle. The capacitor in parallel behaves in the same manner. $R_v$ is a variable resistance representing the needle valve which controls the flow of gas into the

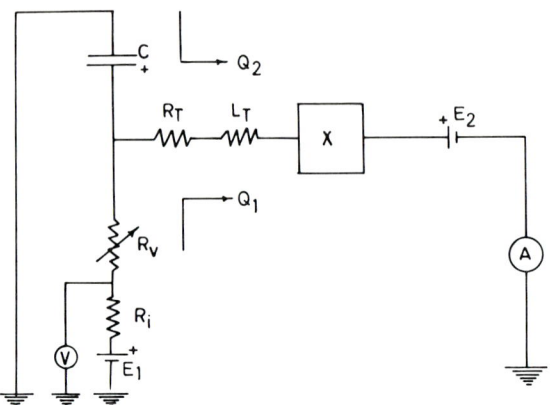

FIG. 27. Electrical analog for bubble formation.

system. $X$ is an element which represents the bubble itself. This element conducts only when the capacitor voltage (chamber pressure) has built up to a certain minimum value $(P_A + h + 2\gamma/r_0)$ and offers a variable resistance once it is conducting. The resistance of this element decreases as the bubble radius increases. The resistance of element $X$ is actually a variable but for the purposes of calculations an average constant value has been taken which was determined experimentally. Further, the sum of the tip resistances and the resistance of element $X$ can be neglected in comparison with the valve resistance $R_v$. The loss in accuracy due to these assumptions is very small. The flow through the system can then be given by

$$Q_t = \frac{E - E_2}{R_v} + \frac{E_o}{R_D} e^{-(t/R_D C)} \tag{164}$$

Equation (165) gives the gas flow rate into the bubble as a function of time.

The bubble volume is calculated by solving Eq. (164) with Eqs. (162) and (163) through numerical methods. First, Eqs. (164) and (162) are solved through trial and error and then Eqs. (164) and (163) are solved by the Runge-Kutta method with the boundary conditions that when $t = 0$, $dx/dt = v = 0$, and $x = 0$. The condition of detachment is taken when $x = r_{fb}$.

The above method of calculation is time consuming and a simpler approximate method has been evolved. This method makes use of the fact that Eq. (164) gives flow rate as the sum of a constant and a variable term. The final bubble volume can then be considered as the sum of the volume due to constant flow rate and the volume by which the capacitor expands before the bubble detaches.

The first calculation of bubble volume is made at average constant flow rate by using Eqs. (54) and (61). As the formation cycle consists of two periods (viz. the formation time during which the flow into the bubble occurs and the weeping time during which no flow into the bubble occurs), the average flow rate must be modified for the weeping periods.

From the bubble size and the average flow rate, a first approximation of frequency (thereby total cycle time), as well as the limits between which the chamber pressure oscillates, is determined. The weeping time is calculated from the wave-form equations of chamber pressure, which are:
During storage:

$$(p - p_o)/(p_H - p_o) = 1 - e^{-(t/R_V C)} \qquad (165)$$

During discharge:

$$(p' - p_h)/(p'_o - p'_h) = e^{-(t/R_D C)} \qquad (166)$$

The flow rate is then corrected by a ratio of the weeping time to the cycle time. Next the discharge time constant ($R_D C$) of the capacitor is determined. If this is shorter than formation time, the entire capacitor expansion is added to bubble volume. If not, the same percentage of the capacitor expansion volume is added because the formation time is of the order of the capacitor discharge constant. The final value thus obtained forms the starting point for the next iterative step. This is continued till the values from two successive iterations are nearly the same.

The predicted values by this method are shown in Fig. 28 where the influence of flow rate on bubble volume for three submergences is shown. The agreement between theory and experiment is good and the theory predicts the variation of bubble volume with submergence satisfactorily.

## B. Bubble Formation at Sintered Disks

Porous plates or sintered disks find extensive use in industry for gas-liquid contacting because they yield large numbers of very small bubbles. Under the usual operating conditions, sintered disks give bubbles with diameters ranging from 0.05 to 0.5 cm. Experiments (B7) on sintered disks reveal that the average bubble volume increases with increasing gas flow rate and liquid viscosity. Increases in surface tension and average pore diameter increase the bubble volume at low flow rates but have relatively negligible effects at higher flow rates.

Bowonder and Kumar (B7) have proposed a model for this phenomenon based on the following assumptions:

The resistance offered by the disk is so large that the bubble formation takes place under essentially constant flow conditions.

A sintered disk contains an extremely large number of "potential sites" for bubble formation, only a fraction of which form the "operative sites."

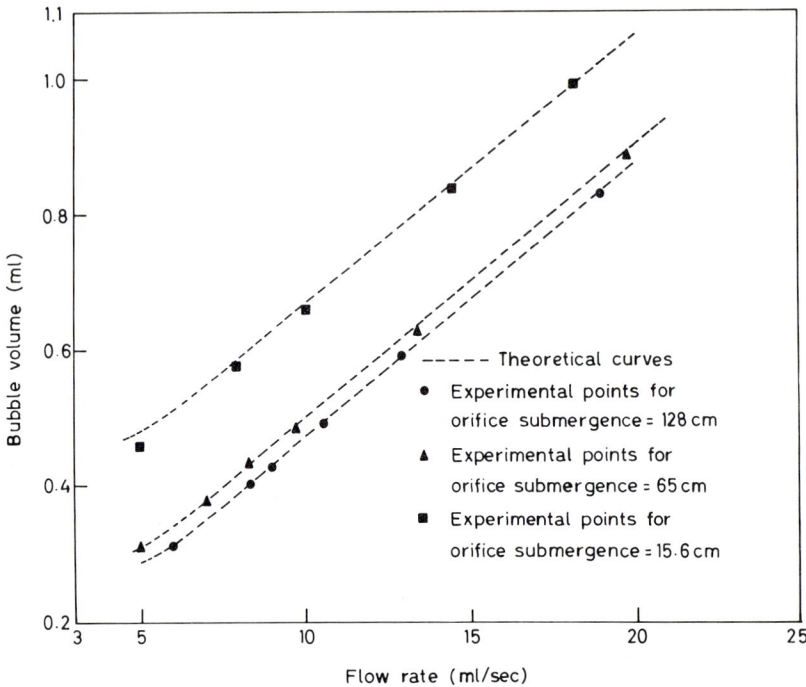

Fig. 28. Influence of flow rate on bubble volume for different orifice submergences.

The number as well as the location of the operative sites adjust themselves as the flow rate varies. This assumption has been experimentally verified.

The flow rate through each operative site is the same, and no flow occurs through the "non-operative sites."

The model further assumes that the number of effective sites is such that the bubbles formed from them as if from single nozzles are just sufficient to cover the whole area when arranged in hexagonal close packing. This assumption has been verified by measuring the average bubble size as a function of flow rate per site and comparing it with the results from an isolated nozzle. The hexagonal close packing arrangement is based on visual observation. The increase of bubble size with flow rate as well as the packing arrangement of bubbles is presented in Fig. 29.

The above conditions may be expressed as

$$\frac{[(3/4\pi)V_{F|Q}]^{-2/3}}{2\sqrt{3}} = n \qquad (167)$$

where $V_{F|Q}$ denotes the bubble volume from an isolated nozzle at flow rate $Q$,

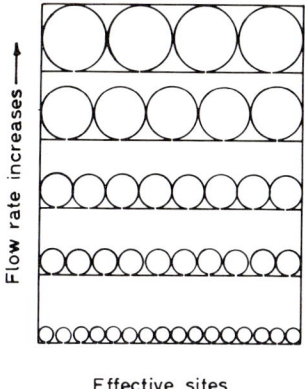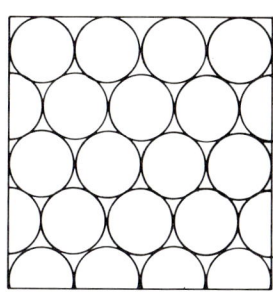

Fig. 29. Effect of flow rate on number of sites and the hexagonal close packing arrangement for bubbles.

and

$$Q = \frac{q}{(\text{Area of the Disk})(n)} \quad (168)$$

where $n$ is the number of operative sites per unit area.

The values of $V_F$ for any $Q$ can be obtained from Eqs. (54) and (61) for constant flow conditions.

As $V_F$ is implicitly expressed in terms of $Q$ in Eqs. (54) and (61), iterative procedure must be used in solving Eq. (167). A value of $n$ is first assumed and $Q$ calculated. Corresponding to this $Q$, $V_F$ is evaluated by Eqs. (54) and (61). This value of $V_F$ is used in Eq. (167) to determine $n$. The calculated $n$ is then compared with the $n$ assumed initially. Depending on the difference between the two, a new $n$ is assumed. This process is continued till the two agree.

A typical set of data for an air-water system on a sintered disk of 3.61 cm$^2$ area is presented in Fig. 30. The increase in average bubble volume and the decrease in effective number of sites with increasing flow rate are evident from Fig. 30. The solid lines correspond to those calculated on the basis of the model. The agreement between the theoretical and the experimental values is very satisfactory.

The model has been tested over a viscosity range of 1 to 604 cp, surface tension from 22 to 72 dyn/cm, pore diameter from 45 to 110 microns, and flow rate per unit area from 1 to 100 cm$^3$/sec.

The success of the model, apart from helping in the evaluation of bubble volume from sintered disks, demonstrates the applicability of the simple "single-site" analysis to an obviously complex situation.

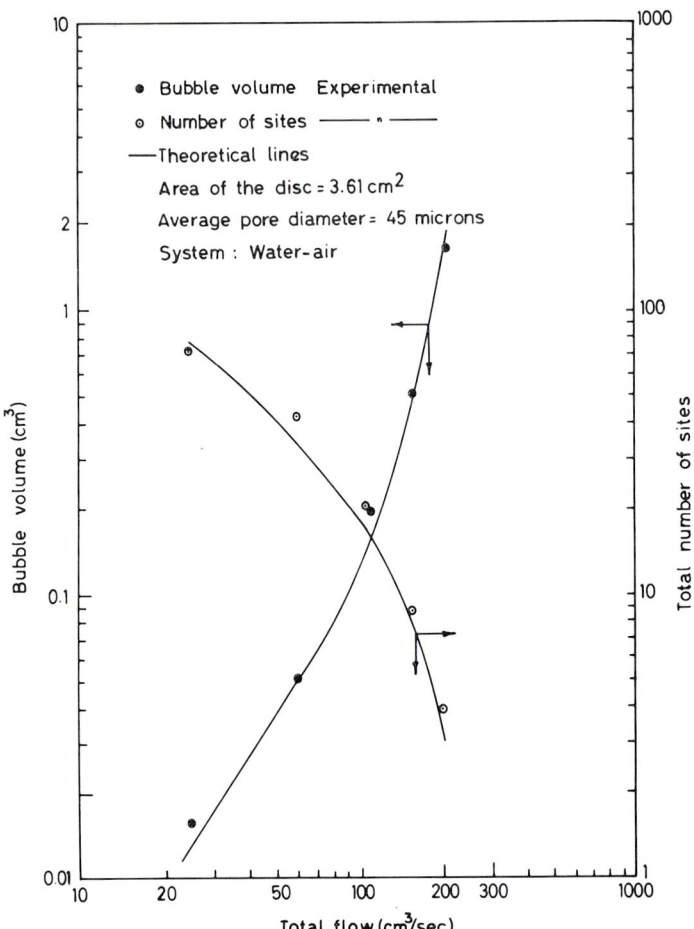

FIG. 30. Variation of bubble volume and the effective sites with flow rate.

## XIX. Concluding Remarks

The single bubble formation at isolated nozzles is now understood both under constant flow and constant pressure conditions, though considerable work still needs to be done in the intermediate region where bubble size is highly influenced by resonance effects. The influence of geometric factors like nozzle geometry and orientation can also be predicted. For multiple orificed systems, a theoretical analysis is still not achieved, though the various influencing factors have been recognized. For a limited region of variables, the equations for single orifices can be applied to multiple nozzled systems also.

The problem of spraying, atomization, and bubble formation in agitated systems still need considerable study, though quite a bit of work has already been reported on them. The formation in the above cases has mostly been studied in the absence of heat and mass transfer and chemical reaction, the presence of which can greatly influence the bubble volume. This has to receive considerable attention if the performance of pertinent industrial equipment is to be adequately explained.

The very important field of bubble and drop formation in non-Newtonian fluids remains virtually untouched, even in the absence of heat and mass transfer and chemical reactions.

## Nomenclature

$a$    Dimensionless index; gas-liquid interfacial area per unit volume of dispersion, $cm^{-1}$ [Eq. (124)]
$A$    Substitution given in the text; area of the noncircular orifice, $cm^2$ [Eq. (107)]
$A_f$    Frontal area of the bubble, $cm^2$
$b$    Dimensionless index
$B$    Substitution given in the text
$c$    Dimensionless index: velocity of sound in the gas, cm/sec
$C$    Substitution given in the text; proportionality constant [Eq. (108); chamber capacitance [Eq. (166)]
$C_1$    Substitution given in the text
$C_2$    Substitution given in the text
$C_D$    Drag coefficient dimensionless
$d$    Diameter of the bubble, drop, droplet, cm
$d_0$    Minimum droplet diameter, cm
$d_m$    Diameter of maximum surviving bubble, cm
$d_u$    Maximum droplet diameter, cm
$d_L$    Linear mean droplet diameter, cm
$d_S$    Diameter of static drop, cm
$\Delta d$    Range of size of droplets, cm
$D$    Substitution given in the text; diameter of the orifice, cm

$D_e$    Effective diameter of a hypothetical equivalent circular orifice, cm [Eq. (107)]
$E$    Substitution given in the text; pressure, $dyn/cm^2$ [Eq. (164)]
$E_0$    Pressure to which chamber charges just before it begins to discharge, $dyn/cm^2$
$E_2$    Pressure that can be supplied by the source when the flow rate is zero, $dyn/cm^2$
$f$    Function [Eq. (155)]; frequency, number/min
$f_\lambda$    Ratio of actual mean drop or bubble diameter to diameter at $\lambda = 0.1$
$F$    Substitution given in the text; drag of the bubble, $gm \cdot cm/sec^2$
$F_B$    Force due to buoyancy in a still liquid, dyn [Eq. (118)]
$F_B'$    Force due to buoyancy in a moving continuous phase, dyn [Eq. (118)]
$F_D$    Drag force due to moving continuous phase; total drag on the sphere, dyn [Eq. (97)]
$g$    Acceleration due to gravity, $cm/sec^2$
$G$    Volume of the liquid, $ft^3$ [Eq. (160)]

| | |
|---|---|
| $h$ | Orifice submergence, cm |
| $H$ | Substitution given in the text |
| $H_g$ | Volume fraction gas hold-up |
| $J$ | Substitution given in the text |
| $k$ | Constant [Eq. (159)]; parameter in Ostwald-de Waele model, dimensions dependent on $n$ [Eq. (96)] |
| $K$ | Orifice constant, $cm^{7/2}/gm^{1/2}$; correction to Stokes Law [Eq. (127)] |
| $L$ | Paddle diameter, cm |
| $m$ | Mass of the bubble, gm |
| $m'$ | Mass of the liquid, gm [Eq. (67)] |
| $M$ | Virtual mass of the bubble, drop, gm |
| $n$ | Parameter in Ostwald-de Waele model, dimensionless [Eq. (96)]; number of effective sites per unit area of the sintered disk [Eq. (167)] |
| $N$ | Substitution given in the text; angular velocity of the stirrer, revolutions sec [Eq. (160)] |
| $N_{Fr}$ | Froude number |
| $N'_{Fr}$ | Substitution given in the text |
| $\bar{N}_{Fr}$ | Modified Froude group |
| $N_{Re}$ | Reynolds number |
| $N'_{Re}$ | Substitution given in the text |
| $\bar{N}_{Re}$ | Modified Reynolds number |
| $N_{We_{Cr}}$ | Critical Weber number of maximum bubble size capable of survival |
| $p$ | Pressure of the liquid [Eq. (155)]; pressure outside a bubble, $gm/cm \cdot sec^2$; chamber pressure at any time $t$ during chamber charging period, $dyn/cm^2$ [Eq. (165)] |
| $p'$ | Chamber pressure at any time $t$ during chamber discharging period, $dyn/cm^2$ |
| $p_h$ | Sum of atmospheric pressure and submergence, $dyn/cm^2$ |
| $p_i$ | Pressure inside a bubble, $gm/sec^2$ |
| $p_H$ | Applied pressure, $dyn/cm^2$ |
| $P$ | Substitution given in the text |
| $\Delta P$ | Pressure difference, $gm/cm \cdot sec^2$ |
| $P'$ | Substitution given in the text |
| $P_1$ | Pressure in the drum, $gm/cm \cdot sec^2$ |
| $q$ | Total flow rate through the sintered disk, $cm^3/sec$ [Eq. (168)] |
| $Q$ | Volumetric flow rate, $cm^3/sec$ |
| $Q_t$ | Flow rate at any time, $cm^3/sec$ |
| $Q_{fb}$ | Volumetric flow rate at the end of the force-balance-bubble stage, $cm^3/sec$ |
| $r$ | Radius of the bubble, drop, cm |
| $r_F$ | Final radius of the bubble, cm |
| $r_{fb}$ | Radius of the force-balance bubble, cm |
| $r_{st}$ | Radius of the maximum stable droplet, cm |
| $R$ | Radius of the orifice, cm; ratio of the area of a noncircular orifice to that of the inscribed circle [Eq. (106)] |
| $R_0$ | Radius of the residual bubble, cm |
| $R_D$ | Discharge resistance, $(dyn/cm^2)/(cm^3/sec)$ |
| $R_T$ | Tip resistance, $(dyn/cm^2)/(cm^3/sec)$ |
| $R_V$ | Valve resistance, $(dyn/cm^2)/(cm^3/sec)$ |
| $Re_o$ | Orifice Reynolds number [Eq. (122)] |
| $Re'$ | Particle Reynolds number; modified Reynolds number [Eq. (98)] |
| $s$ | Distance, cm |
| $t$ | Time, sec |
| $t_c$ | Time of detachment, sec |
| $t_e$ | Time relating to the first stage, sec |
| $T$ | Substitution given in the text |
| $u$ | Velocity of the continuous phase, cm/sec |
| $U_0$ | Fluidizing velocity, cm/sec |
| $v$ | Velocity, cm/sec |
| $v_o$ | Velocity of the flow of gas through the orifice, cm/sec [Eq. (62)] |
| $v'$ | Velocity of the center of the bubble, cm/sec |
| $v'_c$ | Velocity of the dispersed phase through the nozzle, cm/sec [Eq. (127)] |

$v_e$    Velocity relating to the first stage, cm/sec
$v_j$    Velocity of the jet, cm/sec
$v_s$    Superficial gas velocity, ft/sec
$V$    Volume, cm$^3$
$V_0$    Volume of the residual bubble, cm$^3$
$V_t$    Bubble volume at any time, cm$^3$
$V_F$    Final volume of the bubble, drop, cm$^3$
$V'_F$    Bubble volume in a liquid moving with a velocity $u$, cm$^3$
$V_{fb}$    Volume of the force-balance bubble, cm$^3$
$(V_F)_i$    Bubble volume from the inscribed circle, cm$^3$
$(V_F)_{n.c.}$    Bubble volume from the noncircular orifice, cm$^3$
$V_K$    Negative volume resulting from buoyancy, cm$^3$
$V_m$    Excess volume over and above static bubble volume, cm$^3$ [Eq. (109)]
$V_S$    Static bubble volume, cm$^3$ [Eq. (110); static drop volume applying Harkins' correction, cm$^3$ [Eq. (125)]
$V_v$    Volume representing the force experienced by the moving drop, cm$^3$
$W$    Power dissipated by impeller, horsepower [Eq. (159)]
$x$    Distance, cm
$X_n$    Function of the parameter, $n$
$Z$    Substitution given in the text

**GREEK LETTERS**

$\gamma$    Surface tension; interfacial tension, dyn/cm
$\Delta$    Symmetrical "rate of deformation" tensor [Eq. (96)]
$\theta$    Contact angle, degrees; substitution (D8)
$\lambda$    Volume fraction of the dispersed phase
$\mu$    Micron ($10^{-4}$ cm); viscosity of the liquid, poise [Eq. (155)]
$\mu_c$    Viscosity of the continuous phase, poise; centipoise [Eq. (127)]
$\mu_d$    Dispersed-phase viscosity, poise
$\mu_g$    Viscosity of the gas, poise
$\mu_l$    Viscosity of the liquid, poise
$\mu_w$    Viscosity of water, poise
$\mu_D$    Viscosity of the dispersed phase, poise [Eq. (161)]
$\rho$    Density, gm/cm$^3$; density of the continuous phase, gm/cm$^3$ [Eq. (159)]
$\rho'$    Effective mean density, gm/cm$^3$ [Eq. (160)]
$\Delta\rho$    Density difference between the liquid and the gas, gm/cm$^3$
$\rho_c$    Density of the continuous phase, gm/cm$^3$
$\rho_d$    Density of the dispersed phase, gm/cm$^3$
$\rho_g$    Density of the gas, gm/cm$^3$; lb/ft$^3$ [Eq. (123)]
$\rho_l$    Density of the liquid, gm/cm$^3$
$\tau$    Proportionality factor
$\tau$    Shear stress, dyn/cm$^2$ [Eq. (96)]
$\phi$    Substitution given in the text; angle between the orifice plate and the horizontal, degrees; correction factor of Harkins and Brown [Eq. (132)]
$\psi$    Substitution given in the text; Function of diameters, [Eq. (125)

## References

A1. Andrew, S. P. S., *in* "Symposium on Distillation" (P. A. Rottenburg, ed.), p. 73. Inst. Chem. Engrs., London, 1960.
A2. Appel, F. J., and Elgin, J. C., *Ind. Eng. Chem.* **29**, 451 (1937).
A3. Astarita, G., and Apuzzo, G., *A.I.Ch.E.J.* **11**, 815 (1965).
B1. Batchelor, G. K., and Davis, R. M., "Surveys in Mechanics," p. 160. Cambridge University Press, Cambridge, 1956.

B2. Baumgarten, P. K., and Pigford, R. L., *A.I.Ch.E.J.* **6**, 115 (1960).
B3. Benzing, R. L., and Myers, J. E., *Ind. Eng. Chem.* **47**, 2087 (1955).
B4. Blanding, F. H., and Elgin, J. C., *Trans. Am. Inst. Chem. Engrs.* **38**, 305 (1942).
B5. Bloore, P. D., and Botterill, J. S. M., *Nature,* **190** 250 (1961).
B6. Botterill, J. S. M., George, J. S., and Besford, H., *Chem. Eng. Progr. Symp. Ser.* **62**, No. 62, 7 (1966).
B7. Bowonder, B., and Kumar, R., *Chem. Eng. Sci.* In press.
C1. Calderbank, P. H., *Trans. Inst. Chem. Engrs.* (*London*) **34**, 79 (1956).
C2. Calderbank, P. H., *Brit. Chem. Eng.* **1**, 206 (1956).
C3. Calderbank, P. H., *Trans. Inst. Chem. Engrs.* (*London*) **36**, 443 (1958).
C4. Calderbank, P. H., and Rennie, J., *Trans. Inst. Chem. Engrs.* (*London*) **40**, 3 (1962).
C5. Cantor, M., *Ann. Physik* **47**, 399 (1892).
C6. Caswell, B., and Schwarz, W. H., *J. Fluid Mech.* **13**, 417 (1961).
C7. Clay, P. C., *Proc. Acad. Sci. Amsterdam* **43**, 852, 979 (1940).
C8. Coppock, P. D., and Meiklejohn, G. T., *Trans. Inst. Chem. Engrs.* (*London*) **29**, 75 (1951).
C9. Cross, C. A., and Rider, H., *J. Appl. Chem.* (*London*) **2**, 51 (1952).
C10. Crozier, R. D., Ph.D. thesis, Univ. of Michigan, 1956.
D1. Dana, W. L., and Spencer, R. C., *Natl. Advisory Comm. Aeron.* Tech. Notes **424** (1932).
D2. Dana, W. L., and Spencer, R. C., *Natl. Advisory Comm. Aeron.* Tech. Notes **454** (1933).
D3. Danckwerts, P. V., Sawistowski, H., and Smith, W., *in* "Symposium on Distillation" (P. A. Rottenburg, ed.), p. 7. Inst. Chem. Engrs., London, 1960.
D4. Datta, R. L., Napier, D. H., and Newitt, D. M., *Trans. Inst. Chem. Engrs.* (*London*) **28**, 14 (1950).
D5. Davidson, L., and Amick, E. H., *A.I.Ch.E.J.* **2**, 337 (1956).
D6. Davidson, J. F., and Harrison, H., "Fluidised Particles," Cambridge University Press Cambridge, 1963.
D7. Davdison, J. F., Paul, R. C., Smith, M. J. S., and Duxbury, H. A., *Trans. Inst. Chem. Engrs.* (*London*) **37**, 323 (1959).
D8. Davidson, J. F., and Schuler, B. O. G., *Trans. Inst. Chem. Engrs.* (*London*) **38**, 144 (1960).
D9. Davidson, J. F., and Schuler, B. O. G., *Trans. Inst. Chem. Engrs.* (*London*) **38**, 335 (1960).
F1. Fararoui, A., and Kintner, R. C., *Trans. Soc. Rheol.* **5**, 369 (1961).
F2. Fleming, J. F., and Johnson, H. F., *Chem. Eng. Progr.* **49**, 497 (1953).
G1. Goldstein, S., "Modern Developments in Fluid Dynamics." The University Press, Oxford, 1938.
G2. Guyer, A., and Peterhans, E., *Helv. Chim. Acta* **26**, 1099 (1943).
H1. Halberstadt, S., and Prausnitz, P. H., *Z. Angew. Chem.* **43**, 970 (1930).
H2. Harkins, W. D., and Brown, F. E., *J. Am. Chem. Soc.* **41**, 499 (1919).
H3. Harrison, D., and Leung, L. S., *Trans. Inst. Chem. Engrs.* (*London*) **39**, 409 (1961).
H4. Hayes, W. B., Hardy, B. W., and Holland, C. D., *A.I.Ch.E.J.* **5**, 319 (1959).
H5. Hayworth, C. B., and Treybal, R. E., *Ind. Eng. Chem.* **42**, 1174 (1950).
H6. Hinze, J. O., *A.I.Ch.E.J.* **1**, 289 (1955).
H7. Hughes, R. R., Handlos, A. E., Evans, H. D., and Maycock, R. L., *Chem. Eng. Progr.* **51**, 557 (1955).
J1. Jackson, R., *Chem. Engr.* No. 178, CE107 (1964).
J2. Jackson, R. W., *Ind. Chemist* **28**, 350 (1952).
J3. Johnson, A. I., Robinson, D. G., and Michellepis, C. P., *in* "Proceedings of the Symposium on Mechanics of Bubbles and Drops." Am. Inst. Chem. Engrs., New York, 1955.

J4. Johnson, H. F., Jr., and Bliss, H., *Trans. Am. Inst. Chem. Engrs.* **42**, 331 (1946).
K1. Kalyanasundaram, C. V., Ph.D. thesis. Indian Inst. of Science, Bangalore, India, 1967.
K2. Kalyanasundaram, C. V., Kumar, R., and Kuloor, N. R., *Tech. Rept.* No. CE/67-3, Indian Inst. of Science, Bangalore, India, 1967.
K3. Keith, F. W., and Hixson, N., *Ind. Eng. Chem.* **47**, 258 (1955).
K4. Knoll, A., *Kolloid-Z.* **86**, 1 (1939).
K5. Krishnamurthi, S., Kumar, R., and Datta, R. L., *Indian J. Tech.* **2**, 3 (1964).
K6. Krishnamurthi, S., Ph.D. thesis. Indian Inst. of Science, Bangalore, India, 1965.
K7. Krishnamurthi, S., Kumar, R., and Datta, R. L., *Trans. Indian Inst. Chem. Engrs.* **14**, 78 (1961-62).
K8. Krishnamurthi, S., Kumar, R., and Kuloor, N. R., *Ind. Eng. Chem. Fundamentals* **7**, 549 (1968).
K9. Krishnamurthi, S., Kumar, R., and Kuloor, N. R., *Chem. Proc. Eng.* **49**, 91 (1968).
K10. Krishnamurthi, S., Kumar, R., Datta, R. L., and Kuloor, N. R., *J. Sci. Ind. Res. (India)* **21B**, 554 (1962).
K11. Krishnamurthy, P. G., Kumar, R., and Kuloor, N. R., *Brit. Chem. Eng.* **10**, 76 (1965).
K12. Krishnamurthy, P. G., Kumar, R., Datta, R. L., and Kuloor, N. R., *Indian J. Tech.* **2**, 68 (1964).
K13. Krishnamurthy, P. G., Kumar, R., Datta, R. L., and Kuloor, N. R., *Brit. Chem. Eng.* In press.
K14. Kumar, R., *Tech. Rept.* No. CE/67-4, Indian Inst. of Science, Bangalore, India, 1967.
K15. Kumar, R., and Kuloor, N. R., *Brit. Chem .Eng.* **9**, 400 (1964).
K16. Kumar, R., and Kuloor, N. R., *Chem. Tech.* **19**, 78 (1967).
K17. Kumar, R., and Kuloor, N. R., *Chem. Tech.* **19**, 657 (1967).
K18. Kumar, R., and Kuloor, N. R., *Chem. Tech.* **19**, 733 (1967).
K19. Kumar, R., and Kuloor, N. R. In press.
K20. Kurana, A. K., and Kumar, R., *Chem. Eng. Sci.* **24**, 1711 (1969).
L1. Lamont, J. C., and Scott, D. S., *Can. J. Chem. Eng.* **44**, 201 (1966).
L2. Leibson, I., Holcomb, E. G., Cacoso, A. G., and Jacmic, J. J., *A.I.Ch.E.J.* **2**, 296 (1956).
L3. Licht, W., Jr., and Conway, J. B., *Ind. Eng. Chem.* **42**, 1151 (1950).
L4. Littaye, G., *Compt. Rend.* **217**, 340 (1943).
M1. Macmillan, W. P., *J. Imp. Coll. Chem. Eng. Soc.* **13**, 64 (1960-61).
M2. Maier, C. G., *U.S. Bur. Mines Bull.* 260 (1927).
M3. Merrington, A. C., and Richardson, E. G., *Proc. Phys. Soc.* **59**, 1 (1947).
M4. Milne Thomson, L. M., "Theoretical Hydrodynamics," 3rd ed. Macmillan & Co. Ltd., London, 1955.
M5. Minard, G. W., and Johnson, A. I., *Chem. Engr. Progr.* **48**, 62 (1952).
M6. Morse, R. D., and Ballou, C. O., *Chem. Eng. Progr.* **47**, 199 (1951).
M7. Mugele, R. A., and Evans, H. D., *Ind. Eng. Chem.* **43**, 1317 (1951).
N1. Neumann, H., and Seeliger, R., *Z. Physik* **114**, 571 (1939).
N2. Newman, A. O., and Lerner, B. J., *Anal. Chem.* **26**, 417 (1954).
N3. Nukiyama, S., and Tanasawa, Y., *Trans. Soc. Mech. Engrs. Japan* **5**, 1 (1939).
N4. Null, H. R., and Johnson, H. F., *A.I.Ch.E.J.* **4**, 273 (1958).
O1. Orr, C., Jr., "Particulate Technology." Macmillan, New York, 1966.
P1. Padmavathy, P., Kumar, R., and Kuloor, N. R., *Indian J. Tech.* **3**, 133 (1965).
P2. Pars, L. A., "Introduction to Dynamics," pp. 162–165. Cambridge University Press, Cambridge, 1953.
P3. Pilcher, J. M., and Miesse, C. C., *Wright Air Development Center Tech. Rept.* 56 (1957).
P4. Prandtl, L., and Tietjens, O. G., "Applied Hydro- and Aero-mechanics." McGraw-Hill, New York, 1934.
P5. Prausnitz, P.H., *Kolloid-Z.* **76**, 227 (1936).

Q1. Quigley, C. J., Johnson, A. I., and Harris, B. L., *Chem. Eng. Progr. Symp. Ser.* **S1**, 31 (1955).
R1. Ramakrishnan, S., Kumar, R., and Kuloor, N. R., *Chem. Eng. Sci.* **24**, 731 (1969).
R2. Rao, E. V. L. N., M.Sc. thesis. Indian Inst. of Science, Bangalore, India, 1965.
R3. Rao, E. V. L. N., Kumar, R., and Kuloor, N. R., *Chem. Eng. Sci.* **21**, 867 (1966).
R4. Rayleigh, Lord, *Proc. London Math. Soc.* **10**, 4 (1878).
R5. Rennie, J., and Evans, F., *Brit. Chem. Eng.* **7**, 498 (1962).
R6. Rodger, W. A., Trice, V. G., and Rushton, J. H., *Chem. Eng. Progr.* **52**, 515 (1956).
R7. Rodriguez, F., Grotz, L. C., and Engle, D. L., *A.I.Ch.E.J.* **7**, 663 (1961).
R8. Romero, J. B., and Johanson, L. N., *Chem. Eng. Progr. Symp. Ser.* **58**, No. 38, 28 (1962).
R9. Rosin, P., and Rammler, E., *J. Inst. Fuel*, **7**, 29 (1933).
R10. Rowe, P. N., and Partridge, B. A., *Trans. Inst. Chem. Engrs. (London)* **43**, 157 (1965).
R11. Rowe, P. N., Partridge, B. A., Lyall, E., and Ardran, G. M., *Nature* **195**, 278 (1962),
S1. Saradhy, Y. P., and Kumar, R., *Tech. Rept.* No. CE/67-6, Indian Inst. of Science, Bangalore, India, 1967.
S2. Sargent, R. W. H., and Macmillan, W. P., *Trans. Inst. Chem. Engrs.* **40**, 191 (1962).
S3. Satyanarayan, A., Kumar, R., and Kuloor, N. R., *Chem. Eng. Sci.* **24**, 749 (1969).
S4. Schiller, L., and Naumann, A., *Z. Ver. Deut. Ing.* **77**, 318 (1933).
S5. Schurmann, R., *Kolloid-Z.* **80**, 148 (1937).
S6. Schurmann, R., *Z. Phys. Chim.* **143**, 456–74 (1929).
S7. Shankar Srinivas, N., and Kumar, R., *Tech. Rept.* No. CE/67-7, Indian Inst. of Science, Bangalore, India.
S8. Sherwood, T. K., Evans, J. E., and Longcor, J. V., *Ind. Eng. Chem.* **31**, 1144 (1939).
S9. Shinnar, R., and Church, J. M., *Ind. Eng. Chem.* **52**, 253 (1960).
S10. Shulman, H. L., and Molstad, M. C., *Ind. Eng. Chem.* **42**, 1058 (1950).
S11. Siemes, W., and Kaufmann, J. F., *Chem. Eng. Sci.* **5**, 127 (1956).
S12. Silberman, E., *Proc. Fifth Midwestern Conference on Fluid Mech.* (Ann Arbor, Michigan), April 1-2, 263 (1957).
S13. Slattery, J. C., *A.I.Ch.E.J.* **8**, 663 (1962).
S14. Slattery, J. C., and Bird, R. B., *Chem. Eng. Sci.* **16**, 231 (1961).
S15. Spells, K. E., and Bakowski, S., *Trans. Inst. Chem. Engrs. (London)* **28**, 38 (1950).
S16. Subramaniyan, V., and Kumar, R., *Tech. Rept.* No. CE/67-5, Indian Inst. of Science, Bangalore, India, 1967.
S17. Sullivan, S. L., Hardy, B. W., and Holland, C. D., *A.I.Ch.E.J.* **10**, 848 (1964).
T1. Thorogood, R. M., *Brit. Chem. Eng.* **8**, 329 (1963).
T2. Tomita, Y., *Bull. Soc. Mech. Engrs.* **2**, 469 (1959).
T3. Toomey, R. D., and Johnstone, H. F., *Chem. Eng. Progr.* **48**, 220 (1952).
V1. Van Krevelen, D. W., and Hoftijzer, P. J., *Chem. Eng. Progr.* **46**, 29 (1950).
V2. Venugopal, B., Kumar, R., and Kuloor, N. R., *Ind. Eng. Chem. Process Design Develop.* **6**, 139 (1967).
V3. Vermeulen, T., Williams, G. M., and Langlois, G. E., *Chem. Eng. Progr.* **51**, 85 (1955).
V4. Verschoor, H., *Trans. Inst. Chem. Engrs. (London)* **28**, 52 (1950).
W1. Wallick, G. C., Sovins, J. C., and Atterburn, D. R., *Phys. Fluids* **5**, 367 (1962).
W2. Wasserman, M. L., and Slattery, J. C., *A.I.Ch.E.J.* **10**, 383 (1964).
W3. West, F. B., Robinson, A. P., Morgenthaler, A. C., Beck, T. R., and McGregor, D. K., *Ind. Eng. Chem.* **43**, 234 (1951).
W4. Wilcox, J. D., and June, R. K., *J. Franklin Inst.* **271**, 169 (1961).
Y1. Yasui, G., and Johanson, L. N., *A.I.Ch.E.J.* **4**, 445 (1958).
Z1. Zuiderweg, F. J., and Harmens, A., *Chem. Eng. Sci.* **9**, 89 (1958).

# AUTHOR INDEX

Numbers in parentheses are reference numbers and indicate that an author's work is referred to although his name is not cited in the text. Numbers in italics show the page on which the complete references is listed.

### A

Ackoff, R. L., 196(C1), *253*
Ailam, G., 7, 8, *92*
Amick, E. H., 266(D5), 268, 269, 270, 277, 356, *366*
Andrew, S. P. S., 334(A1), *365*
Anscombe, F. J., 137(A1, A2), *181*
Appel, F. J., 347(A2), *365*
Apuzzo, G., 346(A3), *365*
Arabadzhi, V. I., 47, *91*
Ardran, G. M., 259, *368*
Arendt, P., 51, *91*
Arnoff, E. L., 196(C1), *253*
Astarita, G., 346(A3), *365*
Atkinson, A. C., 121(H9), *182*
Atterburn, D. R., 316(W1), 344(W1), *368*
Ayen, R. J., 110, 176(A3), *181*

### B

Bakowski, S., 268, 270 (S15), 321, *368*
Ball, W. E., 130(B1), *181*
Ballou, C. O., 318(M6), *367*
Ballou, J. W., *91*
Bartlett, M. S., 114(B2), *181*
Batchelor, G. K., 334, *365*
Baumgarten, P. K., 260, 318(B2), *366*
Beach, R., *91*
Beale, E. M. L., 104(B3), 118(B3), 121(B3), 128(B3), *181*
Beck, T. R., 347(W3), *368*
Bekryaev, V. I., 73, *93*
Benson, S. W., *181*

Benzing, R. L., 266(B3), 267(B3), 272(B3), 273, 278, 299, *366*
Berg, T. G. O., 25, 26, 68, 70, 80, 82, *91*
Besford, H., 259, *366*
Billingsley, D. S., 206, *253*
Bird, R. B., 346(S14), *368*
Bishoff, K. B., 101(H6), *182*, 188(H2), 193(H2), 210, *253*
Blake, D. E., 2, *91*
Blanding, F. H., 348(B4), *366*
Bliss, H., 347(J4), *367*
Bloore, P. D., 259, 319, 320, *366*
Bôcher, M., 189(B2), *253*
Booth, G. W., 104(B5), *181*
Botterill, J. S. M., 259, 319, 320, *366*
Boudart, M., *181*
Bowonder, B., 359, *366*
Box, G. E. P., 116(B10, B18), 118(B8), 121 (B8, B12), 125(B16), 128(B8), 130 (B12), 137(B8), 147, 152(B14), 154(B14), 155(B7), 156(B7), 157(B19), 159(B7, B11), 160(B11), 161, 164(B11, B17), 168(B9), 169(B16), 171(B13), 173(B15), 175(B16), 178(B12), *181, 182*
Brackett, F. S., 80, 82, *92*
Bradley, R. S., 36, 36, *91*
Branson, L., 25, 38, 39, 79, *95*
Brasefield, C. J., 71, *91*
Bredov, M. M., 67, *91*
Brown, F. E., 335, 337, 342, *366*
Brown, G. G., 82, *92*
Burk, R. E., 104(H7), *182*
Butt, J. B., 130(M4), *183*

## C

Cacoso, A. G., 258(L2), 268, 333, *367*
Calderbank, P. H., 264, 319, 333, 334, 354(C3), 355, 356, *366*
Cantor, M., 321(C5), *366*
Carr, N. L., 113(C1), 133(C1), 156(C1), *182*
Caswell, B., 346(C6), *366*
Chalmers, J. A., *43*, *91*
Chilton, C. H., 189(P1), *253*
Cho, A. Y. H., 47, 48, 75, 79, 81, *91*
Chou, Chan-Hui, 125(C2), *182*
Church, J. M., 354, *368*
Churchill, S. W., 125(W3), *183*
Churchman, C. W., 196(C1), *253*
Clay, P. C., 354, *366*
Cochet, R., 48, 50, *91*
Coehn, A., *66*, *91*
Cohen, E., 38, 39, 79, *91*
Conghran, K. D., 226(S5), *254*
Conway, J. B., 347(L3), *367*
Cooper, W. F., 58, *92*
Coppock, P. D., 272(C8), 273, 278, 299, 300, 302, 342, *366*
Corn, M., 30, 36, *92*
Coutie, G. A., 116(B10), *181*
Cox, D. R., 142(C3, C4), 159(B11), 160(B11), 164(B11), *181*, *182*
Cross, C. A., 321, 356(C9), *366*
Crowe, C. M., 187(S2), *253*
Crozier, R. D., 334, *366*

## D

Dalla Valle, J. M., *81*, *93*
Dana, W. L., 349(D1, D2), *366*
Danckwerts, P. V., 334(D3), *366*
Daniel, J. H., 80, 82, *92*
Datta, R. L., 262, 266(D4), 267(D4), 268, 270, 272, 273, 276, 277(K13), 278, 299, 300, 301, 302, 303, 321, 324(K12), 332, 347(K12), *366*, *367*
Davidson, J. F., 258(D8), 269, 270, 272, 273, 278, 279, 280, 282(D8), 283, 285, 286, 289, 290, 293, 295, 300, 301, 302, 304, 307(D8, D9), 308(D8), 313, 318, 319, 338, *366*

Davidson, L., 266(D5), 268, 269, 270, 277, 356, *366*
Davies, C. N., 6, *92*
Davies, O. L., 126(D1), 133(D1), 155(D2), 156(D2), 159(D2), *182*
Davignon, L., 160, *183*
Davis, R. M., 334, *365*
Dawkins, G. S., 30, *92*
De Maeyer, L., 73, *92*
Deutsch, W., 48, *92*
Dimick, R. C., 79, *96*
Dionne, P. J., 226(S5), *254*
Dodd, E. E., 45, 57, 59, 76, *92*, *94*
Doyle, A. W., 7, 25, 28, 41, 42, 59, 77, *92*, *96*
Draper, N. R., 111(D4), 114(D4), 121(B12), 130(B12), 133(D4), 137(D4), 176(D3), 178(B12), *181*, *182*
Drozin, V. G., 39, 40, *92*
Dubois, J., 74, *92*
Dunskii, V. F., 42, 43, *92*
Duxbury, H. A., 318, *366*

## E

Earhart, R. F., 31, 45, *92*
Eigen, M., 73, *92*
Einbinder, H., 53, *92*
Elgin, J. C., 347(A2), 348(B4), *365*, *366*
Engle, D. L., 355, 356, *368*
English, W. N., 45, *94*
Erjavec, J., 125(K3), 134(K3), 155(K3), 157(K3), *183*
Ermer, D. S., *183*
Evans, D. G., 73, *92*
Evans, F., 333, *368*
Evans, H. D., 258(H7), 266(H7), 268, 269, 270(H7), 272(H7), 349, 356, *366*
Evans, J. E., 347(58), *368*

## F

Fararoui, A., 346(F1), *366*
Fernish, G. C., 25, 26, 68, 80, 82, *91*
Fleming, J. F., 348(F2), *366*
Flood, W. J., 68, 80, 82, *91*
Foster, W. W., 11, 51, *92*
Fowler, J. F., 71, *92*
Franklin, N. L., 159(F2, F3), *182*

# AUTHOR INDEX

Franckaerts, J. F., 106, 140(F1), *182*
Franks, R. G. E., 99(F4), *182*
Froment, G. F., 106, 140(F1), *182*
Frickel, R., 77, *92*
Frost, A. A., 104(F5), *182*
Fuchs, N. A., 13, 15, 36, *92*
Fukada, E., 71, *92*

## G

Gallily, I., 7, 8, *92*
Gaukler, T. A., 25, 26, 70, *91*
Geist, J. M., 82, *92*
George, J. S., 259, *366*
Gerke, R. H., 77, *96*
Germer, L. H., 45, *92*
Gillespie, T., 80, *92*
Göhlich, H., 28, 43, *93*
Goldstein, S., 283(G1), *366*
Gordieyeff, V. A., 38, *93*
Goyer, G., 28, 77, *93, 94*
Graf, P. E., 7, 38, 40, 79, *93*
Groenweghe, L. C. D., 130(B1), *181*
Grotz, L. C., 355, 356, *368*
Gruen, R., 28, 77, *93, 94*
Guest, P. G., *93*
Guttman, I., 121(G1, G2), 128(G1, G2), *182*
Guyer, A., 321(G2), 334(G2), *366*
Guyton, A. C., 82, *93*

## H

Halberstadt, S., 334, *366*
Hamielec, A. E., 187(S2), *253*
Handlos, A. E., 258(H7), 266(H7), 268, 269, 270(H7), 272(H7), 356, *366*
Hansen, J. W., 76, *94*
Hardy, B. W., 263, 268(H4), 269, 270(H4), 282(H4), 304, *368*
Harkins, W. D., 335, 337, 342, *366*
Harmens, A., 334(Z1), *368*
Harper, W. R., 55, 57, 62, 63, 67, *93*
Harris, B. L., 260, 268, 272, 273, 274, 278, 299, 304, *368*
Harrison, D., 265, 319, 320, 344(H3), *366*
Harrison, H., 280, 286, *366*

Hartley, H. O., 116(H1), *182*
Hayes, W. B., 268(H4), 269, 270(H4), 282(H4), 304, *366*
Hayworth, C. B., 334(H5), 340, 343, 347, 352, *366*
Hendricks, C. D., Jr., 7(H13), 8(H13), 25, 38(H13), 39, 75, 79(H13), *93*
Hersh, S. P., *93*
Hertz, H., 68, *93*
Hewitt, G. W., 79, 81, 82, *93*
Hill, W. J., 148(H5), 155(H3), 159(M8), 160(M8), 164(H2), 171(B13), 173(H10), 178(H4), *181, 182, 183*
Himmelblau, D. M., 101(H6), *182*, 188(H1, H2), 192(H1), 193(H1, H2), 210, *253*
Hinkle, B. L., 81, *93*
Hinshelwood, C. N., 104(H7), *182*
Hinze, J. O., 354, *366*
Hixon, N., 347(K3), 348, *367*
Hoffman, T. U., 187(S2), *253*
Hoftijzer, P. J., 267(V1), 279, 286, *368*
Hogan, J. J., 7, 8, 24, 25, 38, 40, 42, 79, *93*
Hohnstreiter, G. F., 79, *96*
Holcomb, E. G., 258(L2), 268, 333, *367*
Holland, C. D., 263, 268(H4), 269, 270(H4), 282(H4), 304, *366, 368*
Hooke, R., 186(H3), *253*
Hopper, V. D., 76, 77, *93*
Hougen, O. A., 105, *183*
Hughes, R. R., 258(H7), 266(H7), 268, 269, 270(H7), 272(H7), 356, *366*
Hunter, W. G., 109(K7), 114(H8), 117(K7), 121(H9), 125(K11), 127(K7), 130(H8), 147, 148(K12), 149(K12), 151(H11, K12), 152(B14), 154(B14), 155(H3, K7), 168(K11), 171(H13), 173(B15, H10, H12, H14), 176(D3, K11), 178(H4), *181, 182, 183*
Hurd, R. M., 8, *93*
Hutchinson, W. C. A., 73, *92*

## J

Jackson, R. W., 321(J2), 356, *366*
Jacmic, J. J., 258(L2), 268, 333, *367*
Jeeves, T. A., 186(H3) *253*

Jex, C. S., 63, *96*
Johanson, L. N., 107, 109, 136, *182*, 318(R8,Y1), *368*
Johnson, A. I., 187(S2), *253*, 260, 268, 272, 273, 274, 278, 299, 304, 321, 348(M5), *366*, *367*, *368*
Johnson, H. F., 334, 335, 340, 347(J4), 348(F2), 352, *366*, *367*
Johnson, R. A., 113(J1), 114(J1), *182*
Johnstone, H. F., 30, *94*, 318(T3), *368*
Jones, C. R., 101(H6), *182*
Jordan, D. W., 35, *93*
June, R. K., 349(W4), *368*

## K

Kabel, R. L., 107, 109, 136, *182*
Kachurin, L. G., 73, *93*
Kallman, H., 51, *91*
Kaufmann, J. F., 272, 278, 300, *368*
Kalyanasundaram, C. V., 335, 339, 340 (K2), 342(K2), 350(K2), 351(K1), 354 (K2), *367*
Keily, D. P. 83, *93*
Keith, F. W., 347(K3), 348, *367*
Kintner, R. C., 346(F1), *366*
Kirkpatrick, S. D., 189(P1), *253*
Kisliuk, P., 45, *94*
Kitaev, A. V., 42, 43, *92*, *94*
Kittrell, J. R., 106(M6), 109(K7), 110(K6), 114(K9), 115(K8), 116(K8), 117(K7, K8, M7), 118(K2, K8, M7), 121(M7), 125 (K3, K11), 127(K7), 128(M7), 134(K3), 142(K8), 143(M5), 144(M5), 146(K5), 147(K5), 148(K12), 149(K12), 151(K12), 155(K3, K7), 157(K3), 159(K4, K10, M8), 161(K10), 167(K4), 168(K11), 169 (K6), 176(K11), *182*, *183*
Kloot, N. H., 142(W4), *183*
Knoblauch, O., 63, *94*
Knoll, A., 321(K4), *367*
Komatsu, S., 186, *253*
Kraemer, H. F., 30, 81, *94*
Krishnamurthi, S., 262, 268, 269, 276, 290, 301, 321, *367*
Krishnamurthy, P. G., 256(K11), 277(K13), 324(K12), 332, 347(K12), *367*
Kruger, J., 28, *94*

Kshemyanskaya, I. Z., 67, *91*
Kuloor, N. R., 256(K11), 258(K18), 261, 262(V2), 264, 266(P1), 268, 269, 270, 272, 276, 277(K13), 280, 282, 285, 286, 288(K16), 292, 295, 296(K19), 301, 302(R1), 310, 313(S3), 314, 315(S3), 316, 319, 320(K17), 321, 324, 325, 332, 334(R3), 335, 337, 339, 340(K2), 342(K2), 344(R3), 345(R3) 347, 350, 353, 354(K2, R3), 356(P1), *367*, *368*
Kumar, R., 256(K11), 258(K18), 261, 262(V2), 264, 266(P1), 268, 269, 270, 272, 276, 277(K13), 280, 282, 285, 286, 288 (K16), 292, 295, 296(K19), 301, 302, 310, 313(S3), 314, 315(S3), 316, 318(S16), 319, 320(K17), 321, 324(K12), 325, 332, 330(S3), 334(R3), 335, 337, 339, 340(K2), 342(K2), 343, 344(K14, R3), 345(R3), 347(K12), 350, 353, 354(K2, R3), 356(P1), 357, 359, *366*, *367*, *368*
Kunkel, W. B., 15, 23, 68, 76, *94*
Kunz, K. S., 186(K2), *253*
Kurana, A. K., 357, *367*

## L

Laby, T. H., 76, 77, *93*
Ladenburg, R., 48, *94*
Laible, J. R., 166(L1), *183*
Laidler, K. J., 100(L2), 104(L2), *183*
LaMer, V. K., 28, 77, *93*, *94*
Lamont, J. C., 263, *367*
Langer, G., 80, *94*
Langlois, G., E., 354, 355, 356, *368*
Langstroth, J. O., 80, *92*
Lapidus, L., 117(L3, L4), *183*
Lapple, C. E., 2, 28, *91*, *96*
Latham, J., 73, *94*
Ledet, W. P., 216, *253*
Lee, W., 221, *253*
Leibson, I., 258(L2), 268, 333, *367*
Lerner, B. J., 264, *367*
Leung, L. S., 265, 319, 320, 354(H3), *366*
Levenburg, K., 116(L5), *183*
Licht, W., Jr., 347(L3), *367*
Lih, M. M., 105, *183*
Lindblad, N. R., 25, *94*
Littaye, G., 349 (L4), *367*

Loeb, L. B., 44, 45, 50, 55, 57, 58, *94*
Longcor, J. V., 347(S8), *368*
Lucas, H. L., 125(B16), 169(B16), 175(B16), *181*
Lyall, E., 259, *368*
Lynch, R. D., 79, 80, 82, *95*

## M

McGregor, D. K., 347(W3), *368*
Macmillan, W. P., 334, *367*
Maier, C. G., 276, 277, 332, *367*
Marquardt, D. L., 116(M1, M2), 118(M2), *183*
Marshall, W. R., 38, 39, 41, *95*
Mason, B. J., 38, 40, 58, 73, *94, 95*
Masters, J. I., 79, *95*
Mathur, G. P., 168(M3), *183*
Matthews, J. B., 38, 40, 58, *95*
Maybank, J., 73, *94*
Maycock, R. L., 258(H7), 266(H7), 268, 269, 270(H7), 272(H7), 356, *366*
Medley, J. A., 59, *95*
Meeter, D. A., 121(G1, G2), 128(G1, G2), *182*
Meiklejohn, G. T., 272(C8), 273, 278, 299, 300, 302, 342, *366*
Merrington, A. C., 334(M3), *367*
Mezaki, R., 106(M6), 113(J1), 114(J1, K9), 115(K8), 116(K8), 117(K8, M7), 118(K8, M7), 121(M7), 128(M7), 130(M4), 142(K8), 143(M5), 144(M5), 146(K5), 147(K5), 148(H5, K12), 149(K12), 151(H11, K12), 159(K4, K10, M8), 160(K4, K10, M8), 161(K10), 167(K4), 173(H12), *182, 183*
Michellepis, C. P., 321, *366*
Mickley, H. S., 165, *183*
Mierdel, G., 48, *95*
Miesse, C. C., 349(P3), *367*
Milne Thomson, L. M., 279(M4), *367*
Minard, G. W., 348(M5), *367*
Mitchell, R. I., 26, *95*
Moffett, D. R., 7, 25, 28, 41, 42, 59, 77, *92, 96*
Molstad, M. C., 321(S10), *368*
Montgomery, D. J., 55, 63, *95, 96*
Moreau-Hanot, M., 48, *95*
Morgenthaler, A. C., 347(W3), 352, *368*

Morse, R. D., 318(M6), *367*
Mugele, R. A., 349, *367*
Myers, J. E., 266(B3), 267(B3), 272(B3), 273, 278, 299, *366*

## N

Nagiev, M. F., 186, *253*
Napier, D. H., 266(D4), 267(D4), 270, 272, 273, 276, 278, 299, 301, 302, 303, 306, *366*
Natanson, G. L., 57, 77, *95*
Naumann, A., 332, *368*
Netter, Z., 210, *253*
Neubauer, R. L., 38, 41, 42, *96*
Neumann, H., 334(N1), *367*
Newitt, D. M., 266(D4), 267(D4), 270, 272, 273, 276, 278, 299, 300, 301, 302, 303, *366*
Newman, A. O., 264, *367*
Nukiyama, S., 349, *367*
Null, H. R., 334, 335, 340, 352, *367*

## O

Orr, C., 81, *93*
Orr, C., Jr., 349(O1), *367*

## P

Padmavathy, P., 266(P1), 270, 319, 356(P1), *367*
Pannetier, J., 160, *183*
Parker, J. H., 45, *94*
Pars, L. A., 304(P2), *367*
Patridge, B. A., 259(R10), *368*
Paul, R. C., 318, *366*
Pauthenier, M., 48, *95*
Pearson, R. G., 104(F5), *182*
Penney, G. W., 79, 80, 82, *95*
Perry, J. H., 189(P1), *253*
Peskin, R. L., 38, 40, *95*
Peterhans, E., 321(G2), 334(G2), *366*
Peters, M. S., 110, 176(A3), *181*
Peterson, J. W., 70, *95*
Peterson, T. I., 104(B5), 117(L3, L4, P2), 119(P2), 128(P2), *181, 183*
Pigford, R. L., 260, 318(B2), *366*
Pilcher, J. M., 349(P3), *367*
Pinchbeck, P. H., 157(P3), 159(F2, F3), *182, 183*

Popper, F., 159(F2, F3), *182*
Prandtl, L., 283(P4), *367*
Prater, C. D., 104(W1), *183*
Prausnitz, P. H., 334(H1, P5), *366, 367*

## Q

Quigley, C. J., 260, 268, 272, 273, 274, 278, 299, 304, *368*

## R

Raco, R. J., 38, 40, *95*
Radnik, J. L., 80, *94*
Rammler, E., 349, *368*
Ramakrishnan, S., 268, 272, 302(R1), 321, 350(R1), *368*
Randall, J. M., 38, 39, 41, *95*
Ranz, W. R., 81, *94*
Rao, E. V. L. N., 334(R3), 335, 337, 342, 344(R3), 345(R3), 350(R3), 351(R2), 353, 354(R3), *368*
Raydt, V., 66, *91*
Rayleigh, Lord, 24, 25, *95*, 349, *368*
Reed, C. E., 165, *183*
Reiner, A. M., 171(H13), *182*
Rennie, J., 333, *368*
Richards, H. F., 66, 68, *95*
Richardson, E. G., 28, *95*, 334(M3), *367*
Rider, H., 321, 356 (C9), *366*
Robertson, A. J. B., 45, *95*
Robinson, A. P., 347(W3), 352, *368*
Robinson, D. G., 321, *366*
Rodger, W. A., 355, *368*
Rodriguez, E., 355, 356, *368*
Rogers, G. B., 105, *183*
Rohmann, H., 48, *95*
Romero, J. B., 318(R8), *368*
Rose, G. S., 67, *95*
Rosen, E. M., 186, *253*
Rosenbrock, H. H., 116(R2), *183*
Rosin, P., 349, *368*
Rowe, P. N., 259, *368*
Rudd, D. F., 99(R3), *183*, 221, *253*
Rushton, J. H., 355, *368*
Russell, A., 30, 31, 32, *95*
Ryce, S. A., 42, *95*

## S

Sachsse, H., 48, *94*
Saradhy, Y. P., 343, *368*
Sargent, R. W. H., 334, *368*
Sargent, W. H., 204, 220, *253*
Satyanarayan, A., 272, 313(S3), 314, 315 (S3), 316, *368*
Sawistowski, H., 334(D3), *366*
Schiller, L., 332, *368*
Schmid, G. M., 8, *93*
Schonland, B. F. J., 13, 43, *95*
Schuler, B. O. G., 258(D8), 278, 279, 280, 269, 270, 272, 273, 278, 282(D8), 283, 285, 286, 289, 290, 293, 295, 300, 301, 302(D8, D9), 304, 307, 308(D8), 313, 319, 338, *366*
Schultz, R. D., 8, 25, 38, 39, 43, 79, *95*
Schurmann, R., 272, 273, 334(S5), *368*
Schwarz, W. H., 346(C6), *366*
Scott, D. S., 263, *367*
Seeliger, R., 334(N1), *367*
Sergiyeva, A. P., 80, *95*
Shabaker, R. H., 105, 139, *183*
Shaffer, R. E., 9, 10, 42, 71, 77, 79, *96*
Shankar Srinivas, N., 347, *368*
Shannon, P. J., 187(S2), *253*
Shapiro, A. R., 77, *96*
Shashoua, V. E., *96*
Shaw, P. E., 63, *96*
Sherwood, T. K., 165, *183*, 347(S8), *368*
Shinnar, R., 354, *368*
Shulman, H. L., 321(S10), *368*
Siemes, W., 272, 278, 300, *368*
Silberman, E., 268, *368*
Silsbee, F. B., *96*
Slattery, J. C., 316(S13, W2), 317(W2), 344(S13, W2), 345, 346(S13, S14, W2), *368*
Sliney, P. M., 7, 25, 28, 41, 59, 77, *96*
Smith, H., 111(D4), 114(D4), 133(D4), 137(D4), *182*
Smith, M. J. S., 318, *366*
Smith, W., 334(D3), *366*
Snavely, E. S., 8, *93*
Solov'yev, V. A., 77, *96*
Soo, S. L., 79, *96*
Sovins, J. C., 316(W1), 344(W1), *368*
Spang, H. A., 115(S2), 116(S2), *183*
Spells, K. E., 268, 270(S15), 321, *368*

# AUTHOR INDEX

Spencer, R. C., 349(D1, D2), *366*
Standel, N. A., 113(J1), 114(J1), *182*
Stasny, R. J., 28, *96*
Steward, D. V., 193(S3), 196, 201, 203, 204(S3), 212, 213, 225, *253*
Storey, C., 116(R2), *183*
Straubel, H., 41, *96*
Subramaniyan, V., 316, 318 (S16), *368*
Sullivan, S. L., 263, *368*
Suzuki, S., 81, *96*
Swanson, C. D., 226(S5), *254*

## T

Tanasawa, Y., 349, *367*
Taneya, S., 69, 71, 77, *96*
Thieme, G. G., 226(S5), *254*
Thodos, G., 168(M3), *183*
Thomas, D. G. A., 71, *96*
Thorogood, R. M., 333(T1), *368*
Tidwell, P. W., 161, 164(B17), *182*
Tietjens, O. G., 283(P4), *367*
Tomita, Y., 316(T2), 344(T2), 346(T2), *368*
Tomura, M., 81, *96*
Toomey, R. D., 318(T3), *368*
Traub, J. F., 186(T1), *254*
Treybal, R. E., 334(H5), 340, 343, 347, 352, *366*
Trezek, G. J., 79, *96*
Trice, V. G., 355, *368*
Tschernitz, J. L., 38, 39, 41, *95* '
Tukey, J. W., 137(A2), *181*

## V

Van Krevelen, D. W., 267(V1), 279, 286, *368*
Van Ostenburg, D. O., 63, *96*
Vela, M. A., 186, *254*
Venugopal, B., 262(V2), 264, *368*
Vermeulen, T., 354, 355, 356, *368*
Verschoor, H., 321(V4), *368*
Vick, F. A., 62, *96*
Viney, B. W., 45, *95*
Vonnegut, B., 7, 9, 25, 28, 38, 41, 42, 59, 77, 92, *96*

## W

Wagner, P. E., 70, 72, *96*
Wallick, G. C., 316(W1), 344(W1), *368*
Ward, S. G., 67, *95*
Warrington, M., 45, *95*
Wasserman, M. L., 316(W2), 31(W2), 344(W2), 345, 346(W2), *368*
Watson, C. C., 99(R3), 109(K7), 110(K6), 114(K9), 115(K8), 116(K8), 117(K7, K8), 118(K8), 125(K11), 127(K7), 142(K8), 155(K7), 159(K10), 160(K10), 161(H10), 168(K11), 169(K6), 176(K11), *183*
Watson, W. K. R., 77, *96*
Wei, J., 104(W1), *183*
Weller, S., 104(W2), *183*
Wells, P. U., 77, *96*
West, F. B., 347(W3), 352, *368*
Westerberg, A. W., 204(S1), 220(S1),*253*
Whitby, K. T., 53, *96*
White, H. J., 48, 49, 51, 79, *96*
White, R. R., 125 (W3), *183*
Whitman, V. E., 79, *96*
Wichern, D. W., 173(H14), *182*
Wiech, R. E., 8, 25, 38, 39, 43, 79, *95*
Wilcox, J. D., 349(W4), *368*
Williams, E. J., 142(W4, W5), *183*
Williams, G. M., 354, 355, 356, *368*
Wilson, H. L., 116(B18), *182*
Woods, D. R., 187(S2), *253*
Wright, T. E., 28, *96*
Wu, S. M., *183*

## Y

Yang, K. H., 105, *183*
Yasui, G., 318(Y1), *368*
York, J. L., 82, *92*
Youle, P. V., 157(B19), *182*

## Z

Zebel, G., 11, 15, 17, *96*
Zuiderweg F., J., 334(Z1), *368*

# SUBJECT INDEX

## A

Adhesion, electrostatic charge and, 30–38
Adjacency matrix
  Boolean, 194–195
  branch in, 213
  graph and, 207
  maximal loops in, 200–202
  paths and other loops in, 205
  for process system, 220
  tearing and, 213–214
  tree of paths in, 213–214
Aerosols
  charge measurement in, 79–81
  Coulomb forces in, 10
  deposition rates in, 28–29
  drop coalescence in, 25–26
  electrostatic effects in, 2, 10–25
  flocculation coefficient in, 15
  flocculation rate in, 17
  floc properties of, 21–23
  monodisperse homopolar, 16
  optimum charge in, 14
  in respiratory tract, 26
  stability of, 15–21
Air, properties of, 87
Alcohol dehydration, 166–167
  confidence region in, 129
  ethanol adsorption constant and, 141, 146
  model of, 107–108
  parameter estimation in, 120
  sums-of-squares surfaces in, 118–119
Algorithm, decomposition, 196–200
  *see also* Decomposition; Steward's algorithm
Ambipolar ion, 49
Analysis of variance
  in model adequacy tests, 131–137
  $F$ statistic in, 133
  regression in, 133–135
Atomic Energy Commission, example of decomposition procedure for, 226–236
Atomization
  charge-to-mass ratio in, 40–42
  corona in, 41–42
  electrostatic, 38–43
  in high vacuum, 38–40
  in liquid-gas system, 348–349
  in liquid-liquid system, 347–348
  of oils, 40–41
  spraying and, 347–349
Avogadro's number, 87, 89

## B

Back discharge, 46–47
Balloelectricity, 55
Billingsley's algorithm, 208
Boltzmann constant, 87, 89
Boolean adjacency matrix, 194–195
Boolean matrix, in process models, 188–189
Boolean multiplication, 189–190
Boolean union, 190–191
  partitioning and, 202
Branch, in equation systems, 213
Brownian motion, 11
Bubble expansion of at orifice, 304–306
Bubble and drop formation, 255–363
  *see also* Bubble formation; Drop formation
  in dispersed phase, 257
  interface surface in, 256
  unified model for, 350–354
Bubble chamber, volume of, 268–270

Bubble formation
  bubble volume and, 257–258
  chamber volume, and 268–270, 356–357
  in constant-flow conditions, 277–304, 327–330
  under constant pressure, 304–316, 330–332
  continuous-phase velocity and, 277
  Davidson and Schuler's model of, 278–279, 285, 289–290, 307–310, 319
  density and, 302–304
  electric analogue in, 357
  experimental equipment for, 265–266
  flow rates in, 276–277, 286, 290–291, 361
  frequency of, 263–265
  in gas fluidized beds, 318–320
  gas properties of, 274–277
  in highly viscous liquids, 289–295, 308–310
  idealized sequence in, 282
  at inclined orifice, 330
  under intermediate conditions, 356–359
  in inviscid fluids, 278–289, 307–313
  Kumar and Kaloor's model of, 280–281, 285, 292–298, 310–318, 320
  liquid density and, 273–274, 302–304
  liquid viscosity and, 272–273
  models of, 278–281, 285, 289–298, 307–320
  at multiple-orificed plates, 333–334
  in non-Newtonian fluids, 316–320
  orifice characteristics and, 267–268
  orifice geometry and, 321–324
  orifice orientation and, 324–332
  in Ostwald-de-Waele fluids, 316
  particle size and, 319
  at sintered disks, 359–362
  submergence effects on, 270–271
  surface tension and, 286–289, 298–301, 307
  system variables in, 267
  Tate's law and, 271
  upward buoyance force in, 282–283
  viscosity and, 278–295, 307–313
  wetting properties, and 271–272
Bubble size
  continuous-phase velocity and, 332–333
  experimental set-up in, 265–266
  factors in, 265–277
  flow rate and, 276–277, 286, 290–291, 361
  orifice geometry and, 321–324
  in stirred vessels, 354–356
  weeping time and, 359
Bubble volume
  cinephotography in, 258–259
  direct methods in, 258–260
  flow rate and, 314, 362
  gamma-ray absorption method in, 259–260
  indirect methods for, 260–265
  liquid properties and, 278
  measurement of, 257–265
  orifice orientation and, 325–326
  rigorous calculations for, 287
  simplified calculations for, 287
  submerged orifice and, 360
  surface tension and, 274–276, 287–289, 298–301, 312–313
  viscosity and, 301–302
  X-ray cinephotography in, 259
Bulk fluid, rupture of, 257
Buoyancy force, bubble volume and, 298–299
Butanol dehydration, 146
Butyl alcohol, dehydration of, 143

## C

Cab-O-Sil, 68
Capacitance technique, in bubble formation measurement, 264–265
Carbon tetrachloride, dispersion of, 41
Carbopol, 71
Carbowax, 68
Chamber volume, bubble formation and, 356–357
Charge
  electrostatic, 2–3; *see also* Electrostatic charge
  in metal-metal contact, 62
  phase change and, 73–74
  surface gradient and, 5
Charged bodies, defined, 2
Charged particles
  collision rate of, 15
  Coulomb's law and, 30–31
  electrostatic force between, 30–35
Charged spheres
  attractive force between, 33
  electrostatic factors in, 32

## SUBJECT INDEX

Charge measurement, techniques of, 75–87
Charge separation, in liquid-gas interface, 55–56
Charge transfer, mechanism of, 61–62
Charge-to-mass ratio, 4, 25, 27
  atomization and, 39, 42
  for electron and proton, 87
  measurement of, 77, 81
Charging
  through condensation, 74
  contact, 46–67
  induction or field, 47–51
  and interface alteration, 56
  metal-to-metal contact, 62
  by visible light, 74
Chemical engineering, models in, 98–99
  see also Model(s)
Chemical plant, modeling of, 99
Chemical reactions, mathematical modeling of, 97–181
Chemical warfare, aerosols in, 14
Cinephotography
  in bubble measurement formation, 263–264
  in bubble volume measurement, 258–259
Cloud
  as aerosol, 10
  charged, 12
  corona charge loss in, 13
  edge velocity of, 11–12
  expansion of, 11–14
  maximum permissible diameter of, 13
  properties of, 11–14
Coalescence, electrostatic charge and, 25–28
Coal-iron contact, charging in, 71
Coehn's rule, 66
Compressive contact, charging in, 67–68
Computer
  in model discrimination and parameter estimation, 178
  in partitioning and disjoint subsystem work, 211
  Sargent's algorithm and, 221
Computer program, for precedence ordering in large-scale equation systems, 237–251
Condensation, charging through, 74
Confidence region
  in model-discrimination designs, 176–177

  in nitric oxide reduction, 168–169
  plotting of in parametric estimation, 129–131
  in reaction-rate modeling, 124–129
Constant-flow conditions, bubble formation in, 277–304
Constant-pressure conditions, bubble formation in, 304–316
Contact charging, 46–47, 62
Contact electrification, 60
Continuous-phase velocity, in bubble formation, 277
Conversion factors, for physical units (table), 84–86
Corona
  atomization and, 41–42
  negative, 45
  point-to-plane, 44
  positive, 45
Coulomb force, in aerosol, 10
Coulomb's law, 6
  interparticle electrostatic force and, 30–31

### D

Data location, in model designs, 168
Davidson and Harrison model, of bubble formation, 319–320
Davidson and Schuler model, bubble formation, 278–279, 285, 289–290, 307–310, 319
Decomposition
  Hanford N-reaction plant examples of, 226–236
  mathematical definition of, 187
  partitioning in, 198–208
  precedence ordering techniques in, 222–226
  tearing in, 211–220
Decomposition algorithms, criteria for, 199–200
Decomposition procedure, for solving large-scale systems of equations, 185–253
Density, bubble formation and, 273–274, 302–304
Dependent variable, in reaction-rate models, 159–160
Deposition, electrostatic charge and, 28–29

Diagnostic parameters
  adaptive model-building with, 147–156
  for hyperbolic models, 147–152
  model discrimination and, 142–147
  for power-function models, 152–156
Dibutyl phthalate, atomization of, 42
Differential models, 101–102
Differentiation, in power-function models, 103
Diffusion, force equation for, 37
Diffusion charging, 51–53
Diffusion rate, in aerosols, 17
Digraph, matrices and, 188–189
Disjoint subsystems, identifying of, 209–211
Dispersion, through submerged orifices, 257
Drag coefficient, in bubble formation, 305
Drop and bubble formation, unified model for, 350–354
Drop formation, 334–343
  see also Bubble formation
  dispersed phase viscosity and, 339
  Hayworth and Treybal model of, 335–336
  Kumar and Kuloor model of, 337–339
  in non-Newtonian fluids, 343–346
  Null and Johnson model of, 337
  other models of, 340–343
  in power-law fluids, 343–345
  in stirred vessels, 354–356
  variables in, 334–335
  for vertically oriented orifices, 346–347
Drop volume
  dispersed-phase viscosity and, 342
  interfacial tension and, 341
  viscosity and, 340–342
  volumetric flow rate and, 341
Dual-site model, 147, 167

### E

Electrical endosmosis, 58, 66
Electrification, contact and static, 60
Electroconstriction, 8
Electroendosmotic experiments, 66
Electron
  charge on, 90
  energy level of, 60–61
  Fermi level of, 61–62
  ionization by, 44
  in solid state, 61
  in thermionic emission charging, 53–54
Electronic probe, in charge measurement, 82
Electron point charge, 34
Electron transfer, through rubbing, 60–62
Electron tunneling, 63
Electrostatic charge
  see also Charge; Particle
  adhesion and, 30–38
  biological processes and, 38
  defined, 2–3
  deposition and, 28–29
  effects of, 7, 25–28
  evaporation and, 28–29
  field intensity and, 74
  in liquid-solid contact, 58–60
  and material properties, 7–8
  maximum stable, 43–46
  measurement of, 75–87
  methods of charging or discharging particles, 43–75
  nomenclature in, 88–90
  powder dissemination and, 43
  in solid-solid contact, 60–72
Electrostatic dissemination techniques, 38–43
Electrostatic effect
  on aerosol properties, 10–25
  on containment or bulk storage, 8
  on material properties, 28–29
Electrostatic field
  high intensities of, 9
  polarization in, 3
Electrostatic force
  between charged particles, 30–35
  comparison with other forces, 35–38
Electrostatics, fundamentals of, 2–6
Empirical model, 102
  response-surface methodology in, 155–159
  techniques for, 156–167
  transformation of variables in, 159–164
  tuning in, 164–167
Endosmosis, 58–59, 66
Energy level, solid state and, 60–61
Equation block, 211
  adjacency matrix and, 214
  tearing methods for, 212–220
Equations
  adjacency matrix and, 200–202
  decomposition methods for, 185–253

## SUBJECT INDEX

information flow among, 187
irreducible sets of, 200–202
large-scale systems of, 185–253
maximal loops and, 200–202
occurrence matrix and, *see* Occurrence matrix
output set for, 196–198
partitioning in, 198–208
subsystems of, 187
systems of, 193–196
Equilibrium charge, 48
Equilibrium vapor pressure, 7
Evaporation, electrostatic charge and, 28–29
Experimental designs, in model making, 168–174

### F

Faraday cage, 79
Fermi-Dirac statistics, 61–62
Fermi level, 61, 63
Field charging, 47–51, 74
Field intensity, of point charge, 34, 74
Flocculation
  aerosol properties in, 21–23
  effective size in, 23
Flocculation coefficient, 15
Flocculation rate, in aerosols, 17
  particle charge and, 20
  particle mass concentration and, 19
  particle number concentration and, 18
Floc diameter, effective size and, 23
Flow rates
  bubble formation and, 286, 290–291
  bubble size and, 361
  bubble volume and, 314, 362
Fluid, rupture of, 257
Fluid-fluid contact, interface in, 256–257
Force balance bubble volume, 312, 317
Force equations, for various mechanisms, 36–37
Force field, particle stability and, 24
FORTRAN
  precedence ordering techniques and, 223
  Steward's algorithm and, 206
Fractional-life methods, in power-function models, 103
Freezing, charging through, 73–74

Froude number, in bubble formation, 305–306
$F$ statistic, in analysis of variance, 113

### G

Gamma-ray absorption, in bubble volume measurement, 259–260
Gas-flow rates, and bubble formation, 276–277
Gas properties, bubble formation and, 274–276
Gas sample collection, in bubble volume evaluation, 261
Graph
  adjacency matrix and, 207
  properties of, 188–190
  maximal loop of, 192
  of reactor process, 219
Ground fog, measurement of, 83

### H

Hanford N-reactor plant
  decomposition examples from, 226–236
  reordered occurrence matrix of, 252
Hayworth and Treybal model, of drop formation, 335–336
Hexyl alcohol, dehydration of, 166
Homopolar charging, absence of in powders, 72
Hougen-Watson model, 105
Hyperbolic models, 105–110
  estimated parameters in, 109–110
  of methane oxidation, 148–152
  reaction rate and, 105–106
  uncertainty in, 125

### I

Independent variables
  plots of in model adequacy analysis, 139–140
  transformations of in empirical modeling, 161–162
Induction charging, 47–51

Information flow
  Boolean adjacency matrix and, 194
  among equations, 187
  maximal loop in, 200–202
Insecticide spray, optimum charge in, 14
Insulators, charging of, 62–63, 66
Integral models, 101–102
Integration, in power-function models, 102–103
Interface, in fluid-fluid contact, 256–257
Interface alteration
  particle charging by, 55–72
  processes and mechanisms in, 56
Intrinsic parameter, in model discrimination, 144–147
Inviscid liquids, bubble formation in, 307–308, 210–313
Inviscid systems, oriented orifice and, 327–330
Ion, ambipolar, 49
Ion concentration, in interface alterations, 57
Ion generator, particle charge and, 53
Ionization, in atomization, 44
Ionization coefficient, 45
Ion mean free path, particle charge and, 49
Ion mobility, factors affecting, 50
Ion transfer, in insulators, 66
Iron-coal contact, charging in, 71–72
Isooctene hydrogenation, parameter estimation in, 115
Iterate column, in occurrence matrix, 218
Iterative techniques, in parameter estimation, 115–116

J

Jet breakage, in droplet formation, 349

K

Kinetic data, in empirical model tuning, 164
Kinetic modeling, response-surface methods in, 156–158
Kumar and Kuloor model, for bubble formation, 280–281, 285, 292–298, 310–318, 320, 325–332
  for drop formation, 337–339

L

Langmuir-Hinshelwood mechanisms, 100
Large-scale systems of equations, decomposition procedures for, 185–253
  see also Equations
Least-squares method
  in model parameter estimation, 111–121
  nonlinear, 115–121
  weighted, 114–115
Ledet's algorithm, 216–218, 221
  tearing and, 225–226
Lee's algorithm, 226
Legendre polynomials, 24 $n$
Light, charging by, 74
Lindemann theory, 100
Linearization method, in parameter estimation, 116
Linear-least-squares method
  in model adequacy tests, 138
  in model parameter estimation, 111–114
Linear models, confidence intervals in, 125–127
Liquid, electrostatic atomization of, 38–43
Liquid breakup, charge nature and magnitude in, 57–58
Liquid density, bubble formation and, 273–274, 302–304
Liquid-gas interface, charge separation and, 55–56
Liquid-liquid contact, interface in, 256–257
Liquid properties, bubble volume and, 278
Liquid-solid contact, particle charge in, 58–60
Liquid viscosity, bubble formation and, 272–273, 289–296, 301–302
Loops, occurrence matrix and, 215

M

Mathematical modeling
  see also Model(s)
  experimental designs, for 168–178
Matrix
  see also Adjacency matrix; Occurrence matrix
  Boolean multiplication in, 189–190

nonzero elements in, 194–195, 203, 210, 229
occurrence and adjacency, 194–196
Maximal loop, 192, 200
  matrix of, 204
  terminal equation and, 208
Mechanisms, models and, 98
Mercury, charge acquired by, 59–60
Metal-metal contact, charging in, 62
Methane oxidation, hyperbolic model of, 148–152
Michaelis-Menten mechanisms, 100
Millikan cell, 76–77
Model(s)
  adequacy tests for, 131–141
  analysis-of-variance techniques for, 131–137
  empirical, 156–157
  experimental designs in, 168–174
  of bubble formation, 278–281, 289–290, 292–298, 302–303, 307–320, 325–332, 350–354
  computer use in, 178
  confidence intervals and regions in, 124–129
  conjecture and experiment in, 168
  data points in, 168
  diagnostic parameters in, 142–156
  discrimination in, 169–173
  disjoint subsystem for, 209
  of drop formation, 337–339
  hyperbolic, 105–110, 147–152
  integral and differential, 101–103
  kinetic, 156–158
  least-squares estimation of parameters in, 111–121
  linear and nonlinear, 125–129
  linearly reducible, 102–110
  mechanisms and, 98–99
  multiple-response parameter estimation in, 192–131
  parameter estimation in, 111–131, 173–178
  power-function, 102–104
  process, 98–99
  reaction-rate, 99–101
  reparametrization in, 121–124
  residual analysis of, 137–141
  response-surface methods in, 155–159
  rival, 169–172
  single- versus dual-site, 106, 147
Model-building, adaptive, 147–156
Model-discrimination
  designs for, 171–173
  with diagnostic parameters, 142–147

## N

Naphthalene oxidation data, 122–123
Naphthalene oxidation system, parameter estimation in, 119–120
Nebulizer, 57
Needle, drop coalescence from, 25–26
Newtonian fluid
  flow index in, 316
  power-law drops in, 345–346
Newton's iteration function, 186
Newton's second law, in bubble formation, 304
Nitric oxide reduction, data location in, 168–169
Nonintrinsic parameters, in model discrimination, 142
Nonlinear least squares, method of, 115–123
Nonlinear models, confidence intervals in, 127–129
non-Newtonian fluids
  bubble formation in, 316–320
  drop formation in, 343–345
Nonzero element
  in occurrence matrix, 194–195, 210
  in product matrix, 203, 229

## O

Occurrence matrix, 193
  adjacency matrix and, 195
  best tear in, 216–217
  disjoint subsystems and, 209–210
  Hanford N-reactor system and, 227, 229–230, 252
  iterate column in, 218
  nonzero elements in, 210
  output variables and, 234
  reduced, 222
  residual, 222, 252
  torn variables in, 217
  for triple-effect evaporator system, 230–231

## SUBJECT INDEX

Oils, atomization of, 40
Orifice
  bubble expansion and, 304–306
  bubble formation and, 267–268, 324–332
  drop formation from, 346–347
  multiple, 333–334
  wetting properties and, 271–272
Orifice geometry, bubble size and, 321–324
Orifice orientation, bubble formation and, 324–332
Orifice submergence, in bubble formation, 270–271, 360
Ostwald-de-Waele fluids, bubble formation and, 316
Output set, 193
  finding of, 196–198
Output variable, 193
Overall plots, in residual analysis of models, 137–138

## P

PACER digital executive program, 187
Parameter estimation
  analysis of variance and, 132
  computer and, 178
  methods in, 170–171, 178
  model discrimination and, 173–178
  multiple-response, 129–131
  weighted least-squares method in, 114–115
Parametric residuals, in model adequacy tests, 140–141
Partial pressures
  feed-component, in hyperbolic models, 147–148
  in model-discrimination design, 173
Particle(s)
  charging or discharging methods for, 43–75
  contact charge of, 46–47
  diffusion charging of, 51–53
  field charging of, 47–51
  interface alteration charge in, 55–72
  Stokes-Cunningham factor for, 11
  thermionic emission charging of, 53–55
Particle charge
  see also Charge; Charging; Electrostatic charge

  back discharge in, 47
  and charge-to-mass ratio, 27
  contact charging and, 46–47
  container diameter for, 10
  flocculation rate and, 20
  ion solubility and, 49–50
  measurement of, 75–87
Particle concentration, spacing and, 21
Particle mass concentration, flocculation rate and, 19
Particle number concentration, 18
Particle size
  bubble formation and, 319–320
  charge and, 3–4
Particle spacing, concentration and, 21
Particle stability, electrostatic charge and, 24
Particle surface gradient, 5
Particulates, electrostatic phenomena with, 1–90
Partitioning
  Boolean union in, 202
  computation times for, 224
  methods of, 202–208
  of system equations, 198–208
Pauli exclusion principle, 60
Pentane isomerization
  parameter estimates for, 113
  rate models for, 158
Phase change, charging by, 73–74
Photoelectric cell counter, in bubble-frequency measurements, 263
Photography, see Cinephotography
Photoionization, charging through, 75
Physical units, conversion factors for (table), 84–86
Planck's contant, 87–88
Pneumatic conveying contact, charging in, 68
Point change, force and field identity due to, 34
Polarization
  between charged spheres, 33
  in electrostatic field, 3
Powdered milk, charging in, 69–70
Powders
  electrostatic dissemination of, 43
  homopolar charging in, 72
  pneumatically conveyed, 68
Power-function models, 102–104

diagnostic parameters for, 152–156
multicomponent, 104
Precedence ordering
  computer program for, 237–251
  decomposition techniques for, 222–226
Predicted-value plots, in model analysis, 138–139
Pressure pulse technique, in bubble-frequency meaurements, 264
Process exploitation, response-surface methods in, 157–159
Process flow sheet, information in, 188
Process models, 98–99
  information flow in, 188–196
Process system
  adjacency matrix for, 220
  design parameters in, 220
  input-output relationships in, 219
Process units, tearing in, 219–222
Propositions, in symbolic logic, 189
Propylene hydrogenation, 139–140, 163
  discrimination design in, 174–175
Proton, mass of, 89

## Q

Quadrupole mass spectrometer, 77

## R

Ramakrishnan model, of bubble formation, 295–298
Rau and co-workers model, of drop formation, 337–339
Rayleigh instability, 24–25
Reachability matrix, 192
Reaction driving force, in empirical modeling, 163
Reaction order, in empirical modeling, 161
Reaction rate
  in empirical model tuning, 165–166
  in hyperbolic models, 105
  in response-surface methodology, 155–156
Reaction-rate models, 99–101
  confidence intervals and regions in, 124–129
  diagnostic parameters in, 151
  transformation of variables in, 159–161
Reactor process, graph of, 219

Regression, in analysis-of-variance techniques for model adequacy, 133–137
Relaxation time, defined, 9
Reparametrization, in chemical models, 121–124
Reservoir method, in bubble volume evaluation, 260
Residual analysis, in model adequacy tests, 137–141
Residual plots, in model analysis, 139
Respiratory tract, aerosols in, 26
Response surface
  in empirical modeling, 155–159
  in process exploitation, 157–159
Reynolds number, in bubble formation, 306, 316
Riot control, aerosols in, 14
Rolling contact, charging in, 70–71
Rotameter, in bubble volume measurement, 263
Rubbing
  electron transfer through, 60–62
  of powders, 68–70
Runge-Kutta method, in bubble volume calculations, 317

## S

Sargent's algorithm, 220–221, 226
Satyanarayan model, of bubble formation, 310–318
Shock waves, charging through, 75
Silica, charging of in pneumatic conveyors, 68–69
Simultaneous equations, partitioning in, 198–205
  see also Equations
Single-site model, 106, 109, 147
Sintered disks, bubble formation at, 359–362
Sliding contact, charging in, 69
Soap-film flow meter, 262
Solidification, charging through, 73–74
Solid-solid contact
  charge in, 60–72
  compressive, 67–68
  pneumatic, 68–69
  rolling, 70–71
  sliding, 69–70
  solid state structure in, 60–61

Solid-solid reaction, in empirical modeling, 160
Solid state structure, theory of, 60–61
Spray electrification, 55
Spraying and atomization, 347–349
  see also Atomization
Standard air, properties of, 87
Static electrification, 60
Steepest-descent method, in parameter estimation, 116
Steward's algorithm, 205, 216, 221, 223
  output set of, 196–198, 224
  second phase of, 206
Stokes-Cunningham factor, 11
Stokes terminal velocity, bubble velocity and, 325
Stirred vessels, drop and bubble size in, 354–356
Streaming potential, in liquid-solid contact, 58
Submerged orifices, dispersion through, 257
Submergence, in bubble formation, 270–271, 360
Sums-of-squares, surfaces parameters in, 118–119
Surface states, charging and, 71–72
Surface tension
  bubble formation and, 271–272, 286–289, 298–301, 307, 327
  bubble volume and, 274–276, 300, 312–313
  in constant-flow conditions, 274
  orifice geometry and, 323–324
  pressure due to, 7
  vapor pressure and, 7
Symbolic logic, propositions in, 189
Systems equations
  branch in, 213
  Hanford N-reactor plant examples of, 226–236
  partitioning of, 198–208
  representation of, 193–196
  tearing in, 211–220
System variables, for triple-effect evaporator system, 234–235

T

Tate's law, 271
Tearing
  algorithms in, 219–222
  concept of, 211–212
  as decomposition process, 211–222
  for large-scale systems of equations, 212–220
  Ledet's algorithm and, 225
  loops in, 215
Terminal equation, maximal loop and, 208
Thermal equilibrium, electron and, 61
Thermionic emission charging, 53–55
Time-residual plots, in model adequacy tests, 140
Torn variables, in occurrence matrix, 217
Tree, in adjacency matrix, 213–224
Triboelectric series, table of, 64–65
Triboelectrification, 60
Triple-effect evaporator, in Hanford N-reactor example, 230–235
Truth values, in symbolic logic, 189
Tunnelling, electron, 63

U

Union, Boolean, 190–191
Universal constants (table), 87

V

Vacuum, atomization in, 38–40
Van der Waals force, equation for, 37
Vapor pressure, equilibrium, 7
Variables, transformation of in empirical modeling, 159–164
Viscosity
  bubble formation and, 272–273, 289–290
  bubble volume and, 301–302
Viscous liquids
  bubble formation and, 272–273, 289–296
  oriented orifices in, 327–330
Volumetric flow rate, bubble volume and, 260–263

W

Water drops, coalescence of, 26
  see also Drop formation

X

X-ray cinephotography, in bubble volume measurement, 259

Z

Zeta potential, in liquid-solid contact, 58

DOES NOT CIRCULATE